THE BOMB IN THE WILDERNESS

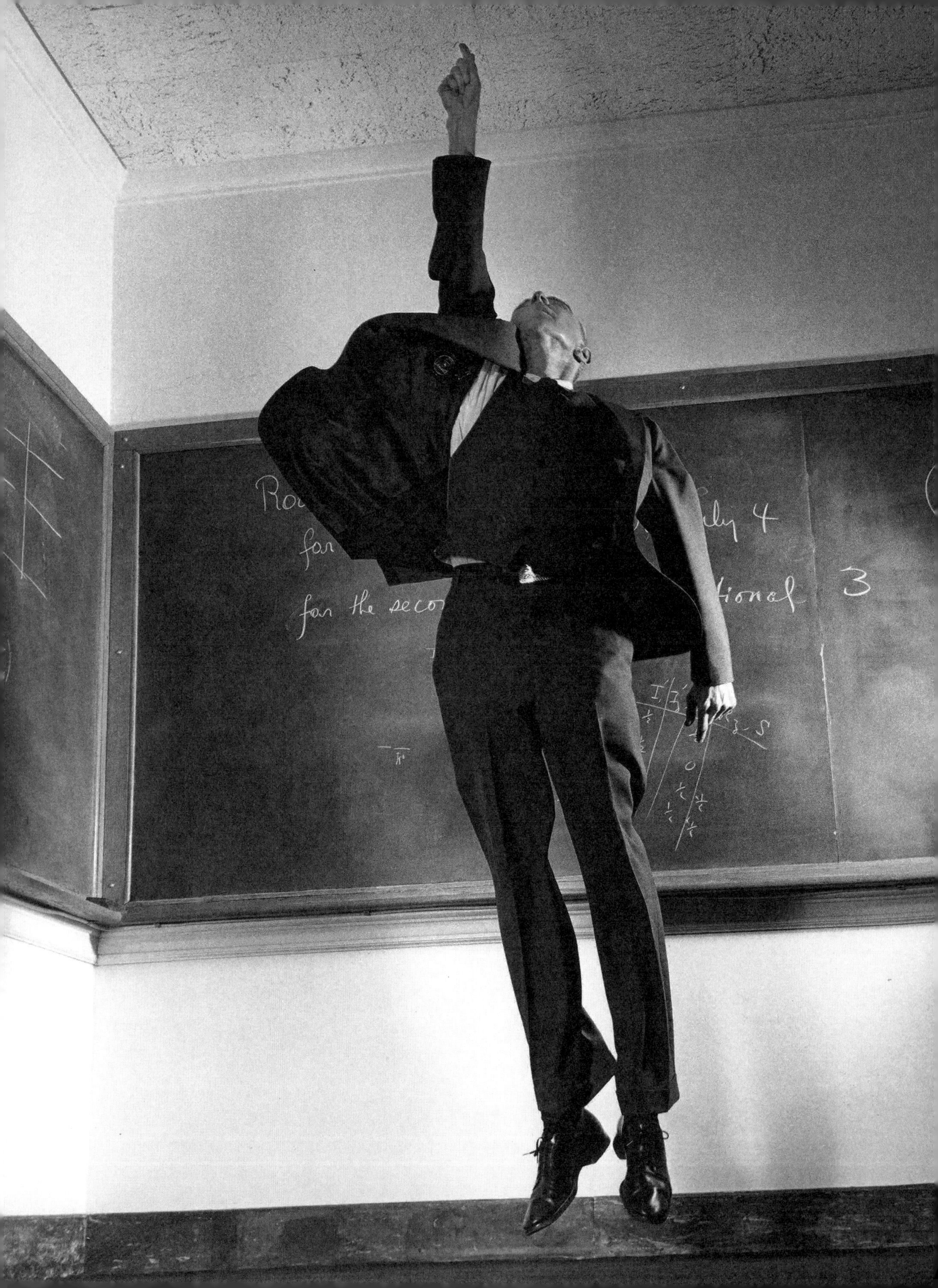

THE BOMB IN THE WILDERNESS

Photography and the Nuclear Era in Canada

JOHN O'BRIAN

UBCPress

29 28 27 26 25 24 23 22 21 20   5 4 3 2 1

Printed in Canada on FSC-certified ancient-forest-free paper (100% post-consumer recycled) that is processed chlorine- and acid-free.

Library and Archives Canada Cataloguing in Publication

Title: The bomb in the wilderness : photography and the nuclear era in Canada / John O'Brian.
Names: O'Brian, John, 1944– author.
Description: In title, the "o" appears as a bomb. | Includes bibliographical references and index.
Identifiers: Canadiana (print) 20200262459 | Canadiana (ebook) 20200262599 | ISBN 9780774863889 (softcover) | ISBN 9780774863896 (PDF) | ISBN 9780774863902 (EPUB) | ISBN 9780774863919 (Kindle)
Subjects: LCSH: Nuclear weapons—Canada. | LCSH: Nuclear energy—Canada. | LCSH: Photography—Social aspects—Canada.
Classification: LCC U264.5.C3 O27 2020 | DDC 355.8/251190971—dc23

Canada

UBC Press gratefully acknowledges the financial support for our publishing program of the Government of Canada (through the Canada Book Fund), the Canada Council for the Arts, and the British Columbia Arts Council.

This book has been published with the help of a grant from the Canadian Federation for the Humanities and Social Sciences, through the Awards to Scholarly Publications Program, using funds provided by the Social Sciences and Humanities Research Council of Canada.

Printed and bound in Canada by Friesens
Set in Garamond Premier Pro and Jost by Michel Vrana
Copy editor: Deborah Kerr
Proofreader: Alison Strobel
Indexer: Stephen Ullstrom
Interior and cover designer: Michel Vrana
Cover image: Robert Del Tredici, *Stanrock Tailings Wall, Elliot Lake, Ontario.*

The University of British Columbia
2029 West Mall
Vancouver, BC V6T 1Z2
WWW.UBCPRESS.CA

Brenda and David McLean Canadian Studies Series

UBC Press is proud to publish the Brenda and David McLean Canadian Studies Series. Each volume is written by a distinguished Canadianist appointed as a McLean Fellow at the University of British Columbia, and reflects on an issue or theme of profound import to the study of Canada.

W.H. New, *Borderlands: How We Talk about Canada*
Alan C. Cairns, *Citizens Plus: Aboriginal Peoples and the Canadian State*
Cole Harris, *Making Native Space: Colonialism, Resistance, and Reserves in British Columbia*
John F. Helliwell, *Globalization and Well-Being*
Julie Cruikshank, *Do Glaciers Listen? Local Knowledge, Colonial Encounters, and Social Imagination*
Sherrill Grace, *On the Art of Being Canadian*
R. Kenneth Carty, *Big Tent Politics: The Liberal Party's Long Mastery of Canada's Public Life*
Tina Loo, *Moved by the State: Forced Relocation and Making a Good Life in Postwar Canada*

For Celia, Mila, Felix, and Emma,
my grandchildren

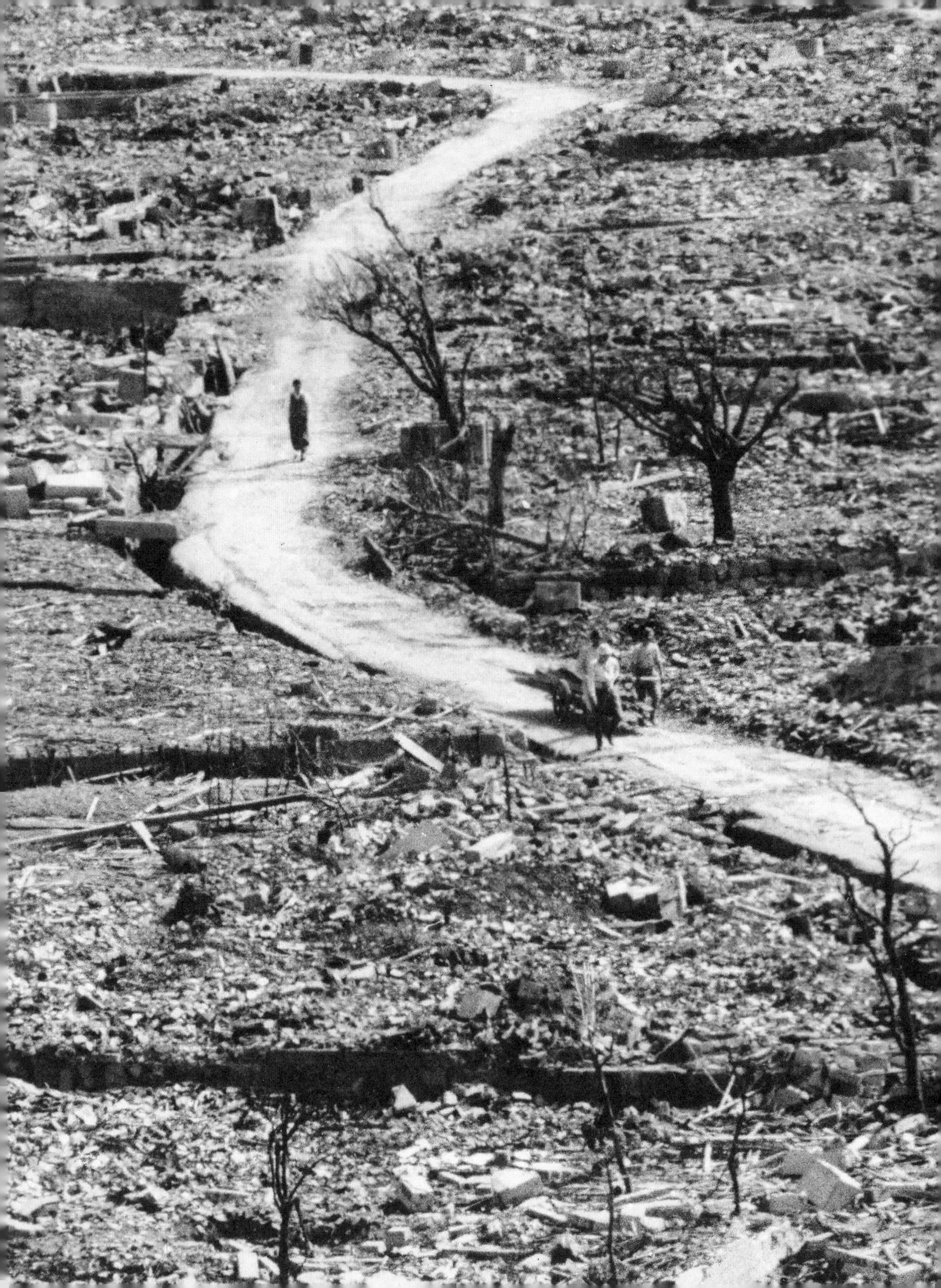

He'd been shown photographs of Hiroshima ...
A silence deeper than dreaming lay over this island of burnt life.

– Dennis Bock, The Ash Garden, *2001*

CONTENTS

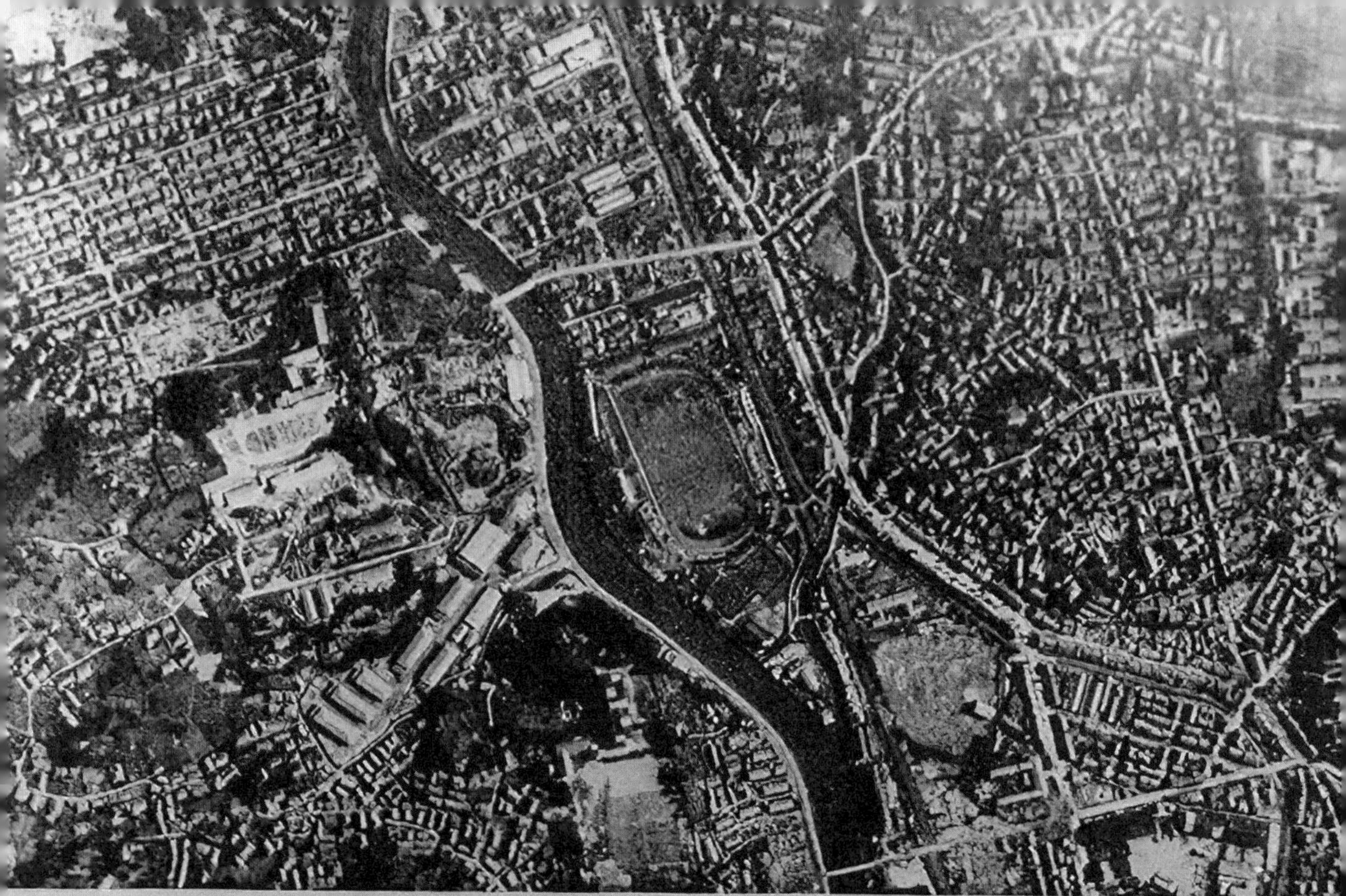

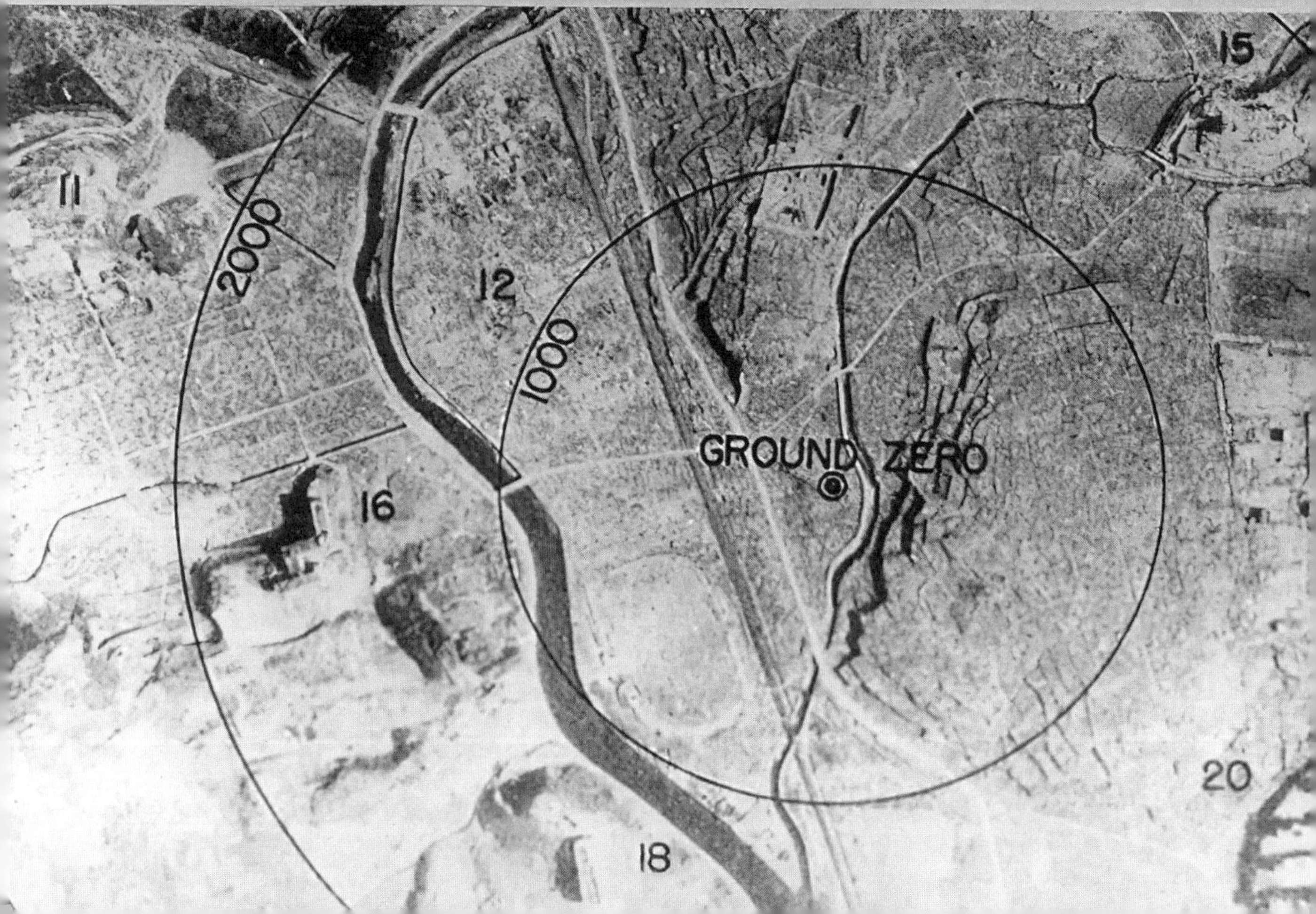
15
11
2000
12
1000
GROUND ZERO
16
20
18

PREFACE

Radioactive Footprints

DURING THE SECOND WORLD WAR, CANADA WAS A MEMBER OF THE MANHATTAN Project, the consortium that developed the atomic bombs dropped on Hiroshima and Nagasaki. The impact of Canada's nuclear programs has been felt ever since the destruction of those cities in 1945. The extent of the programs – the reach of Canada's nuclear footprint – has been global. This book focuses on what photography can demonstrate (or not demonstrate) about the size and shape of the footprint. Some of the revelations are striking. The photographs discussed range from those released by military authorities, to the documentary images of Montreal artist Robert Del Tredici, to the atomic ephemera of consumer culture. The book contends that photography is one of the primary ways, if not the primary way, that nuclear episodes and activities are represented and remembered. Photographs provide a link between the nuclear past and the nuclear present, and play a mediating role in shaping a public image of events. They help to show that the military atom and the civilian atom, notwithstanding disclaimers to the contrary by governments and corporations, are intimately connected. A central focus of the book is on nuclear risk: Do photographs alert viewers to nuclear threat, numb them to its dangers, or by some strange calculus do both at the same time?

| **Figure P.1**: Nagasaki 1945 – Before and After, 1945.

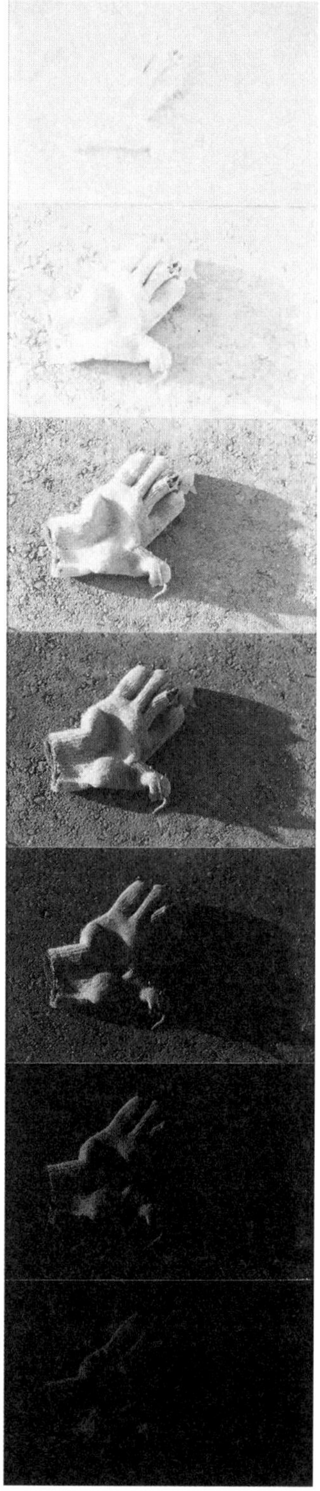

I started paying attention to issues of nuclear risk and representation in the 1980s, when I was writing about postwar visual art. You cannot discuss Barnett Newman and Jackson Pollock, or Joyce Wieland and Roy Kiyooka, without also talking about Cold War anxieties. Kiyooka, a Japanese Canadian artist branded an enemy alien during the Second World War, knew more than most about Cold War tensions. When Kiyooka represented Canada at the 1970 Osaka world's fair, he arrived in Japan while the fairgrounds were still under construction and took photographs of gloves discarded by workers on the site. Mixed with concrete dust and rainwater, some gloves had turned to "stone." [**Figure P.2**] "The scattered gloves haunted the site with the specters of Hiroshima and Nagasaki," one observer writes. "With the fiery disappearance of bodies marked in the silent twists of their folds [they offered] traces of not only those vaporized but also the *hibakusha,*" victims of the bombings who survived.[1] Several vertical sequences of photographs from the *StoneDGloves* series offer a visual parallel to an atomic explosion, where the white of the flash turns to black ash. A single glove shown in a sequence of seven photographs progressing from light to dark is scarcely visible at the extremes. The over-exposed print at the top is almost white, the under-exposed print at the bottom almost black.

Like most of my generation, those born during the Second World War, I was personally caught up in nuclear events long before the 1980s. When the threat of nuclear conflict between the United States and the Soviet Union heightened at the beginning of the Korean War in 1950, I was a first-grade student living just outside New York City. At the time, my father, a career officer in the Royal Air Force, was stationed nearby at West Point (I had my tonsils removed in the military hospital there and can still taste the ether used to knock me out), and I recall doing duck-and-cover drills in class that sent me scrambling beneath my desk. [**Figure P.3**] My father talked sparingly about his experiences in the RAF, skirting questions about his participation

Figure P.2: Roy Kiyooka. *StoneDGloves*. 1970.

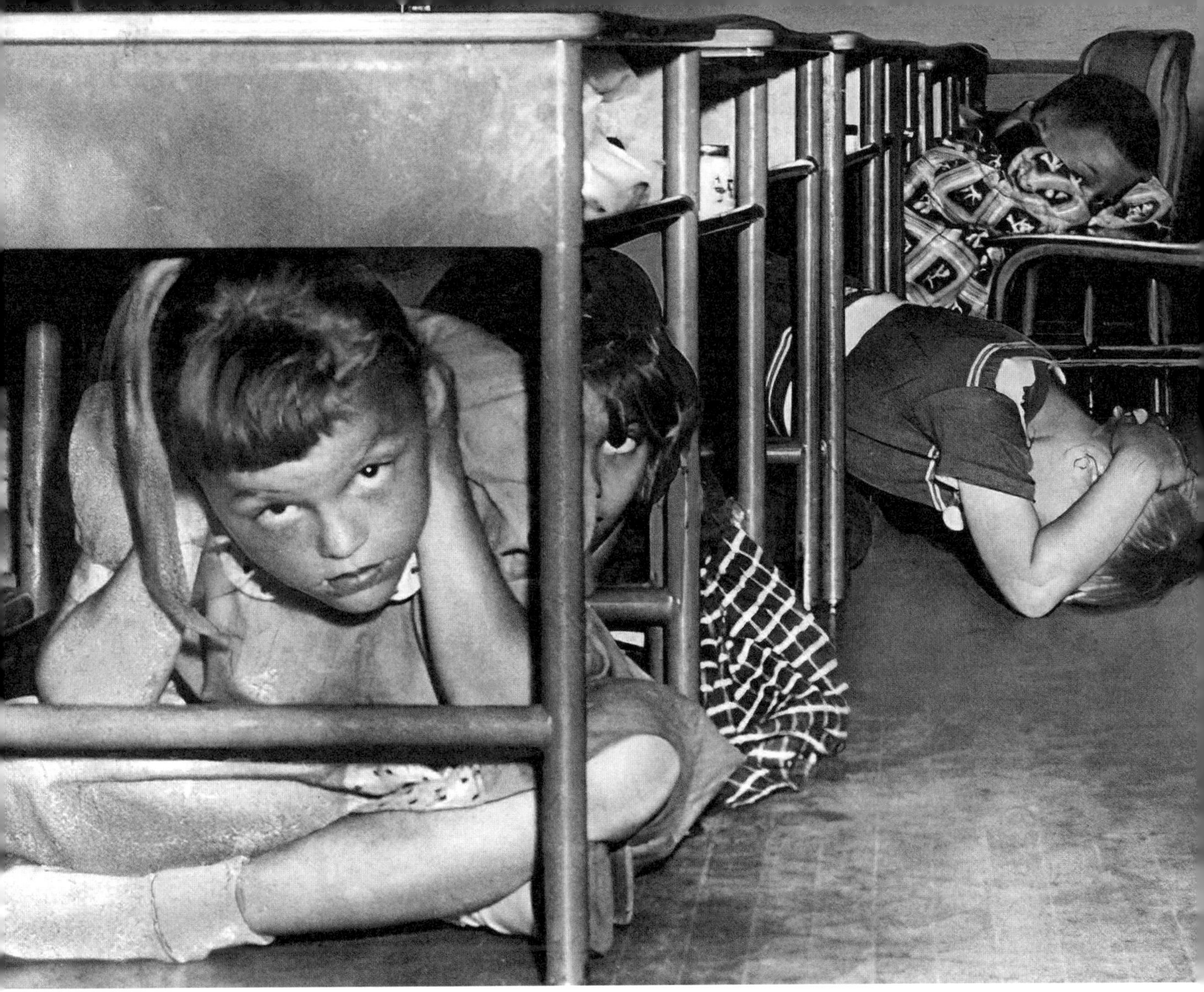

| **Figure P.3**: Unknown. *Duck-and-Cover Drill in School*. c. 1950.

in the Battle of Britain and his postwar assignments to the United States and Germany. When he died, I learned more from his obituaries in British and Canadian newspapers than I did from any conversation I had with him.[2] One of his postings was as commanding officer of RAF Leuchars in Fife, Scotland, the first RAF supersonic base equipped with fighter planes capable of intercepting incoming Soviet bombers armed with nuclear weapons. We lived in staff quarters just beyond the perimeter of the base, and I can still remember the sound of aircraft taking off and landing. I could determine whether they were coming or going by the telltale pitch of their engines, like the tick-tock of the clock in the crocodile's stomach in *Peter Pan*. But I cannot remember forming any idea of what the planes were for.

In a similar way, when I was at Trinity College School in Port Hope, Ontario, as a boarder in the late 1950s, and Eldorado Mining and Refining (now Cameco Corporation) was processing uranium ore on the town's waterfront, I did not think to ask what the factory

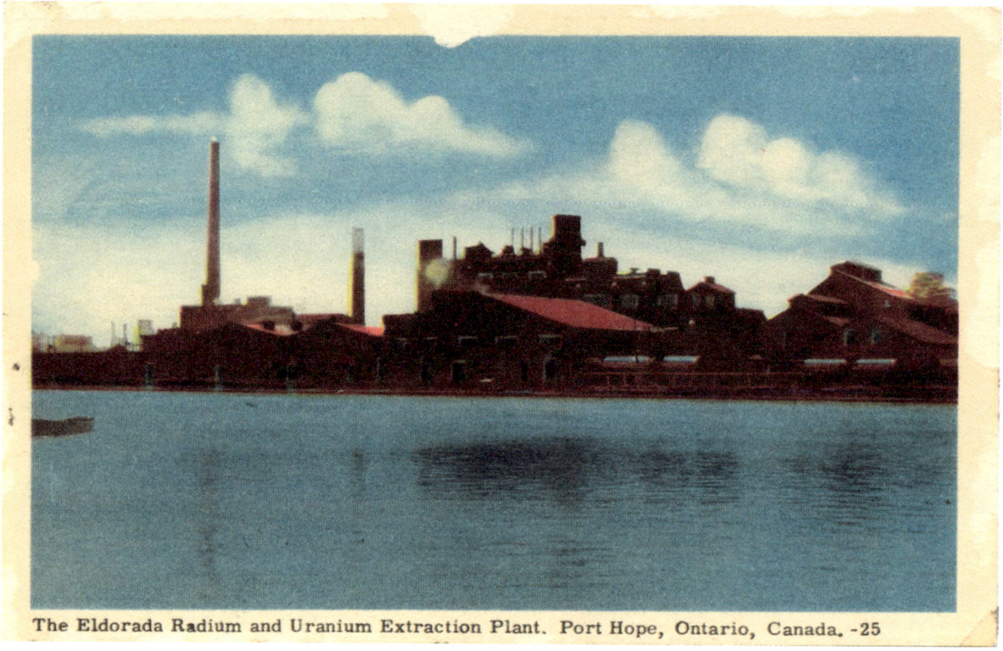

Figure P.4: Postcard. *The Eldorado Radium and Uranium Extraction Plant, Port Hope, Ontario, Canada*. c. 1950.

was doing. I failed to notice, for a second time, that I was on the periphery of a nuclear site. As the "oldest extant nuclear facility in the world," Eldorado was producing uranium dioxide and uranium hexafluoride for shipment to the United States under military contract.[3] The company was a pioneer in the secret as well as the material sides of the atom, in what the fiction writer Philip Pullman calls "uranium atomcraft."[4] The factory appears on a postcard from about 1950 that glows with unnatural blues and reds so intense they might well have been manufactured in an experimental corner of the plant. [**Figure P.4**] In later chapters, I will have more to say about atomic postcards and how they function to domesticate nuclear threat.

Canada's first nuclear town, as Port Hope once named itself, was also releasing radionuclides into the environment.[5] The spread of contamination eventually resulted in the most costly environmental cleanup by a municipality in Canadian history, a cleanup still ongoing, sounding a public alarm about the problem that radioactive waste presents for society. During the construction of St. Mary's school in the town, Eldorado donated sandy fill for the cafeteria, gymnasium, and parking lot. In 1975, when the fill was found to contain levels of radon gas that were as high as those in a uranium mine, the school was evacuated and shut down. This signalled that much of the town was toxic.[6] In the photograph *St. Mary's Elementary School,* Port Hope, Ontario, taken from a high angle by Robert Del Tredici after the school had been decontaminated and reopened, children run in the playground as if the school had never been exposed to the gas produced by the radioactive fill. [**Figure P.5**] The height

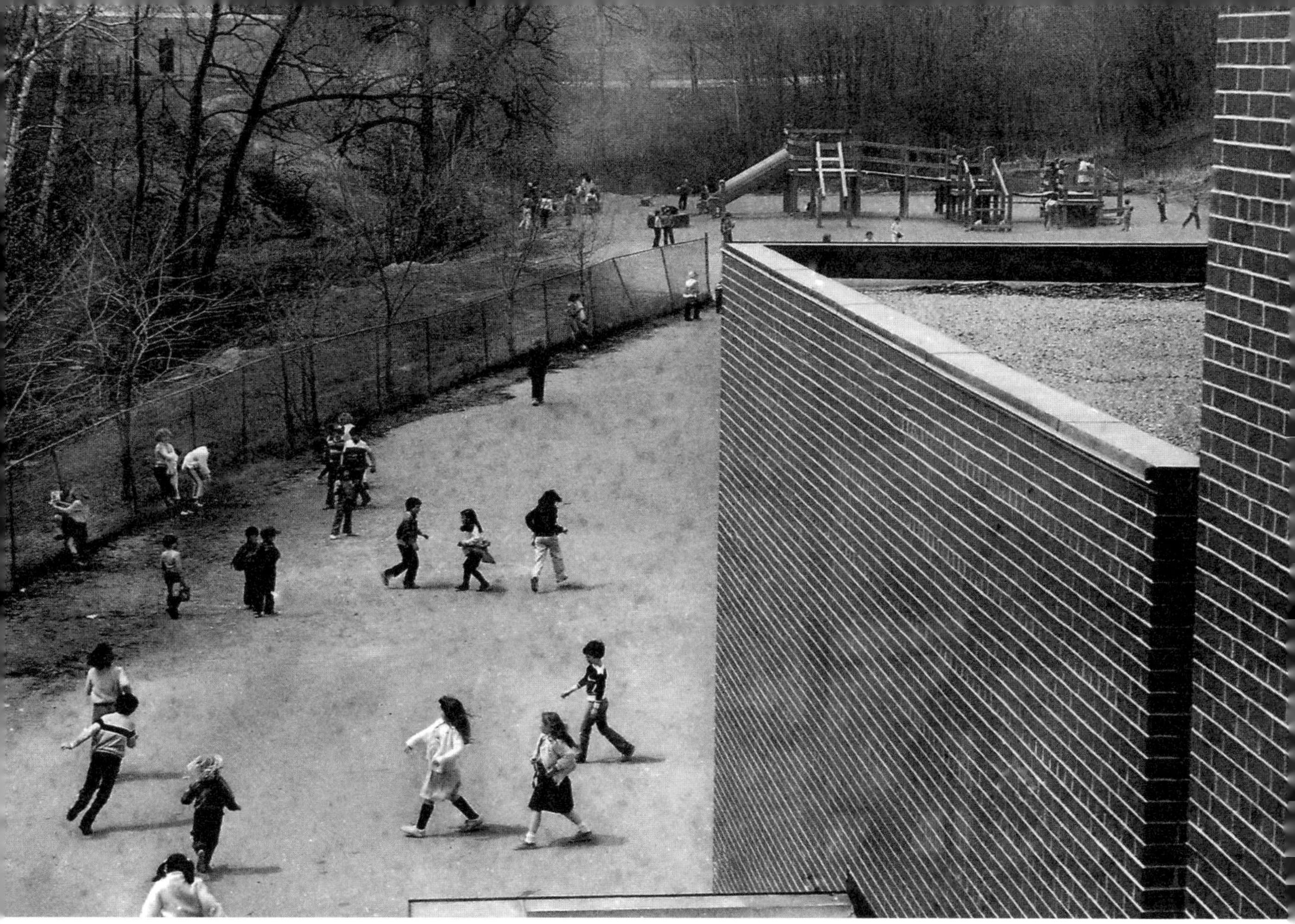

| **Figure P.4**: Robert Del Tredici. *St. Mary's Elementary School.* 1982.

from which the photograph is shot allows Del Tredici to bifurcate his composition. On one side is the hard, immobile brick wall of the school building, on the other the movements of children at play. The inorganic presses against the organic. For those who are aware of the contamination, the image underscores the memory of what happened. The individuals in the school had been exposed to radiation, exposed to an act of "slow violence," to use a term coined by the environmental scholar Rob Nixon.[7]

My childhood brushes with nuclear activities lay dormant for decades. Unlike the time traveller in Chris Marker's 1962 film *La Jetée* or the cultural theorist Paul Virilio, who was obsessed with bunker ruins, I was not haunted by recurring images of disaster.[8] I did not cover my face with my hands to block out premonitions of catastrophe, as the woman does on the viewing platform of the airport terminal in *La Jetée*. [**Figure P.6**] But if a lack in the past makes a claim on the present, as the artist Mary Kelly insists, it was only a matter of time before images of the bomb and nuclear energy caught up with me.[9] Part of what interests Kelly in her feminist practice and me in my research on nuclear representation is the gap between generations. This gap also interests the political theorist Hannah Arendt, who distinguishes between her pre-nuclear generation and that of Kelly and me in her book *On*

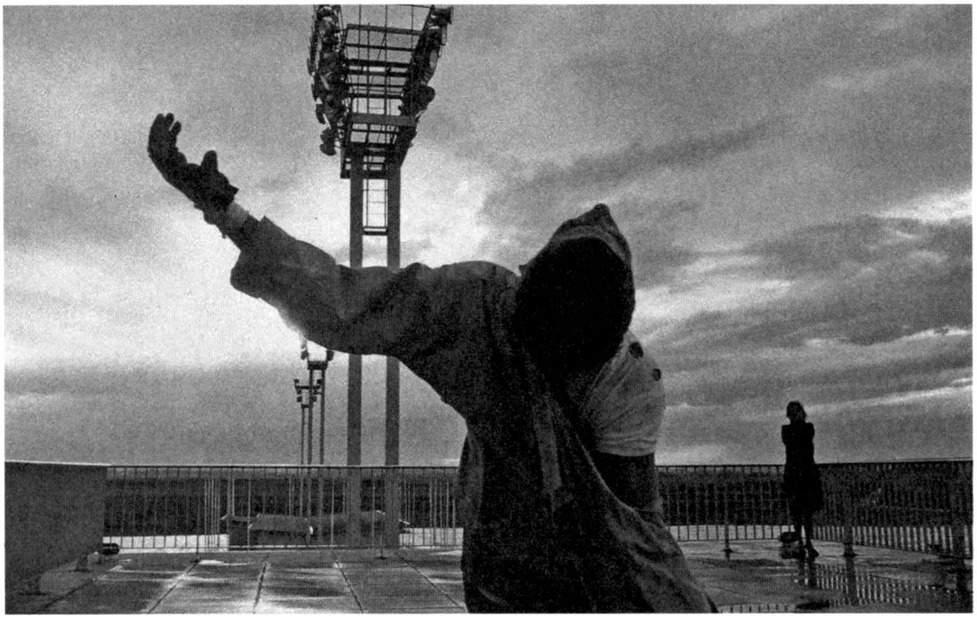

| **Figure P.6**: Publicity still from *La Jetée*. 1962.

Violence. "The possibility of doomsday," Arendt observes, was our generation's "first decisive experience in the world."[10] Those born during the Second World War – who were alive during the bombing of Hiroshima and Nagasaki – are psychologically disposed to return to sites of nuclear catastrophe.

In my travels through the nuclear era, I have learned that atomic flashpoints should not be emphasized at the expense of the everyday activities of the nuclear industrial complex. The nuclear complex has seeped into almost every corner of Canadian life.[11] By bringing the words "bomb" and "wilderness" together in the title of this book, I want to draw attention to Canada's nuclear history. Canada is not terra nullius, a place where nothing happens, though it is often described that way. The First Nations have inhabited the land for thousands of years. Their occupation includes the territory around Port Radium and Great Bear Lake, the traditional lands of the Sahtu Dene people, from which Eldorado shipped uranium ore to its refinery in Port Hope. Like nuclear facilities in Los Alamos, New Mexico, the Port Hope plant has not only been an environmental polluter because of its mass production of uranium dioxide and uranium hexafluoride, but it has also been a major employer in the town. Society's relationship to nuclear weapons and nuclear energy is rarely straightforward.

This relationship is also explored in *Through Post-Atomic Eyes,* an anthology that I edited with Claudette Lauzon.[12] It brings together the nuclear imagery of artists and photographers with the writings of scholars who specialize in quantum physics, military geography, anthropology, big data, art history, performance studies, and forensic architecture.[13] These artists and scholars share our conviction that atomic culture since Hiroshima and Nagasaki has been conditioned by matters of visuality. At the heart of the two volumes lies this question: What might it mean to see the world through post-atomic eyes that have been transformed by Hiroshima and Nagasaki?

Photography has the capacity to shape and reshape perceptions of the nuclear enterprise by revealing hidden activities and risks. *StoneDGloves* by Kiyooka and *St. Mary's Elementary School* by Del Tredici both help to alter nuclear perceptions. I am not suggesting that by throwing light on what has been obscured, by revealing the disfigurements in modern life since the development of the atomic bomb, photography can repair what has been misshapen.[14] Bringing nuclear issues to light cannot mend what has been damaged. Nor am I suggesting that photography can operate outside history and the nuclear events it portrays; photography, like the bomb, is an agent of history. To build more powerful weapons, the United States and other countries with the bomb developed faster cameras that aided the scientific measurement of blast yields. Better cameras led to better photographs, and better photographs led to better bombs. Photography has consistently haunted the nuclear imagination by disclosing what has been masked.

THE BOMB IN THE WILDERNESS

Photography and the Nuclear Era in Canada

JOHN O'BRIAN

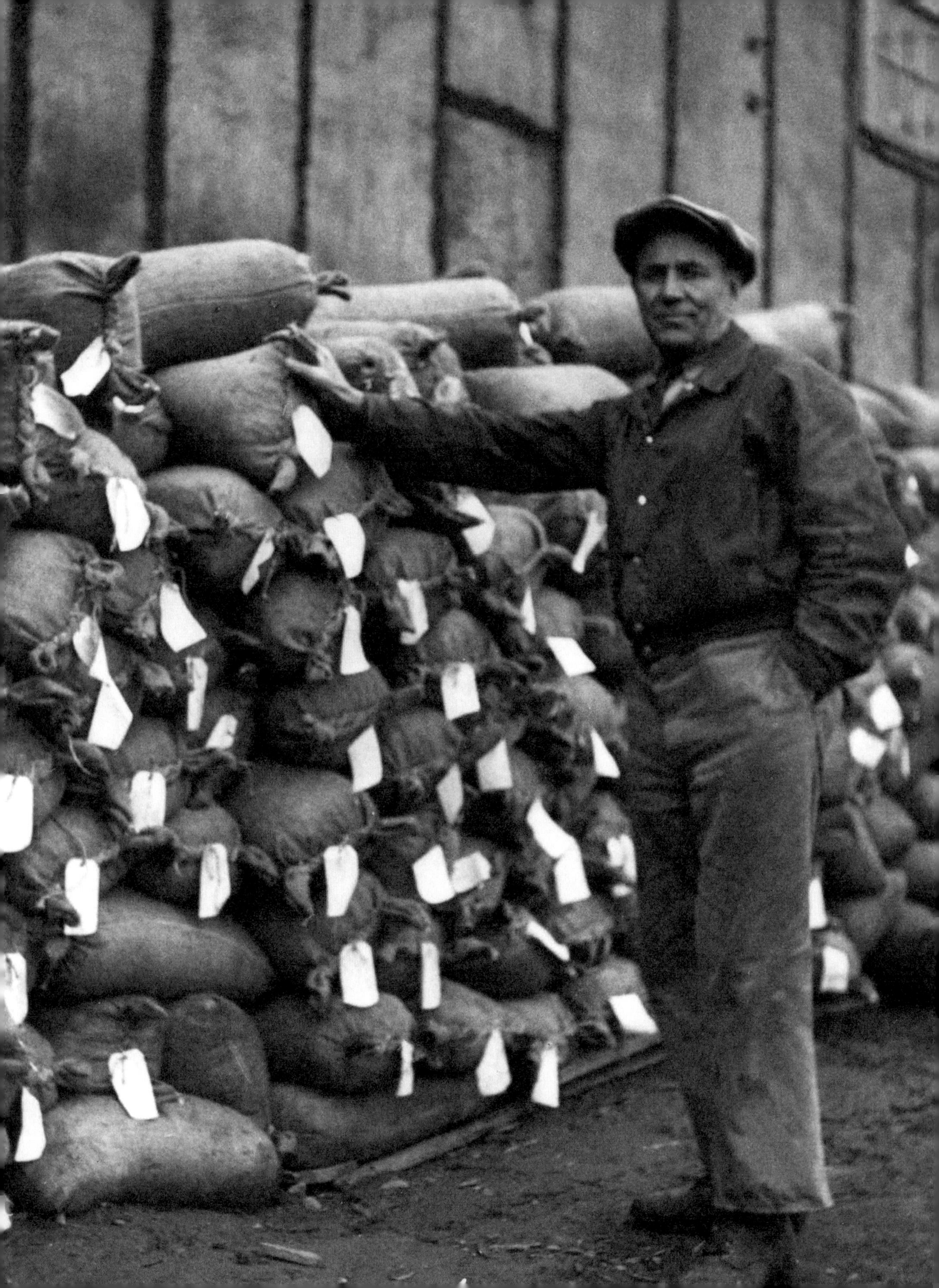

INTRODUCTION

Still Photographs

PHOTOGRAPHS MARK EVENTS.[1] CONSIDER IMAGES OF HIROSHIMA AND Nagasaki, Bikini Atoll and the Nevada Test Site, Chernobyl and Fukushima, Port Radium and Port Hope, Chalk River and Bruce Power. The bombings of Hiroshima and Nagasaki are part of our collective memory less because of books written about the devastation they caused, or because of witness accounts provided by the hibakusha, than because of a visual archive of emblematic photographs.[21] The enriched uranium and plutonium in the atomic bombs came from ore mined at Port Radium in the Northwest Territories that had been refined at Port Hope in Ontario.[3] Photographs exist of both Canadian sites. The filmmaker Richard Finnie photographed an employee of Eldorado Mining and Refining, which owned the Port Radium mine, resting his hand on sacks of pitchblende concentrate rich in uranium.[4] [**Figure I.1**] The image catches the eye not only because of the subject matter but also because of the repetitive pattern produced by the white labels on the burlap sacks. The sacks are waiting to be transported across Great Bear Lake and then south to Alberta for further shipment by rail eastward to Port Hope. The Sahtu Dene people, living in the village of Déline on the western shore of the lake, were employed as carriers of the ore without knowing what it was for. Some slept on the sacks.

Figure I.1: Richard Finnie. *Sacks of Pitchblende Concentrate, Port Radium.* 1939.

Where nuclear activities have occurred, photographers have often been there to frame them. They have been there to record what happens and sometimes to assist in what happens. During the Cold War, they worked primarily in black-and-white rather than colour. This was for professional, technical, and aesthetic reasons, and is reflected in the large majority of examples reproduced in this volume. The Swiss American photographer Robert Frank wrote that "black and white is the vision of hope and despair."[5] It offers nuanced gradations of tone and texture, and carries more emotional weight than colour. It has gravitas. But black-and-white was also chosen for gendered reasons. Especially in the early decades of the Cold War, colour was coded as feminine.[6] The masculinization of nuclear technologies called for black-and-white.

An extensive literature discusses the nuclear era, especially as it relates to the United States, the only nation so far to have used atomic weapons in war.[7] New books and articles appear daily, many containing photographs.[8] But few reflect on nuclear optics, on the mediating role of photography in forming an image of the bomb and commercial nuclear energy. Two exceptions are *The Nuclear Culture Source Book,* edited by Ele Carpenter, which focuses on contemporary visual art, and *Invisible Colors: The Arts of the Atomic Age,* by Gabrielle Decamous, which brings historical and contemporary nuclear art together.[9] Even fewer books reflect on Canada's role as a nuclear power, never mind providing a critical perspective on it. *Canada and the Cold War,* by Reg Whitaker and Steve Hewitt, is one such inquiry, and Peter C. van Wyck's *Highway of the Atom* is another.[10] Both have contributed to this study. The former addresses significant aspects of the Cold War in the public life of Canada, using press photographs as visual documentation; the latter brings archival information and story together to produce narratives about Port Radium, Déline, and the Sahtu Dene people. It includes photographs as well, some by the author.

Whitaker and Hewitt note that Canada, as a member of the Manhattan Project, the bomb-building consortium led by the United States, constructed the first nuclear reactor located outside the United States. The reactor was "born violent," its purpose to produce plutonium for nuclear weapons.[11] It became operational in August 1945 at Chalk River, Ontario, a few weeks after the Hiroshima and Nagasaki bombings. The heavy water technology employed in the design was transferred in following decades to the CANDU reactor program, and Consolidating Mining and Smelting Company of Canada in Trail, British Columbia, produced some of the heavy water required to run the reactors. Chalk River was also the site of another first: the first major reactor accident in the world, on December 12, 1952, when the NRX (National Research Experimental) reactor experienced a partial meltdown. (Future president Jimmy Carter, a naval officer with the US nuclear submarine program, participated in the cleanup and suffered radiation poisoning.) In addition, Chalk River Laboratories pioneered the development of medical isotopes as well as the management of radioactive waste. In 1974, at Pinawa, Manitoba, it built one of the earliest test facilities for high-level nuclear waste disposal, though the site was never used.[12] More recently, in 2012, Bruce Power's group of reactors on the eastern shores of Lake Huron in Ontario became the largest nuclear plant in the world.[13]

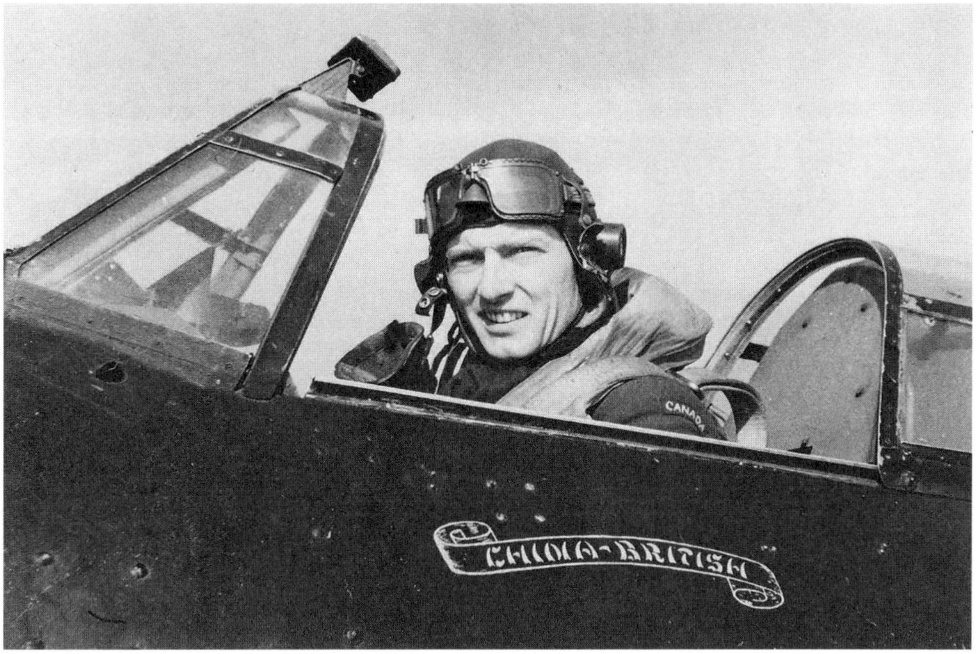

Figure I.2: Royal Air Force. *Peter O'Brian Seated in the Cockpit of a Hawker Hurricane.* 1940.

Van Wyck is more skeptical than Whitaker and Hewitt about the ability of language and photography to convey adequately what occurs during nuclear events. He questions whether it is possible to grasp what happened at Port Radium and Déline, either by visiting the mine and village or by looking into the past. His rhetorical strategy in *Highway of the Atom* is to unite "theory" with "story."[14] Testimony and narrative can register events forcefully, he says, but they cannot always clarify matters. He quotes the philosopher Jacques Derrida: "There is no testimony that does not at least structurally imply in itself the possibility of fiction, simulacra, dissimulation, lies, and perjury."[15] In other words, stories are slippery. Van Wyck, like Derrida, insists on the fragmentary nature of testimony and distrusts claims to certainty. The writer and scholar Julie Salverson, who accompanied van Wyck on his travels along the atomic highway, also distrusts claims to certainty.[16] She expresses her doubts and anxieties in an atomic memoir, *Lines of Flight.* The spectre of nuclear winter haunts its pages.

The plot lines of intersecting photographic and verbal inscriptions on nuclear matters tend to emerge from the underbrush of history as much as from the examined spaces of major atomic events. Photography, like language, does some of its most telling work in the recesses of historical space and time. Susan Sontag argues in her critical writings that photography fails to provide the narrative structure or interpretive coherence offered by language. Photographs need to be supplemented with texts and captions to produce understanding, she insists, whereas language can speak to disagreement and build understanding without photographs.[17] On its own, she would argue, photography does not have the expressive power to convey the pain of the

Sahtu Dene people, the hibakusha in Hiroshima and Nagasaki, or the survivors of Fukushima Daiichi.[18] But there are also constraints and contingencies to language. Shortly after publishing *Precarious Life: Powers of Mourning and Violence* in 2004, Judith Butler wrote a response to Sontag's ideas on photography. Her essay "Photography, War, Outrage" questions Sontag's separation of the work of photography from the work of writing. When Sontag "faults photography for not being writing," on the grounds that it lacks narrative continuity, she fails to grasp how it works.[19] Photographers build interpretation by choosing one scene over another, Butler suggests, by framing what they are looking at and producing visual narratives. The chosen frame carries their purpose, and that purpose is implemented through the frame. Butler cautions against privileging language over photography, a position that strikes me as correct.

Take two photographs of pilots in the cockpits of their planes, one well known, the other little known. [**Figs. I.2 and I.3**] Now imagine that W.G. Sebald, the author of *Rings of Saturn* and *Austerlitz* as well as of accounts of bombing campaigns conducted during the Second World War, came across the less-known example in an English thrift store.[20] During an interview about photography, Sebald mentioned that he often visited thrift stores and spent time sifting through boxes of discarded photos.[21] Some of the prints he purchased on his thrift store visits are published in his books, along with snapshots he took himself. Sebald had a sophisticated understanding of photography as a medium, commenting that many photographs lead a "nomadic existence" until they are "rescued."[22] He recognized that photography is inextricably bound up with modernity as well as with language. It is a phenomenon that inhabits almost every aspect of modern life, from birth to death, war to peace, art to advertising, theoretical physics to applied physics, uranium mining to atomic-bomb building (my list, not Sebald's).

Sebald put great store in his subjective response to photographs as well as in the stories that emerged from them. "Thinking hypothetically" was his term for the process.[23] So it is reasonable to ask how he might have responded to the lesser-known image. Photographs of air force pilots sitting in the cockpits of their planes are a dime a dozen in thrift stores, and Sebald might have skipped over it and continued to look through the box of loose prints. Like so many genres of photography – *cartes-de-visites*, wedding pictures, snaps of family reunions – photographs of posed military personnel are all the same and all different.[24] For Sebald to have paused over the image, a generic picture of a pilot wearing a leather helmet at the controls of his plane, something would have had to catch his eye: the inscription "China-British" painted on the fuselage, perhaps, which was the kind of detail that led the author into literary detours – Sebald's literary detours in his novels are celebrated – or, maybe, the label "Canada" sewn on the shoulder of the uniform. As a chronicler of war, Sebald might have wondered about the message conveyed by the words China, British, and Canada in close proximity to one another. He would not have wondered, however, about what kind of

| **Figure I.3**: US Air Force. *Paul Tibbets Waving from the Cockpit of the Enola Gay*. 1945.

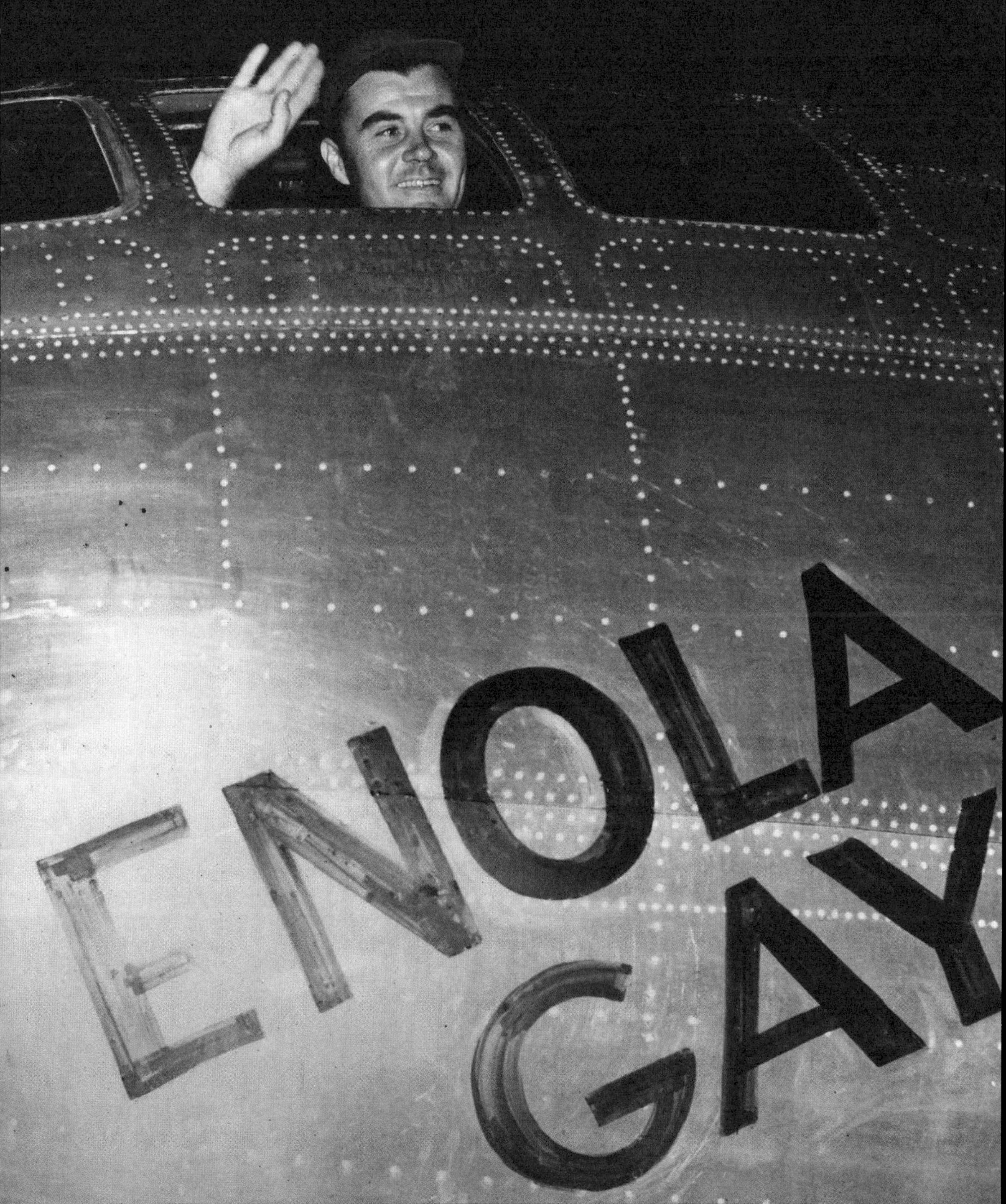
ENOLA
GAY

photographer took the picture. The sharp focus, the positioning of the camera with the sun behind it, and the tight composition announce that it was a professional.

Unless it was written on the back of the print, Sebald could not have discovered why it was taken or the name of the pilot. The absence of this information might have been the beginning of a story for him. He might have seen the picture as representing a missing person, someone no longer alive. Like Roland Barthes in his last book, *Camera Lucida,* Sebald was emotionally susceptible to the stories that arise from photographs at the same time as he was conscious – acutely conscious – of their unreliability.[25] He did not privilege photography over language, or vice versa.

The little-known photo was taken in 1940, shortly after the Battle of Britain, and the well-known one in 1945, a few days before the end of the Second World War. The aircraft are a Hawker Hurricane fighter plane, photographed in daylight, and a B-29 bomber, shot at night using a flash that lights up the words "Enola Gay", the name of the pilot's mother. The primary weapons on board are, respectively, a machine gun and an atomic bomb. The planes are piloted by my father, a Canadian officer in the Royal Air Force, and Paul Tibbets, an American officer in the US Air Force. They are on the same side of the conflict, one in the European theatre (notwithstanding the "China-British" inscription indicating that construction of the aircraft was supported by expatriates) and one in the Pacific theatre.[26] When the pictures were taken, both pilots had been assigned their targets: the German Luftwaffe, on the one hand, and the Japanese city of Hiroshima, on the other. The difference between the targets and the weapons used to destroy them is the difference between one era and another, between the pre-nuclear era and the post-Hiroshima era. The industrial technologies of the pre-nuclear era could be more or less controlled; the post-atomic technologies of nuclear destruction could not. As Rey Chow observes in *The Age of the World Target,* the atomic bomb makes a target of everyone.[27]

All statements about nuclear events are inherently unstable, Derrida declares in his article "No Apocalypse, Not Now (Full Speed Ahead, Seven Missiles, Seven Missives)." He argues that nuclear activities and threats are always rhetorical in structure. Drawing on Plato, he refers to nuclear conflict as a *pharmakon,* something that has the ability to both injure and cure.[28] Radiation can be beneficial as well as harmful; it can cause cancer and it can fight cancer. Regardless of which side of the radiological equation is being discussed, the other side is always present, as demonstrated in Chris Lund's photograph *Radium Treatment of Lip Cancer, Toronto General Hospital,* where the tools of the cure look as brutal as the disease. [**Figure I.4**] In this close-up image of a cancer patient's face during an operation, the upper and lower lips are taped down, while the gloved hands of the surgeon hold instruments of repair. A label in the lower left corner reflects the high cost of radium at the time: "Caution, Radium, Do Not Throw Away." The two sides of the equation are joined together in the photograph. The positive and the negative, as Derrida says, cannot be separated.

A 1955 photo-essay in *Life* magazine called "Science Tackles Radiation Peril" offers another example of the pharmakon at work.[29] [**Figure I.5**] An overview of the physiological effects

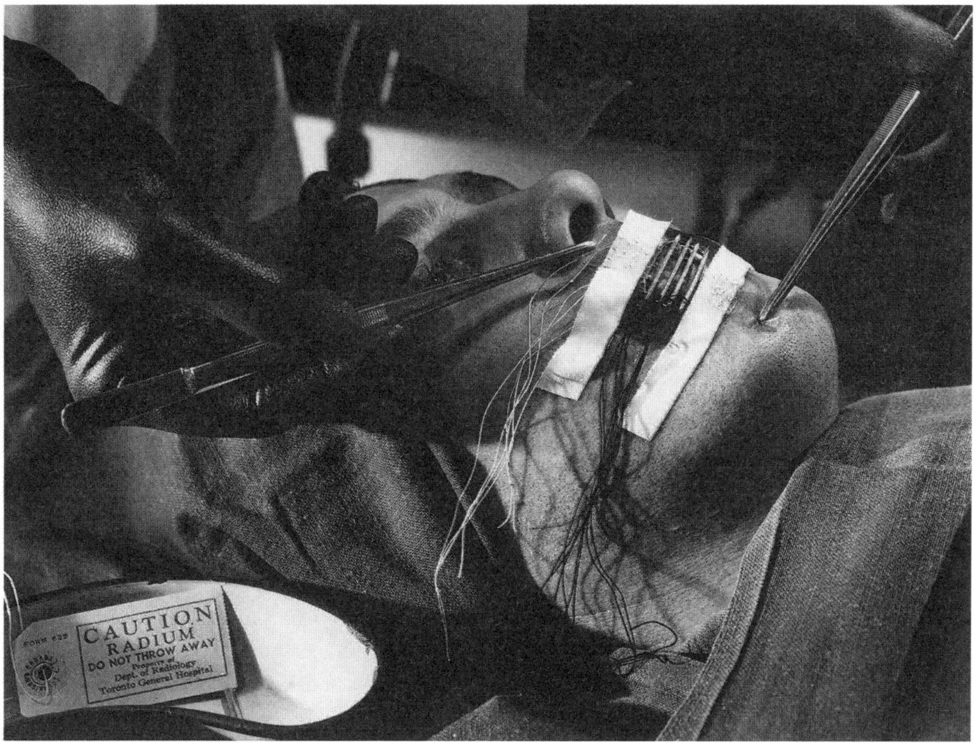

Figure I.4: Chris Lund. *Radium Treatment of Lip Cancer, Toronto General Hospital.* 1949.

of atomic weapons testing, it incorporates statements solicited from scientists who knew about Exercise Desert Rock in the 1950s, the program in which American and Canadian soldiers were ordered to advance across a simulated battlefield toward ground zero during nuclear detonations at the Nevada Test Site. They were not issued with protective gear, so they absorbed large doses of radiation.[30] In "Science Tackles Radiation Peril," Alfred H. Sturtevant, a geneticist, claims that "the last H-bomb test alone probably produced more than 70 human mutations which are likely to produce large numbers of defective individuals in the future," whereas Gioacchino Failla, a physicist, states that "the question of how many H-bombs can safely be tested is irrelevant. To remain free, we must develop powerful nuclear bombs. We must continue our test program."[31] The most striking image in the photo-essay is of another scientist, Theodore Tahmisian, who studies the effects of radiation on grasshoppers. With his deeply shadowed eye sockets, he seems hypnotized by the insects crawling around him, as if he might soon be overwhelmed by them, reinforcing the trope of the mad scientist overcome by his research. The photograph of Tahmisian and his grasshoppers occupies almost half the page.

The statements by Sturtevant and Failla, which seem strongly at odds with one another, prove to be interdependent. They are joined together in the rhetorical structure that, as Derrida asserts, conditions all nuclear discourse. Failla's argument that atomic weapons are beneficial

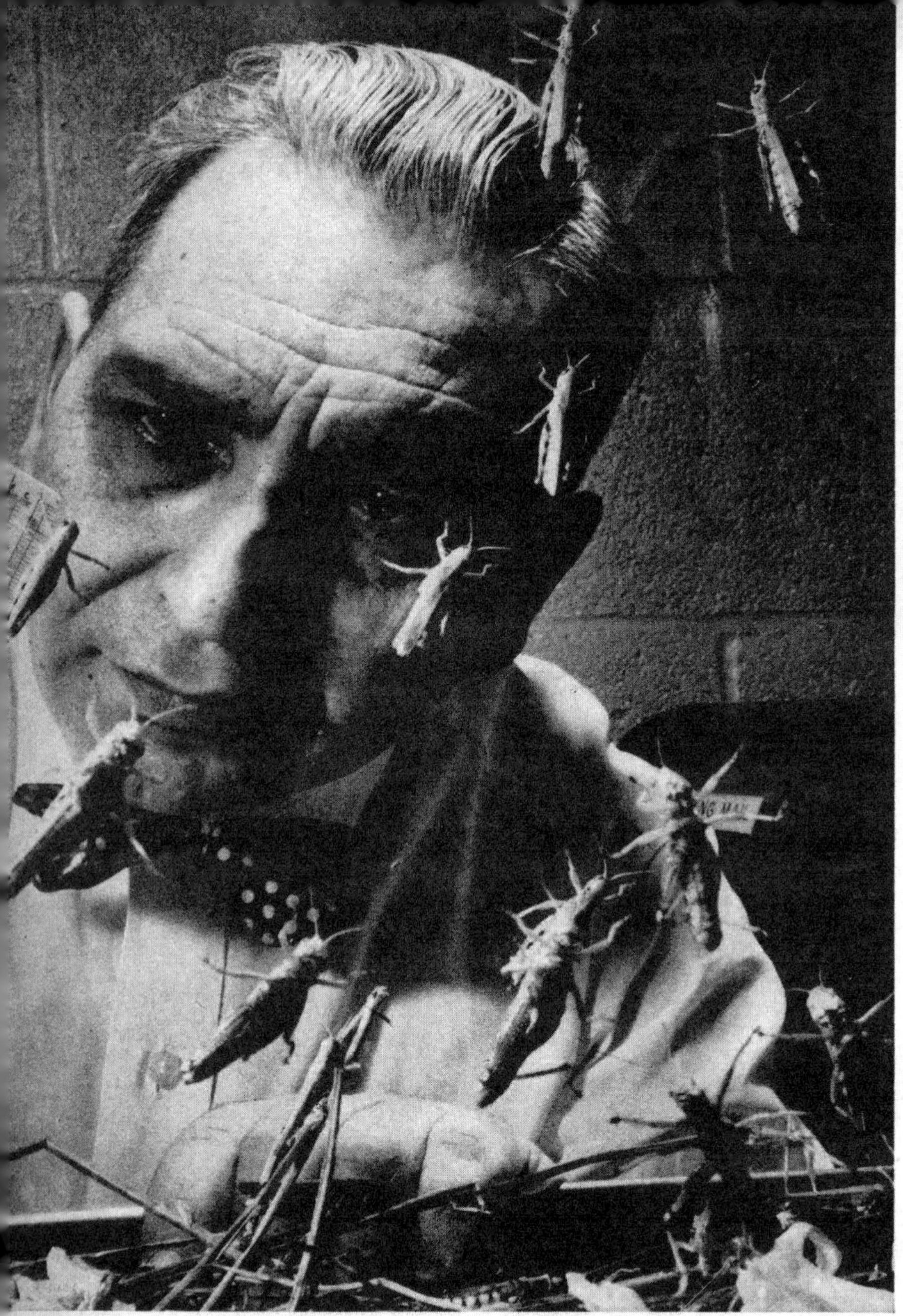

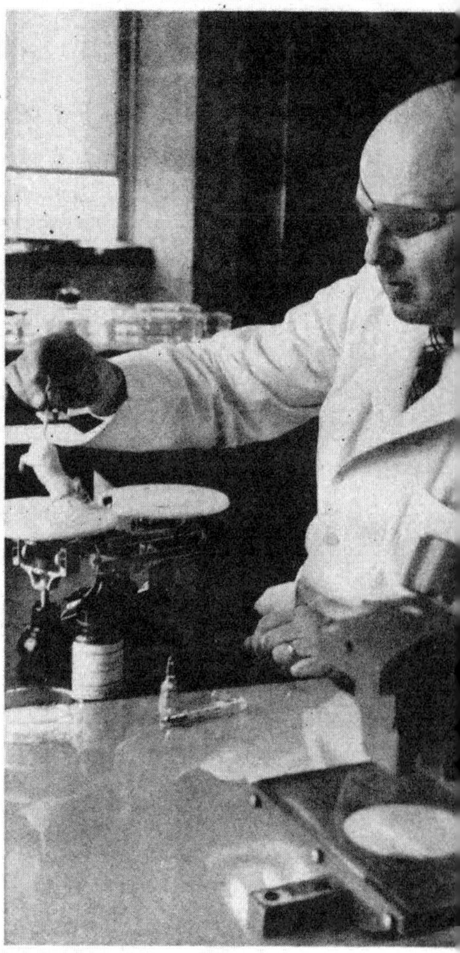

HOPEFUL DRUG is about to be injected into a mouse being weighed at Columbia University laboratory. Drug, beta-mercaptoethylamine, is one of several which raise animals' resistance to radiation.

← **PROVOCATIVE DISCOVERY** was made by Theodore Tahmisian, who found that the effects of radiation on grasshopper eggs (produced by colony, *left*) do not show up while the eggs are in hibernation.

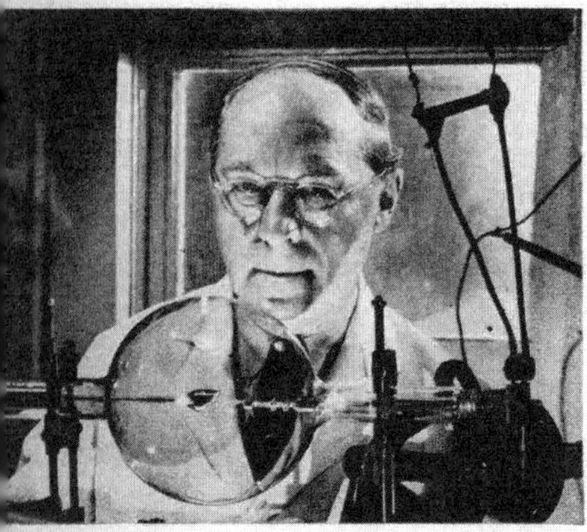

NOBEL PRIZE-WINNER Hermann J. Muller, a geneticist: "Atomic warfare may cause as much genetic damage, spread out over future generations, as the direct harm done to the generation exposed."

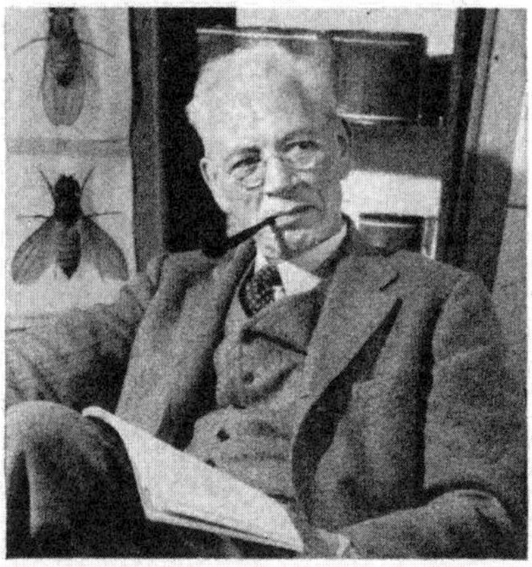

GENETICIST Alfred H. Sturtevant says, "The last H-bomb test alone probably produced more than 70 human mutations which are likely to produce large numbers of defective individuals in the future."

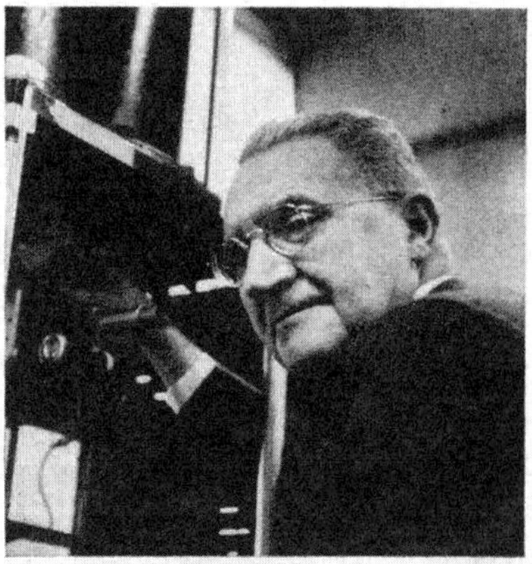

PHYSICIST Gioacchino Failla: "The question of how many H-bombs can safely be exploded is irrelevant. To remain free we must develop powerful nuclear bombs. We must continue our test program."

CONTINUED ON NEXT PAGE

is predicated on their capacity to harm. They extend the possibility of "freedom" and political stability, he maintains, through their ability to annihilate enemy armies and enemy states. They are a deterrent. Sturtevant's comment that nuclear weapons are destructive is also predicated on their capacity to harm. They can cause human mutations and can also annihilate armies and states. Deterrence might fail, he says, and the world could be destroyed.[32] Nuclear weapons, according to Derrida, are "the absolute *pharmakon*."[33]

Methodologically, this book occupies a position somewhere between the critical history offered by Whitaker and Hewitt in *Canada and the Cold War* and the deconstructive stories narrated by van Wyck in *Highway of the Atom*. Much of my research has been undertaken in institutional archives in Canada and elsewhere, and I am indebted to the historical documentation and photographs provided by them.[34] But I am less likely than Whitaker and Hewitt to see nuclear events in the past as retrievable. In a frequently cited passage, Walter Benjamin writes that "articulating the past historically does not mean recognizing 'the way it really was.'"[35] Determining the way it was is an impossibility. There is no single version of the past, he is saying, only multiple versions. In another passage, Benjamin observes that because history tends to break down into images, "there can be no photographic image, no force of arrestment, which does not tell us of the relation between images and history, photography and memory, and space and violence."[36] Photography, history, memory, and violence are part of the same continuum. They cannot be pulled apart. If Sebald and Barthes focused on subjective memories produced by photographs, the sociologist Maurice Halbwachs concentrated on shared memories. He coined the term "collective memory," suggesting that memory resides in group consciousness more than in individual consciousness.[37] Collective memory exists outside the individual; photographs of Hiroshima and Nagasaki are part of a shared consciousness.

This book began as a series of questions. In what ways, I wanted to know, has photography contributed (or not) to an understanding of Canada as a nuclear nation? How has Canada's engagement with the ongoing nuclear era been visualized? What sort of image, for example, is a US military photograph of Test Grable taken at the Nevada Test Site in 1953, in which Canadian troops were exposed to radiation? [**Figure I.6**] In what ways do photographs make the invisible visible? Has photography helped or hindered our comprehension of the nuclear era and nuclear risk? What are its limitations in representing nuclear events? How are the technologies of photography and nuclear fission connected? How do photography and language work together to produce nuclear narratives and construct public memory?

I use these questions to guide the chapters of this book, which are organized as intersecting case studies. In the next chapter, for example, I examine the activities of the nuclear industry and ask if photography can make visible its operations. Robert Del Tredici's book *At Work in the Fields of the Bomb* is the centrepiece of the chapter. Combining text and image, it traces the intricate links between the nuclear industry's far-flung activities. Del Tredici's

| **Figure I.5**: Page from "Science Tackles Radiation Peril," *Life*. 1955.

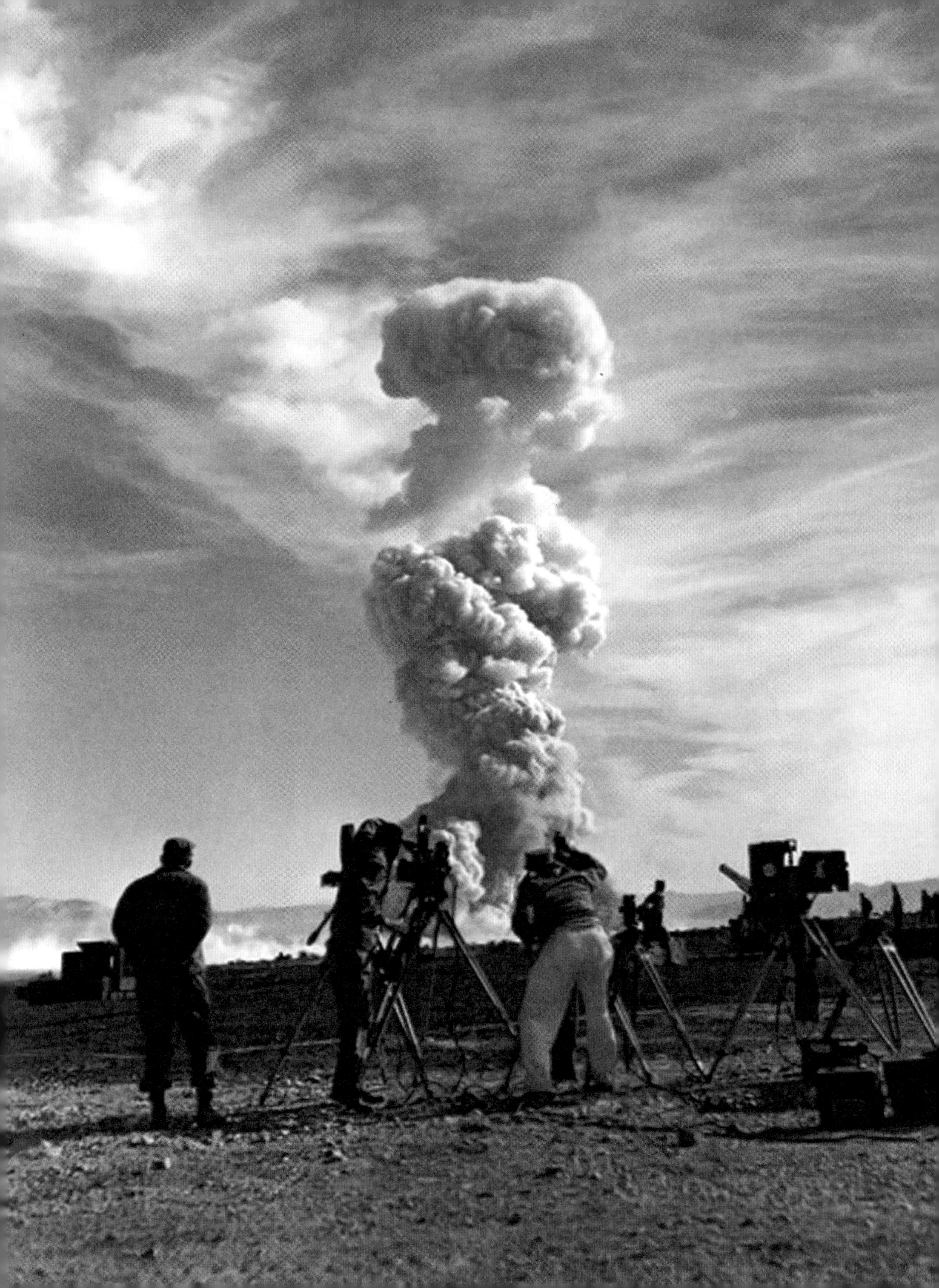

revelations also inform Chapter 2, which focuses on radiological accidents that occurred at the Chalk River nuclear complex and elsewhere, most of which have been minimized by authorities.

Few aspects of the nuclear environment, I argue, have escaped the camera's gaze. The camera possesses the power of deferred sight, the uncanny ability to make seeable what cannot always be seen at the time of an event.[38] The nuclear optics of the visible and the invisible are addressed by Akira Mizuta Lippit in *Atomic Light (Shadow Optics)*, which contends that some things will always remain "profoundly unseen."[39] Still, the physicist Ernest Rutherford used strips of film in his radiation experiments at Montreal's McGill University in the early 1900s to differentiate between invisible alpha and beta particles, producing research that earned him the Nobel Prize.[40] He also developed the concept of radioactive half-life, defined as the time it takes for one-half of the radioactive nuclei of a given element to disintegrate. Engineers and miners at Port Radium used film to find uranium by means of radioactive emission, which left its trace on photosensitive surfaces, rendering visible what otherwise could not be seen.[41] The physicist J.H. Webb at Eastman Kodak discovered the existence of atomic fallout as an atmospheric phenomenon when he tracked radiation that had damaged film in contaminated packaging in Indiana back to Alamogordo, New Mexico, the site of the first nuclear test.[42]

Issues of visibility and invisibility are central to the relationship of photography and nuclear technologies. Albert Einstein's 1905 mass–energy conversion equation, $E^2 = p^2c^2 + m^2c^4$, where "p" is the momentum of the particle, is fundamental not only to nuclear science but also to photography. Because light carries energy, when silver bromide grains in film are struck by light the electric charge of silver shifts to produce indexical images. Photography can provide, as no other technology can provide, quantitative and qualitative documentation on nuclear tests; it is an indispensable means of analyzing the effects of explosions. Photography is "the basic instrument for accurate measurements of movement as a function of time," writes Peter Kuran, director of several documentaries on the nuclear era, expressing himself in language that is strikingly similar to that of Eadweard Muybridge when he described his photographic experiments on motion in the late nineteenth century.[43] Muybridge's experiments pointed the way toward motion pictures, notes his biographer Rebecca Solnit: "He is the man who split the second, as dramatic and far-reaching an action as the splitting of the atom."[44]

But there are limits to photography, as I have mentioned. Because the medium focuses on the here-and-now, it is ill-suited for recording the slow destruction caused by radiation. It cannot readily depict changes in the atmosphere or in the human body that occur over years and decades. The temporality of radiation and its effects, Julia Bryan-Wilson notes, take place in a different register than photography.[45] Photography also privileges certain kinds of information at the expense of others. It helped make the residue of dust and debris that forms the mushroom cloud a dominant image of the atomic age, to the point that it has sometimes

| **Figure I.6**: US Military. *Photographing Test Grable*. 1953.

eclipsed the threat of the bomb itself. The caption at the bottom of a page torn from a book, kept and stamped by the Toronto Public Library, accompanies a photograph of the Nagasaki mushroom cloud. It states, "The power of the atomic bomb is beyond belief."[46] That is all. It makes no mention of the death and destruction that occurred below the mushroom cloud or of the invisible fallout dispersed across a region and beyond. Photographs of the mushroom cloud have diverted attention away from the real and present dangers posed by nuclear weapons.[47]

The nuclear photographs in this book range from images released by civil and military authorities, to works by Canadian artists, to the visual ephemera of brochures and postcards. Aesthetic hierarchies of photography matter less than content in my choice of images. Atomic photography is found in advertising, fine art, scientific experiments, journalism, medical procedures, civil defence manuals, corporate reports, family snapshots, and popular culture. In 1966, Imperial Oil published a brochure on nuclear nomenclature in cooperation with Atomic Energy of Canada Limited to emphasize its involvement with nuclear energy. [**Figure I.7**] The cover image represents the spent fuel pool of an atomic reactor. When fuel rods are removed from reactors, they are too hot to be moved far and must be placed in cooling pools at least twenty feet deep to protect workers from the radiation being emitted. Taken from directly above the pool, the photo shows symmetrically arranged dark shafts tilting inward. It is so abstract as to be virtually unreadable. The degree of abstraction aestheticizes the dangers posed by the rods, which are highly radioactive and need to be sequestered in fuel pools for up to ten years.[48] In the book, I do not reproduce other images as abstract as the cover of the Imperial Oil brochure, but several of them also aestheticize nuclear issues, while purporting to address arms proliferation, radioactive contamination, and environmental risk.

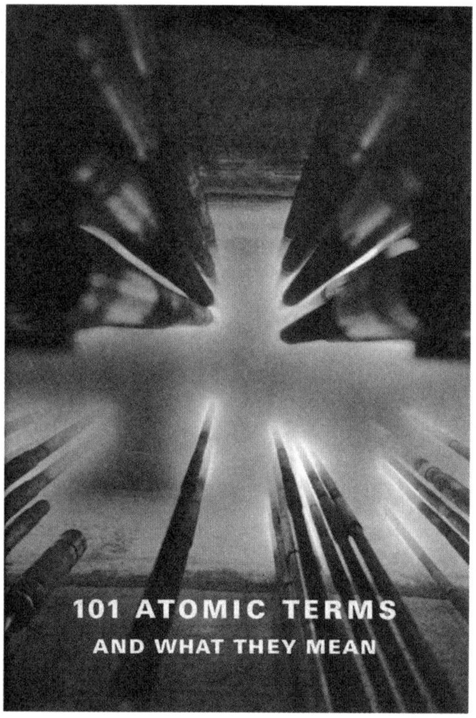

Figure I.7: Imperial Oil Limited. Cover of *101 Atomic Terms and What They Mean*. 1966.

The nuclear shadow that fell across the world following the bombings of Hiroshima and Nagasaki appeared to lift after the Cold War ended in 1989, only to darken again in recent years. The world has now entered a second nuclear age characterized by a growing proliferation of nuclear capabilities, a rising threat of terrorism, and a reconfigured arms race.[49] Although Canada does not possess nuclear weapons of its own, it remains deeply implicated in global nuclear affairs.[50] In the absence of photographs and other lens-based media such as film and video, nuclear phenomena risk being forgotten. I was astonished when the Nuclear Security Summit convened by President Barack Obama in Washington, DC, in early April 2016 faded from public consciousness within weeks, despite dire warnings issued about nuclear terrorists. "If these madmen ever got their hands on a nuclear bomb," the president declared, "it would change the world."[51] His warning might have lodged more sharply in the collective consciousness had a video of a drone releasing a dirty bomb – a weapon combining radioactive materials with conventional explosives – not been withheld from public view. The video was screened behind closed doors and seen only by senior political and military figures, some of whom remarked on its horrific content. In the absence of images, the warning did not register strongly in the public sphere.

When he was a boy in the early 1970s, Arctic sovereignty expert Michael Byers watched the contrails of B-52 bombers above his grandfather's farm in Stoughton, Saskatchewan, as the planes flew north to their advanced staging positions in the Canadian High Arctic.[52] A decade earlier, Robert McNamara, President John F. Kennedy's defense secretary, had informed Congress that in addition to the B-52s flying north, nuclear-tipped Bomarc-B missiles were being supplied to Canada. The missiles would function as decoys in the event of a Soviet attack.[53] They would draw fire away from the United States and turn Canada into a powder monkey (my description, not McNamara's). Canada occupied the main geographical zone between the two superpowers of the Soviet Union and the United States. Spatially, it was the filling in a nuclear sandwich.

Until recently, historians have generally portrayed Canada as little more than a condiment in the sandwich. The arms race, they have contended, was a bipolar contest between the United States and the Soviet Union, in which Canada played a bit part. C.P. Stacey, the official historian of the Canadian Army in the Second World War, comments in *Canada and the Age of Conflict* that "there is no space here to tell in detail the story of Canada's comparatively small involvement in the more-than-epoch-making international drama of atomic energy."[54] Canada had no part in the bombs dropped on Japan, he inaccurately declares.[55]

Stacey's unwillingness to acknowledge Canada's part in building the bomb contradicts the historical evidence.[56] His refusal discounts the country's expertise in nuclear physics and engineering at the end of the Second World War, not to mention its large reserves of uranium, and suggests that had Canada wished to build its own nuclear arsenal it would not have been able to do so. That is incorrect. The atomic complex at Chalk River, Ontario, as part of the Manhattan Project, finished constructing an operational nuclear reactor in August 1945 named the ZEEP (Zero Energy Experimental Pile). A photograph taken from an elevated angle shows a white building housing the ZEEP on the left and a brick building housing the NRX on the right. [**Figure I.8**] The Ottawa River, needed to cool the reactors, flows directly behind the buildings. As the first reactor built outside the United States, the ZEEP was intended to produce fissile material for use in bombs, specifically plutonium.[57] The heavy water concurrently manufactured by Consolidated Mining and Smelting in the West Kootenays – heavy water is used to moderate nuclear fission – was also intended to help

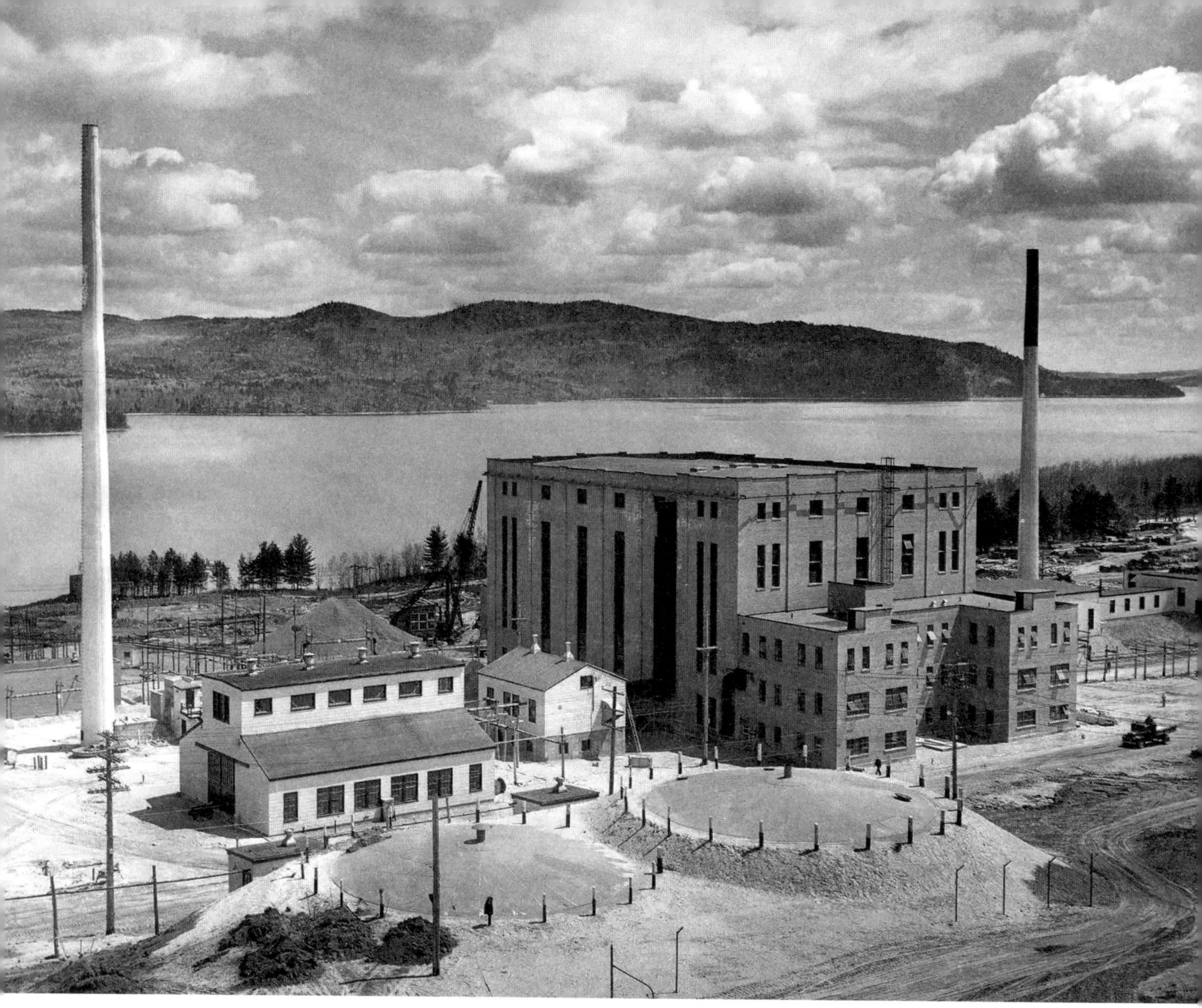

| **Figure I.8**: *ZEEP and NRX Nuclear Reactor Buildings, Chalk River.* 1947.

produce bombs. The clandestine British Columbia operation, titled Project 9, was guarded by an armoured vehicle called *Seabiscuit,* equipped with four machine guns, which looks as if it were manufactured during the Prohibition era.[58] [**Figure I.9**] A photograph of the Chalk River complex taken by Associated Press, in which the foreground is raked by jagged shadows and the buildings are overhung with heavy clouds, shows employees in trench coats walking down a dirt road. [**Figure I.10**] Shot in the slanting light of late afternoon, the image looks noirish, hinting at danger and undisclosed secrets.

In the early 1940s, Canadian scientists played a decisive role in persuading President Franklin D. Roosevelt to mobilize American resources for the purpose of developing a nuclear device, and in the 1950s Canada hosted a delegation from Israel, which had been sent to gather information on how to construct a nuclear reactor of its own.[59] When Prime Minister David

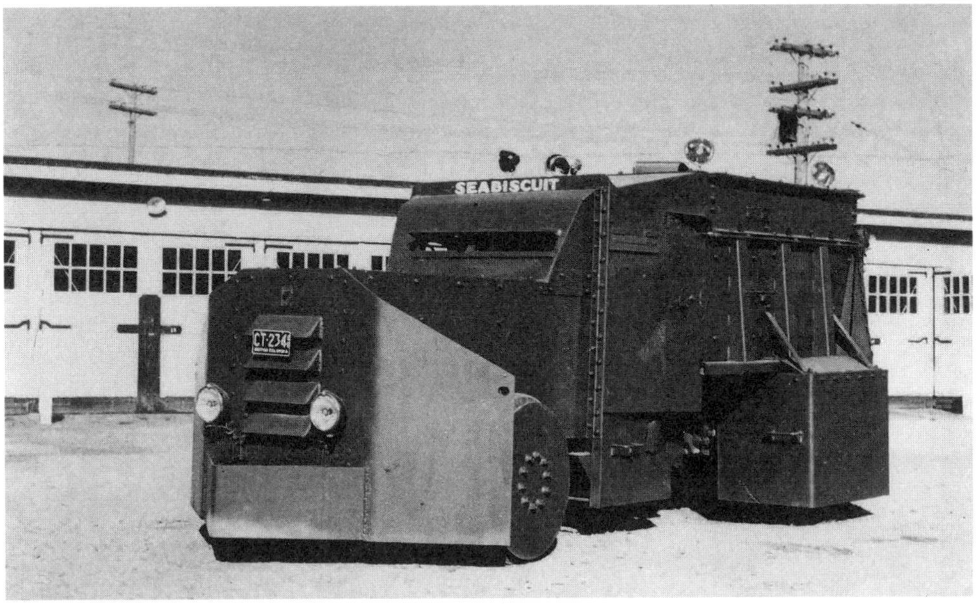

Figure I.9: *Seabiscuit*. c. 1942–45. **Figure I.10**: *ZEEP and NRX Nuclear Reactor Buildings, Chalk River*. 1945.

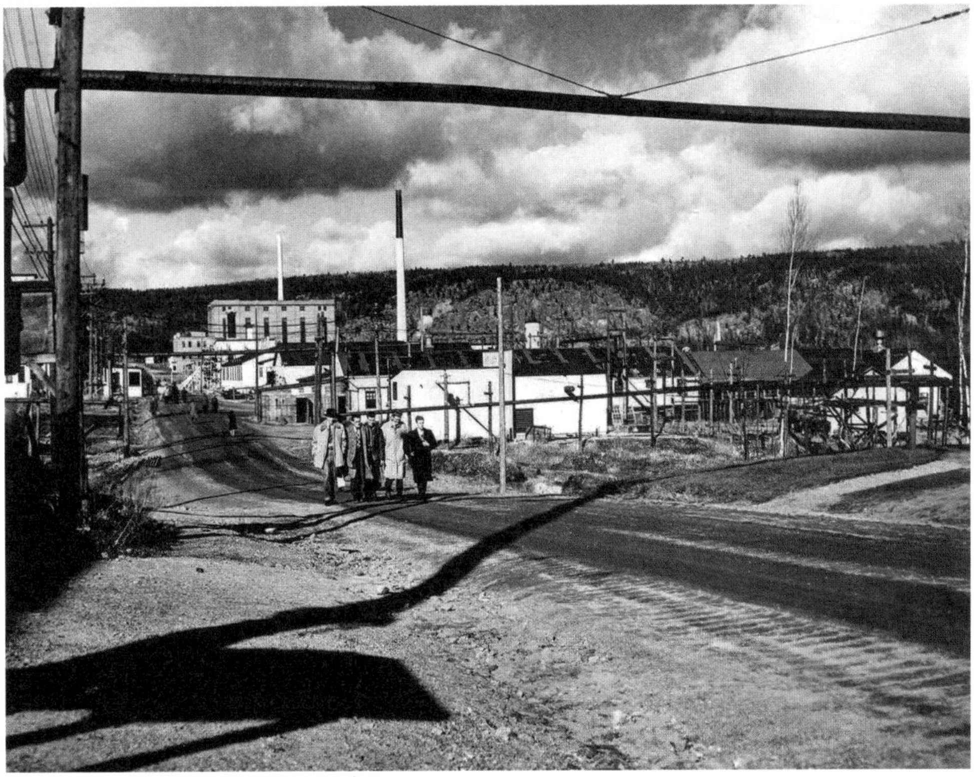

Ben-Gurion announced to the Knesset in 1960 that Israel was building a reactor, he said it would "be similar to the reactor which the Canadian government helped to construct in India, with the difference that our reactor is of smaller capacity."[60] The Dimona reactor in the Negev desert is a near copy of Canada's second reactor, the NRX. Built for the Israelis by the French to produce plutonium for nuclear weapons, it helped provide Israel with atomic hegemony in the Middle East.[61] Because of its nuclear engineering capabilities and its reserves of uranium ore, Canada was a consequential atomic power at the end of the Second World War. It has remained so.[62]

At the beginning of the nuclear era, the Soviet Union had no doubts about Canada's nuclear expertise. On September 5, 1945, a month after the bombings of Hiroshima and Nagasaki, a cipher clerk named Igor Gouzenko, working in the Soviet embassy in Ottawa, defected to Canadian authorities with a stack of incriminating documents revealing the existence of a Soviet atomic spy ring in Ottawa.[63] Although Soviet agents were pursuing him, Gouzenko's request for asylum for himself and his family was almost denied. Prime Minister William Lyon Mackenzie King, with characteristic caution, wished to hand him back to the Soviets to avoid the glare of an international incident. The prime minister was persuaded to do otherwise by his closest advisers, and information from Gouzenko's interrogation and the 109 documents he brought with him were shared with the United States and Britain. Eighteen months later, the spies, which included Canadian member of Parliament Fred Rose and British nuclear physicist Alan Nunn May, were arrested and jailed. May was sentenced to ten years of hard labour.[64]

Gouzenko's defection is sometimes said to mark the beginning of the Cold War.[65] To protect him, Canadian authorities provided him with a new identity and a suburban house in Mississauga, Ontario. There, he wrote an autobiography, *This Was My Choice,* and a best-selling novel about life in the Soviet Union, *The Fall of a Titan.*[66] In 1954, when his novel was published, Gouzenko began to appear on television and in public wearing a hood. [**Figure I.11**] In the first shot of the defector with his head covered, Gouzenko sits in a television studio, "somewhere in Canada" according to the caption, looking as if he is waiting to perform in a horror movie.[67] The photo is a striking portrait, even if Gouzenko's face cannot be seen.[68] The curve of his head echoes the curve of the microphone above him, and although the hands on the clock at upper left are unreadable, we might be justified in thinking they are pointing to midnight.

Gouzenko explained that the hood was to throw Soviet assassins off his scent. It did nothing of the sort, of course, advertising rather than obscuring the wearer and drawing journalists, photographers, and television broadcasters to him. Gouzenko's decision to hood himself for the photograph is a classic case of antiphrasis – one could say, of atomic antiphrasis – where the presence of the hood conveyed an opposite message to the one being declared. By making

| **Figure I.11**: International News Photo. *Soviet Defector Igor Gouzenko Wearing a Hood.* 1954.

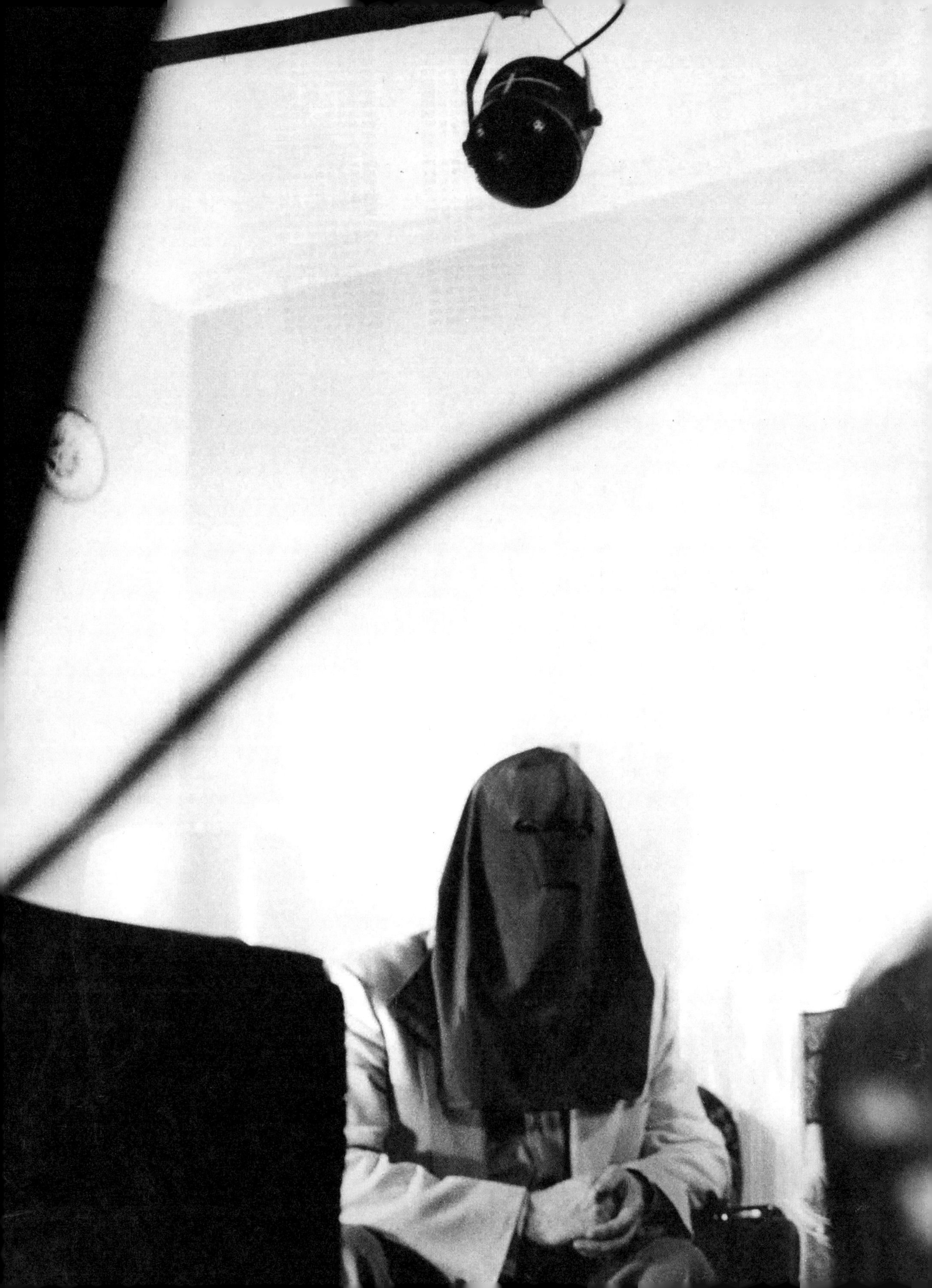

of himself a spectacle for the camera and marketing his defection, by turning himself into a moving target with a bag on his head, Gouzenko became even more of a celebrity than he already was. He became hypervisible. The defector is remembered more for the hooded photographs than for the books he wrote or the television shows he appeared on. "Photographs (and the little words underneath)," observed the photojournalist W. Eugene Smith, "are molders of opinion."[69] They stick in the mind as privileged moments that are more easily revisited than texts or moving images. They represent a sliver of stopped time rather than a stream of information that keeps moving. Because they stay still, they are less likely than moving images to be forgotten.[70] The staying power of photographs is central to my arguments in this book.

The historian Donald Creighton, in *The Story of Canada,* briefly alludes to Eldorado's extraction of uranium at Port Radium in the Arctic, "which made possible the federal government's atomic pile at Chalk River near Ottawa," but like Stacey he neglects to mention Canada's participation in the Manhattan Project.[71] He also neglects to mention that the Sahtu Dene people were involved in carrying the uranium ore.[72] "Before the mine, you never heard of cancer," hunter and trapper Paul Baton comments, "now lots of people have died of cancer."[73] Cindy Gilday, chair of the Deline Uranium Committee, is blunter. She points her finger directly at the nuclear colonialism exercised by Eldorado and the Government of Canada. "In my mind it's a war crime that has been well hidden," she states. "The Dene were the first civilian victims of the war, and are the last to be addressed."[74] Declassified documents support her assessment – the company and the government withheld crucial health and safety information from the Dene. Some of the Dene ore carriers, as mentioned earlier, slept on the burlap sacks during the long barge journey across Great Bear Lake to the transfer point where the uranium was shipped south and then east for processing.[75] Because there were so many cancer deaths among the men, Déline became known as "the village of the widows."

Creighton also fails to mention that Canada exported refined uranium ore from Port Hope to the United States for weapons-grade enrichment and use in the Hiroshima bomb as well as in the making of plutonium for the Nagasaki bomb.[76] Instead, he observes, in an act of painterly displacement, that the Group of Seven depicted the Precambrian Shield in their paintings a quarter of a century before uranium and other minerals were discovered there.[77] Creighton's decision not to mention the Manhattan Project, the Dene, and Canada's export of uranium to its southern neighbour is like choosing not to mention that before-and-after photographs of Hiroshima look different. [**Figs. I.12 and I.13**] The contrast can be seen even in postcards. Before the attack, streets are lined with buildings, telephone poles are strung with wires, and streetcars move along tracks. After the attack, there are no buildings, no wires, no tracks.

Creighton's failure to remark on Canada's participation in the production of the atomic bomb, like Stacey's silence on the matter, strikes me as a form of blindness. What is the point of having a field of scholarship called Canadian history, Ian McKay asks, if the real action is so often conceived as happening elsewhere?[78] Although Creighton's book came out in the middle of the Cold War, it minimized Canada's complicity in the global proliferation of

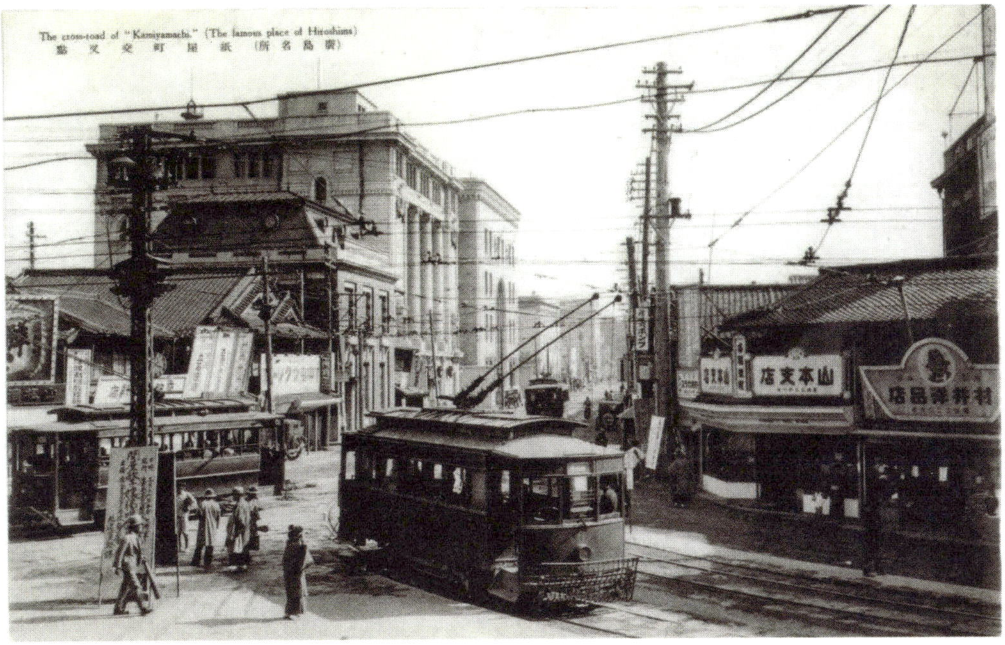

Figure I.12: Postcard. *The Cross-road of "Kamiyamachi" (The Famous Place of Hiroshima)*. c. 1940.
Figure I.13: Postcard. *Hachobori Districts Overlooked from the Roof of the Chugoku Newspaper Office*. 1945.

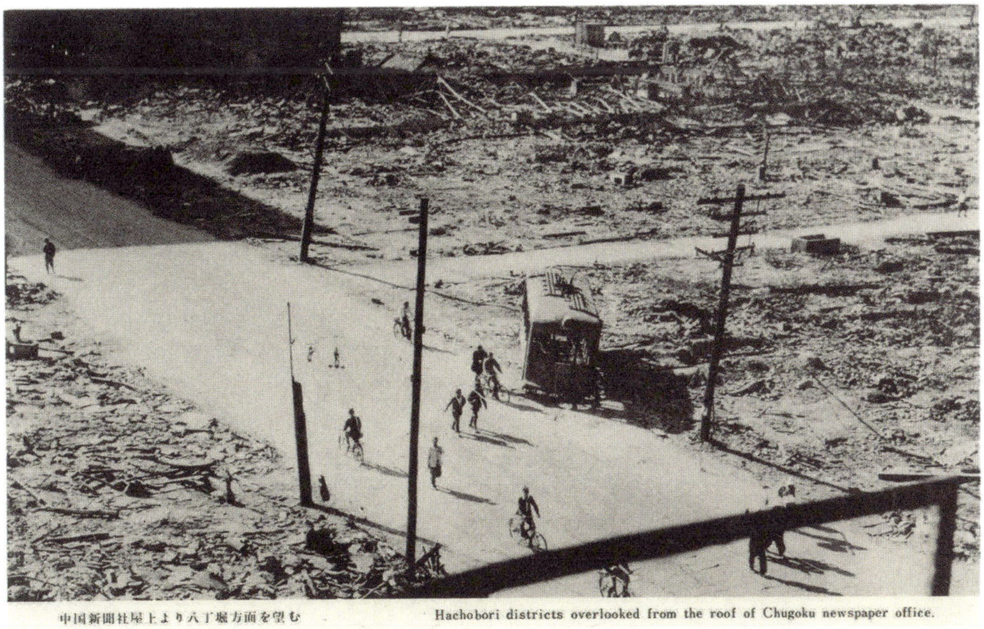

nuclear weapons. Impressed by Foreign Minister Lester B. Pearson's leadership in organizing the first major peacekeeping mission in the history of the United Nations during the Suez Crisis of 1956–57, for which he was awarded the Nobel Peace Prize, as well as by Canada's role in the UN missions to Congo and Cyprus in 1960 and 1964, Creighton imagined Canada as a nation of peacekeepers, even though it meant masking the country's military commitments and nuclear expertise.[79] He was not alone in promoting the peacekeeping fiction. Many Canadians romanticized Pearson as an ambassador of peace, conveniently forgetting that he was as much a cold warrior as a peacekeeper.[80] Nostalgia for Canada's role as peacekeeper has persisted into the twenty-first century, notwithstanding attempts by the Conservative government under Stephen Harper to rebrand the country as a warrior nation of heroes.

After the war, Prime Minister King opted not to pursue a nuclear weapons program, choosing instead to steer a wobbly course between building and banishing the bomb.[81] Canada did not join the nuclear club, whose membership is restricted to nations with atomic weapons, but it did remain a major supplier of uranium to the nuclear programs of the United States and Britain. By the late 1950s, uranium ore was its fourth-largest export after wheat, pulp, and paper. Canada has developed some of the richest uranium mines in the world, including the Gaertner pit at the Key Lake Mine, as seen in Del Tredici's photograph of an immense bucket used for extracting the ore. [**Figure I.14**] The ore is richest where the earth is blackest, Del Tredici observes. Twelve thousand tons of uranium were produced in 1959, more than by any other nation, most of which went toward the fabrication of nuclear weapons.[82]

In addition, Canada turned over its coastlines and the adjacent landmass to the United States as proving grounds for nuclear missile and submarine testing; it partnered in the construction of three early warning radar systems, including the Distant Early Warning Line, or DEW Line, to sound the alarm against Soviet attack; it signed the North American Air Defense Agreement (NORAD) in 1958 with the United States, which placed the air forces of the two countries under joint command; it accepted US Bomarc-B missiles with nuclear warheads onto its military bases; and it exported Canadian nuclear expertise around the globe in the form of the CANDU reactor.[83] Because the CANDU, short for Canada Deuterium Uranium, is an efficient producer of plutonium, some countries, including India and Pakistan, transformed their reactors, which were intended to generate electric power, into mini bomb factories. Taiwan also developed a nuclear weapons program under the guise of Canada's nuclear-generating exports, but it was terminated in 1988 under pressure from the United States.[84] Between Prime Minister King's renunciation of Canadian nuclear weapons and his government's assistance in the production of nuclear weapons by others, there lay a radioactive middle ground. Gordon Edwards, founder of the Canadian Coalition for Nuclear Responsibility, concludes that since the Second World War, "Canada has played a major role in fostering the proliferation of nuclear weapons throughout the world."[85]

The chapters in this volume, bookended by a preface and an epilogue, address distinct nuclear themes. They function as intersecting case studies and are organized thematically rather than chronologically. An atomic timeline at the back of the book provides key dates.

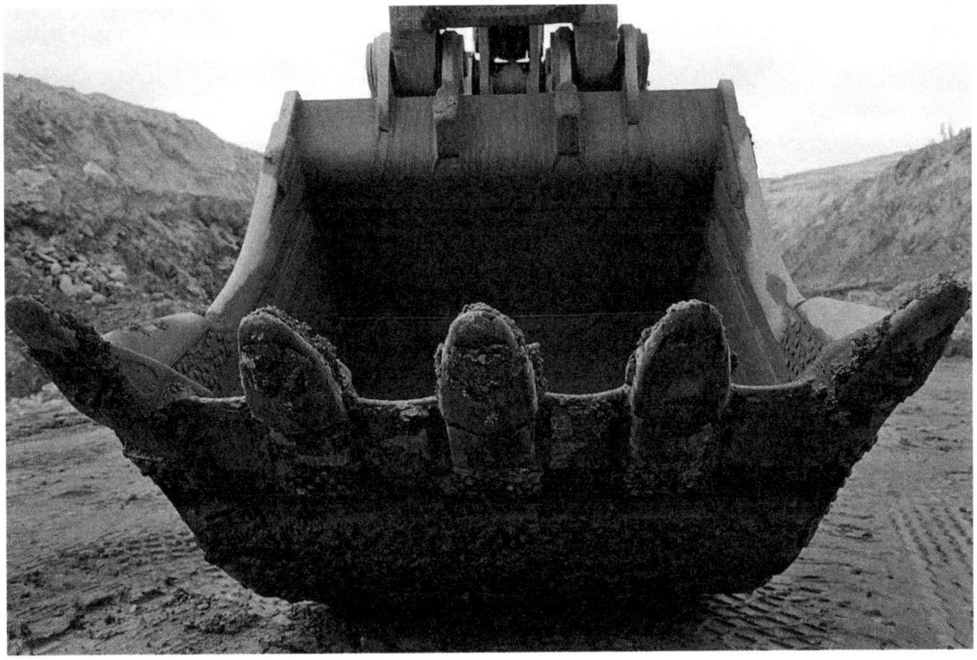

Figure I.14: Robert Del Tredici. *Uranium Shovel (Gaertner Pit, Key Lake Mine, Northern Saskatchewan, Canada).* 1986.

This introduction addresses the principal themes as well as providing an initial response to the questions posed earlier. Chapter 1, as mentioned above, concentrates on the nuclear industry and how Robert Del Tredici's 1987 photobook *At Work in the Fields of the Bomb,* which in my opinion sets the standard for such books, helps to make the activities of the industry visible. It shows how the bomb's far-flung component parts – uranium mining and refining, plutonium production, reprocessing, weapons assembly, testing, research and development, reactor design and construction, radioactive waste management – are intricately connected. Chapter 2 examines the nuclear complex at Chalk River and the radiological accidents there and at other sites around the world. It reflects on how photography and what I am inclined to call the "Godzilla effect" – a metaphor for how nuclear technologies have threatened civilization by outrunning the ability of experts to control them – have shaped perceptions of radiation risk. Chapter 3 discusses military and civil defence, especially the ways government leaders, in Canada and elsewhere during the 1950s and 1960s, advised citizens that a nuclear conflict could be defended against while preparing for a war they knew could not be won. With the development and testing of thermonuclear weapons in the early 1950s, more sophisticated and more powerful than the bombs dropped on Hiroshima and Nagasaki, the possibility of surviving an atomic attack became fantasy.

"Cloud Shadows," Chapter 4, observes that in the Cold War era every part of the globe, including the furthest reaches of Canada, was haunted by the image of the mushroom cloud.

It presented a spectacle of mass destruction, largely through photographs, that made it the dominant icon of the atomic age and its most extravagant figurehead. The spectacle attracted not only producers of popular culture, but also artists who tried to make it strange again. Chapter 5, "Atomic Soldiers," examines discrepancies between official and vernacular memory in the representation of nuclear events. Canadian military personnel were participants in more than two dozen nuclear weapons tests in Nevada, the South Pacific, and Australia from the end of the Second World War until atmospheric testing was discontinued in 1962. In 1957, Operation Plumbbob was the longest and most elaborate series of tests conducted in the continental United States, and it is notorious for the decision by US and Canadian military commanders to use foot soldiers as guinea pigs during operations. The relationship of photography to state power and mass protest is the main concern of Chapter 6. It is ironic that when French secret service agents sank the Greenpeace ship *Rainbow Warrior* in Auckland Harbour in 1985, during a clandestine bombing attack codenamed Opération satanique, the hole ripped in the vessel led to the drowning death of a photographer. The sinking of the ship and the killing of the photographer were a public relations disaster for France. Chapter 7 relates the images of modern and contemporary art photographers to a 1965 book of poems by George Bowering called *The Man in Yellow Boots*. The threat of nuclear war, a recurring theme in the poems, made everyone a bulls-eye. The Epilogue reflects on radioactive waste disposal from the production of atomic weapons and nuclear energy as well as on plans under way to build a new generation of reactors, small modular reactors (SMRs).

Photography shapes the speed and forms of knowledge accessible to nuclear scientists, and the search for scientific knowledge shapes photography. The technologies of photography and nuclear fission are intimately (perhaps even lethally) connected. At the height of the Cold War, Marshall McLuhan noted that the two technologies were responsible for creating social environments in which time speeded up and space was compressed. That was the case for Canada as much as it was for nations in the nuclear club. Photography was producing "a world of accelerated transience," he wrote, and nuclear fission was doing the same.[86] Photography occupies a pivotal place in the representation and understanding of nuclear weapons and nuclear energy.

1 Nuclear Industry

ON AUGUST 25, 1984, THE FRENCH CARGO SHIP *MONT LOUIS* COLLIDED WITH the German ferry *Olau Britannia* in the English Channel and sank off the coast of Belgium. A grainy photograph, taken at a distance with a telephoto lens, shows the *Mont Louis* listing sharply to starboard before going down. [**Figure 1.1**] Without the shot, there would be no public visual record of the vessel sinking. With a deadweight capacity of five thousand tonnes, it was transporting 350 tonnes of uranium hexafluoride from Canada to the Soviet Union, where the gas was to undergo enrichment before being sent on to France for use in light-water reactors.[1] The circulation of uranium and its products is a global enterprise. The uranium hexafluoride on board the *Mont Louis* came from ore mined and refined in Canada, but according to a press report, "no one knew anything" about the ship's contents before the accident.[2] Asked about the risk caused by the collision, the nuclear scientist Yves Lenoir stated that if "the gas came into contact with sea water radioactivity would be released into the atmosphere."[3] Beaches in Belgium and France were cleared of people, while the *Olau Britannia,* with 935 passengers on board, continued to its destination of Sheerness, Kent. The ferry experienced only minor damage, it was reported, and no one on board suffered from radiation exposure. That was the official story, at any rate; it was not reported whether all the lost containers of uranium hexafluoride were recovered.

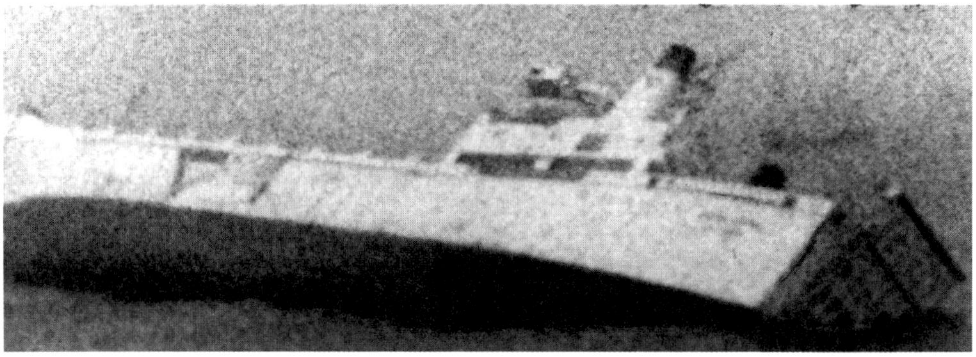

Figure 1.1: Syndication International. *Atom Wreck Peril.* 1984.

Compared to the 1917 collision of the *Mont-Blanc* with the *Imo* in Halifax Harbour, Nova Scotia, which resulted in thousands of photographic images, the collision of the *Mont Louis* with the *Olau Britannia* left a small visual imprint. The Halifax accident produced the largest man-made explosion before the invention of the atomic bomb. The *Mont-Blanc* was preparing to sail for Europe, loaded with a cargo of benzol, picric acid, and gun cotton for use in the First World War, when it collided with the *Imo,* caught fire, and exploded. Its detonation killed two thousand people.[4] The shank of the ship's anchor, weighing half a tonne, landed three kilometres away from the blast. Gertrude Pettipas, a witness of the three-kiloton explosion, reported seeing "a blinding sheet of fire shoot a mile high in the air. It seemed to cover the whole sky ... The house swayed and rocked, and the doors and windows were blown to pieces."[5] Pettipas's report resembles witness accounts of the Hiroshima blast, which was five times the size and killed fifty times more people. (The impact of atomic bombs is measured in megadeaths.)[6] Hatsuyo Nakamura, the widow of a tailor, described the intense brightness of the atomic detonation: "Everything flashed whiter than any white I had ever seen, and [I] seemed to fly into the next room over the raised sleeping platform, pursued by parts of [my] house."[7]

The descriptions by Pettipas and Nakamura are profoundly visual. Two wars, two explosions; two catastrophes, two witnesses. Thousands of photographs exist of each catastrophe. The Halifax explosion killed Canadian civilians accidently, the Hiroshima explosion Japanese civilians deliberately. In the attack on Hiroshima by the *Enola Gay,* piloted by Paul Tibbets, the nuclear weapon on board was produced in part with Canadian expertise and resources. The *Globe and Mail* reported after the bombing that "Canadian science and Canadian uranium played a large part in this epochal achievement."[8] The newspaper was correct about the science and the uranium but wrong about the achievement. The destruction of Hiroshima was the opening of Pandora's box. This chapter is about how Robert Del Tredici's 1987 photobook *At Work in the Fields of the Bomb* helps to make the contents of the box visible. The constituent parts of the nuclear industry – uranium mining and refining, scientific research and development, reactor design and construction, weapons assembly and testing, toxic waste disposal – are intricately related. At the centre of it all is uranium, the mother element that fuels nuclear energy. It is the foundational ingredient in nuclear weapons.

The dropping of the Little Boy bomb on Hiroshima on August 6, 1945, was preceded three weeks earlier by the Trinity test in the New Mexico desert. It was followed three days later by the dropping of the Fat Man bomb on Nagasaki. The detonations announced the beginning of the nuclear era. The level of devastation produced by the bombs demonstrated that the human species now possessed the capacity to annihilate itself. The moment has been identified by some observers as marking the start of the Anthropocene, a term coined by scientists Paul J. Crutzen and Eugene Stoermer to denote our present epoch, in which human activities are significantly altering the geology and ecosystems of the planet.[9] By now, we have internalized the bomb so thoroughly that it has become ingrained in our DNA. The increasing destruction being perpetrated during the Anthropocene is a call, van Wyck comments, to do something about it.[10]

"The earliest weapons were hands, nails and teeth," wrote the Roman poet Lucretius. "Next came stones and branches wrenched from trees, and fire, and flame."[11] Then, centuries later, came nuclear weapons. Had Lucretius known about Little Boy, Fat Man, and Castle Bravo, the code name for the 1954 thermonuclear detonation in the South Pacific – had he known about artificially produced fission and fusion – he might have said they were stolen from the sun. He might also have said that technology had become so powerful we could no longer predict or control its effects. Castle Bravo produced double the megaton yield anticipated by its makers, resulting in lethal radiation poisoning of the residents of the Marshall Islands. The *Daigo Fukuryū Maru* (Lucky Dragon 5), a Japanese fishing boat located outside the test zone, was also inundated with deadly white ash from the explosion.[12] All of its crew became sick and one died. By 1958, the United States had conducted twenty-three atmospheric tests at Bikini, "for the good of mankind and to end all world wars," officials declared.[13] Because of contamination, the residents of Bikini Atoll were forced to leave their island, as recorded in a picture of an uprooted family departing by boat, taken by the Japanese photojournalist Hiromitsu Toyosaki. [**Figure 1.2**] For "the good of mankind," they had become nuclear refugees. In an essay accompanying his photographs, Toyosaki calls them the "hidden and forgotten *hibakusha*."[14] As much as the survivors of Hiroshima and Nagasaki, they are victims of the bomb.

In some cases, no photographs exist of nuclear events and activities — or none for which permission to publish is granted. To guard against terrorist attacks, Bruce Power in Ontario employs a championship-winning SWAT team at its nuclear facility on Lake Huron.[15] "Robust security forces protect nuclear energy stations and other major nuclear organizations against intruders," a 2015 bulletin issued by the Canadian Nuclear Association states. The security forces "routinely conduct threat and risk assessments so that they can protect Canadians from radiation exposure that could result from sabotage." In addition, the tactical teams work with the International Atomic Energy Agency to keep Bruce Power's nuclear materials out of the hands of those who should not have them. Many of the measures used by the security forces are "not publicized," the bulletin states, which is to say, they are kept secret.[16] For four years

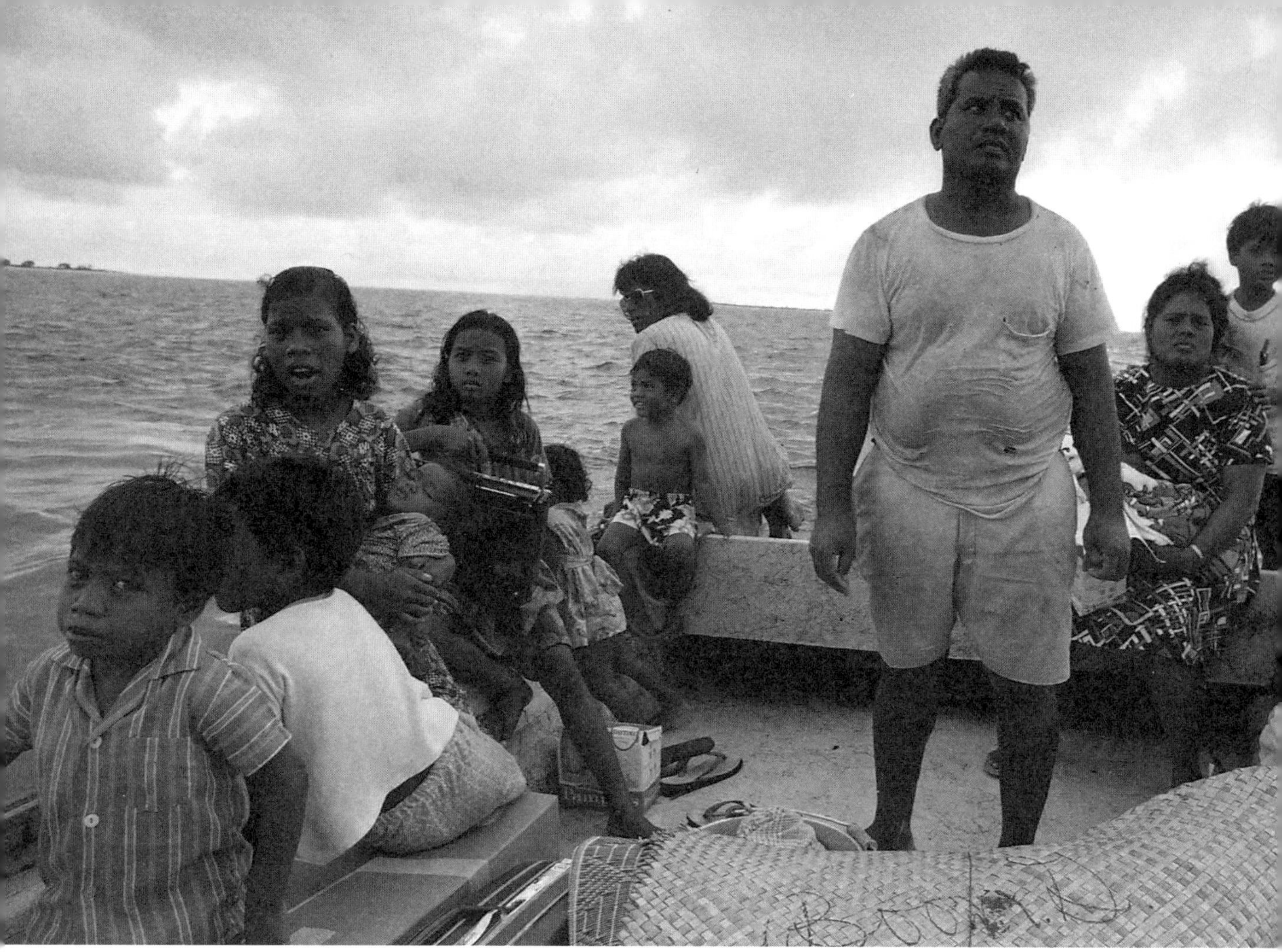

| **Figure 1.2**: Hiromitsu Toyosaki. *Bikini Atoll: Contamination Forces Islanders to Move On.* 1978.

straight, from 2008 to 2011, the Bruce Power team won first prize at the US National SWAT Championships. We are told that in the Bus Assault Competition, the team had to pick off targets in a school bus and rescue dummies inside. In the Countersniper Challenge, team members had to run a prescribed distance and hit targets located more than 250 yards away. The Bruce Power website shows the 2015 SWAT team posing for a photograph, an image I was refused permission to publish for "security reasons," even though it is readily available on the Internet. In it, the team members wear bulletproof vests and carry assault weapons.[17] What is disquieting about the image are the smiles on their faces. Military and paramilitary personnel generally present themselves poker-faced, to emphasize the seriousness of their work. What, then, are we to make of a grinning SWAT team, looking as if it has just won a reality TV contest? What kind of "reality" is being communicated in the shot? The men look more like actors trying to entertain viewers than paramilitaries trained to kill terrorists. The SWAT team photograph offers a mixed message.

A ten-part series of staged photographs by Carole Condé and Karl Beveridge, *No Immediate Threat,* also offers a mixed message. But here it is intentional. The 1986 series examines the problems facing workers in the nuclear industry by following the life of a fictional worker, developed from conversations between the artists and families in the town of Kincardine, Ontario. Kincardine is dominated by Bruce Power and the eight nuclear reactors that make up its site on the shores of Lake Huron, representing the largest concentration of nuclear power in the world.[18] In one of their montaged photographs, Condé and Beveridge portray a bedroom in which a couple is having a disagreement. [**Figure 1.3**] A haz-mat suit in a cruciform shape hangs behind them on a chain-link fence, and two nuclear cooling towers are visible through the window. Turned away from his wife, who stands with a hand on her hip, the worker says, "Sure you pick up radiation, but I've never gone over my dose limit. You don't see any two-headed kids around here, do you?" He is defending the plant that provides him with a job, whereas his wife worries that the price might be too high. Similar conversations occur among families at nuclear sites everywhere, including the Pickering plant near Toronto, one of Canada's oldest and biggest nuclear complexes. Two of its eight reactors have been permanently closed because of their age. The rest were slated for closure but have been given an extended licence to operate until 2024.

Hazards produced by advanced technology on the scale of nuclear devastation, climate change, and species extinction have exposed a fault line between past industrial society and present technoscientific society.[19] In industrial society, the positives and negatives of technological advancement can be evaluated more or less accurately; in the technoscientific era, they cannot. At the time of the Trinity test in Alamogordo, scientists debated whether the engineered explosion would ignite the atmosphere and destroy the world. For months, they had been concerned about ignition temperatures. Enrico Fermi, who built the first nuclear reactor in a squash court at the University of Chicago, and who knew that the risk of atmospheric ignition was minimal, took bets from fellow physicists that the chain reaction would not destroy the planet.[20] On orders from J. Robert Oppenheimer, the test went ahead, and the world changed in an instant, even though the atmosphere remained intact. The negatives of technological advancement had come to exceed any kind of measurable risk. Risk had turned into threat.

Beginning with the Manhattan Project, the threats associated with atomic weapons and nuclear energy have been consistently minimized by technoscientific-military-diplomatic authorities, as Derrida calls them in "No Apocalypse, Not Now."[21] In many countries, including Canada, the United States, and Japan, scientists have been muzzled by political and corporate masters, and their research has been kept under wraps.[22] In her 1962 book *Silent Spring,* biologist Rachel Carson argued that there are limits to technological progress. She called for a ban on the production of DDT, a pesticide that was destroying the natural world, she said, as surely as nuclear weapons had destroyed Hiroshima and Nagasaki. "Chemicals are the sinister and little-recognized partners of radiation," she declared.[23] In technoscientific circles, Carson's message was first met with denial, then with hostility.[24] Robert White-Stevens, a biochemist at the conglomerate American Cyanamid, drew from the Gospel of Matthew in the New Testament to assert that Carson wished to return us to the Dark Ages, where "insects

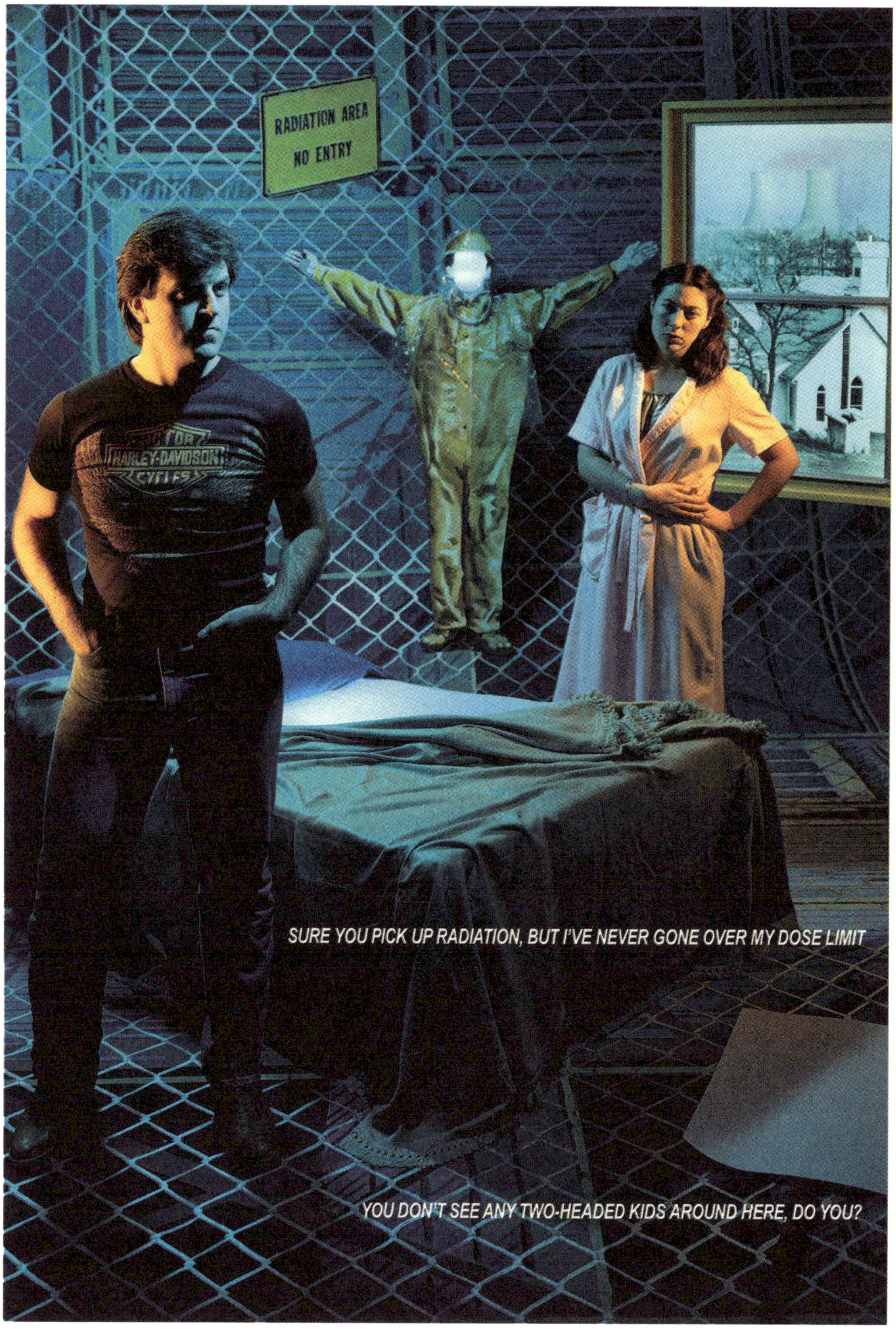

Figure 1.3: Carole Condé and Karl Beveridge. *No Immediate Threat*. 1985–86.

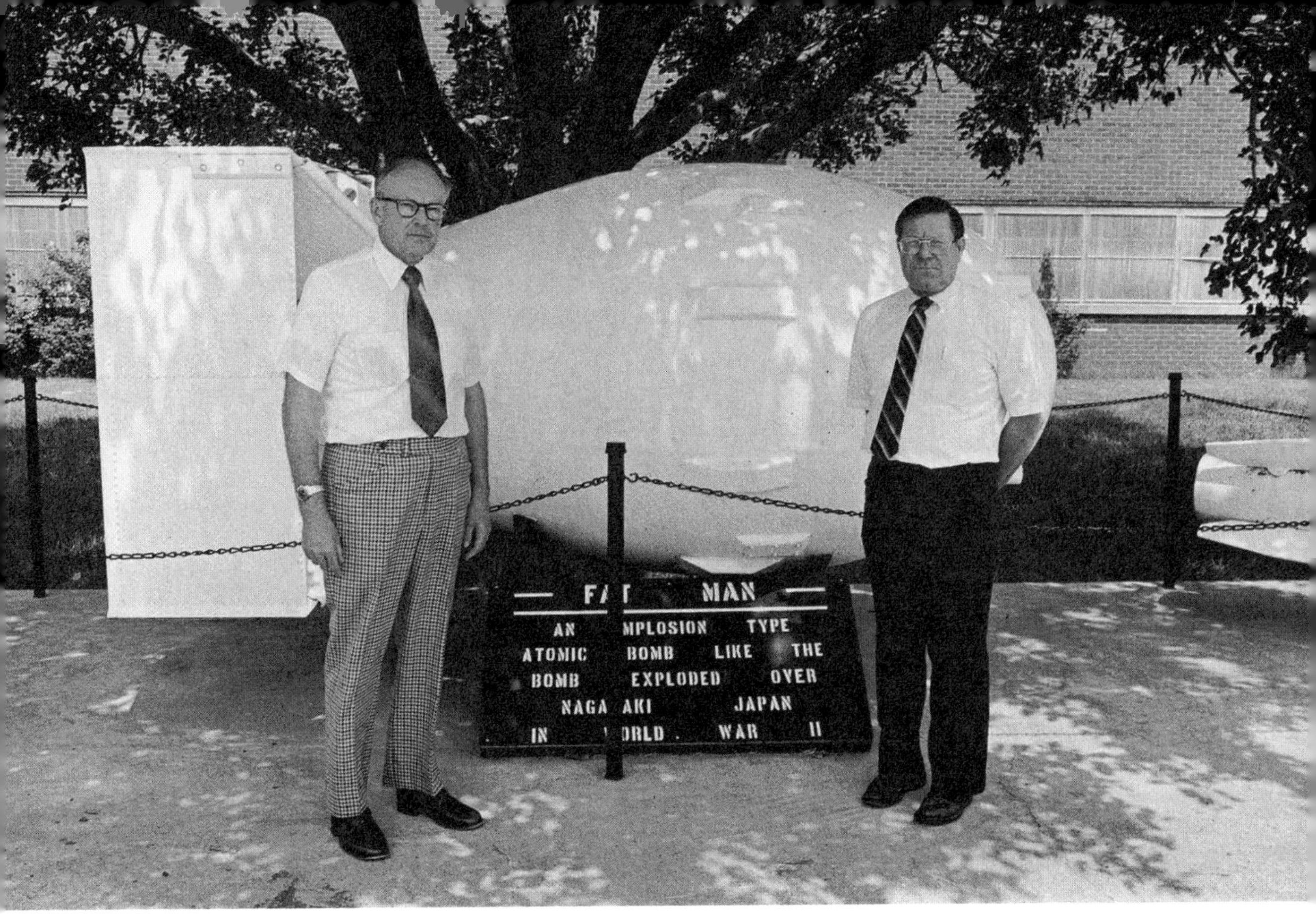

| **Figure 1.4**: Robert Del Tredici. *Duplicate "Fat Man."* 1982.

and diseases and vermin would once again inherit the earth. [She is] a fanatic defender of the cult of the balance of nature."[25] White-Stevens and the corporation he worked for thought that Carson's ideas, like the balance of nature she was defending, must be discredited by all means possible. That meant lying about the effects of pesticides.

Photography can help to tell us about nuclear threat by making it visible through representation. Del Tredici's photobook *At Work in the Fields of the Bomb* aims to reveal what nuclear authorities have kept hidden since the birth of the bomb. Using a combination of words and images, Del Tredici wants to show "the nuclear arsenal at its source," as demonstrated in his photograph *Duplicate "Fat Man,"* (Plate 9), which pictures a replica of the Nagasaki bomb on display outside the Pantex Nuclear Weapons Final Assembly plant in Amarillo, Texas.[26] [**Figure 1.4**] The two public relations employees whom Del Tredici persuaded to pose for him look uneasy in their roles. They are not in a position to reveal anything but approved talking points about the Pantex plant located behind them and the dummy bomb. But Del Tredici is under no such constraints. Focusing on the nuclear state, and producing a "violation" of documentary photography-as-usual, as John Roberts puts it, the photobook discloses more about the technoscientific-nuclear complex and its borderlands, its interconnected networks of uranium processing and weapons production plants, its missileers and bomb crews (cogs in the nuclear machine), its pervasive invisibilities and sacrifice zones, and its radiation-damaged citizens and flak-catching public relations officers than any other book concentrating on nuclear issues.[27] It renders the nuclear complex and its mass-production cycle "culturally visible."[28] The book achieves this by putting the human body in the nuclear machine and by countering the forces of abstraction, as much as it is possible to counter abstraction when confronted with the calculus of the nuclear cycle. It concretizes and historicizes the nuclear complex in a way that compares in a moral sense with James Agee and Walker Evans's *Let Us Now Praise Famous Men* (1941), which gave a human face to the plight of tenant farmers in the United States during the Depression.[29]

While Del Tredici was working on his book, he came up with the idea of forming an international organization of nuclear photographers for the purpose of exhibiting work and educating the public.[30] The Atomic Photographers Guild was established in 1987 with eight other founding members: Carole Gallagher, Yoshito Matsushige, Berlyn Brixner, Kenji Higuchi, Harris Fogel, Dan Budnik, James Lerager, and Günter Zint.[31] They held a group exhibition later that year at the New York Open Center, timed to coincide with the first Global Radiation Victims Conference in New York. Subsequent exhibitions were held in Montreal in 1988, Berlin in 1990, Hiroshima and Nagasaki in 1994, the Diefenbunker Museum near Ottawa in 2001, and Rio de Janeiro in 2012 as well as in various European and North American cities, culminating in the largest show yet, *Life and Death in the Nuclear Age,* at the Rocky Flats Museum in Arvada, Colorado, in 2012.[32] Conceived as an association with a fluctuating membership and an irregular program of exhibitions, the guild continues to mount shows.[33]

In an essay published in 1995 to mark the fiftieth anniversary of the development of the bomb, Del Tredici writes that the guild photographers share "a documentary bent and a hunger for unseen evidence ... an eye for innuendo, a taste for paradox, and the ability to walk among conspiracies and phantoms."[34] Over the years, they have produced a body of nuclear "counterimages," as member Blake Fitzpatrick calls them, which have provided a critical alternative to mushroom cloud imagery.[35] Mike Davis compares the achievement of Del Tredici and the guild to that of the photographers of the Farm Security Administration, in an article called "The Dead West: Ecocide in Marlboro Country."[36] Fitzpatrick's *The Radioactive Autograph Album,* from 2002, presents a narrative sequence of photographs and texts tracing the path of radioactive materials that migrated from the Eldorado refinery on the hands of plant manager Marcel Pochon. [**Figure 1.5**] In contrast to mushroom cloud imagery, Fitzpatrick's project is about how minute amounts of radiation become visible. Pochon collected autographed photos of film stars, which he kept in an album.[37] Uranium was found on a number of the pictures, which had to be removed, leaving behind an eerily vacant album made up of blank pages and captions to the missing portraits. The movie stars Donna Reed and Greer Garson are named but not seen. The contaminated photographs are now housed in Port Hope's Low Level Radioactive Waste Management Office.

At Work in the Fields of the Bomb was published in 1987 by Harper and Row in the United States, George G. Harrap in Britain, and Douglas and McIntyre in Canada. Its introduction was written by journalist and disarmament advocate Jonathan Schell, who observes that Del Tredici "has pitted his camera against all forms of the bomb's invisibility."[38] The book appeared a year after the Chernobyl meltdown, guaranteeing it immediate interest, and two years before the fall of the Berlin Wall.[39] It measures 28 × 22 centimetres and is composed of photographic plates and textual material, including "Field Notes." Del Tredici chose to place the 106 plates together, up front, relegating interviews and other details to the back of the volume. He wanted to establish that it was primarily a photographic work. His previous book, *The People of Three Mile Island,* which documents the 1979 nuclear accident near Harrisburg, Pennsylvania, integrated photos with text. According to the author, this caused confusion among readers about whether it was a book of photographs-with-text or a book of text-with-photographs. His solution to that ambiguity in *At Work in the Fields of the Bomb* was to group the plates at the beginning.[40]

There is no mistaking the volume for a coffee-table book, the dominant genre in photobook production. The cover, just white lettering on a black ground, lacks an image; it is not trying to ingratiate itself with anyone. [**Figure 1.6**] The captioned plates are followed by notes and interviews with each individual photographed. The plates are printed on semi-gloss stock, the notes and interviews on matte. The majority of images were shot in the United States, but eight were taken in Japan, seven in Canada, and half a dozen at sites in England, Germany, Sweden, and the Soviet Union. A schematic map indicates the location of each picture.

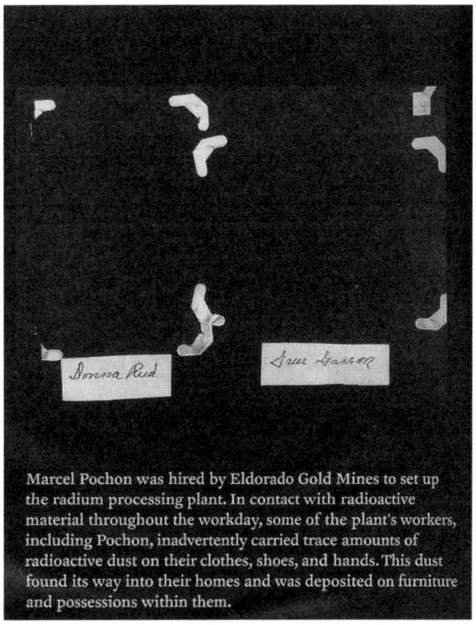

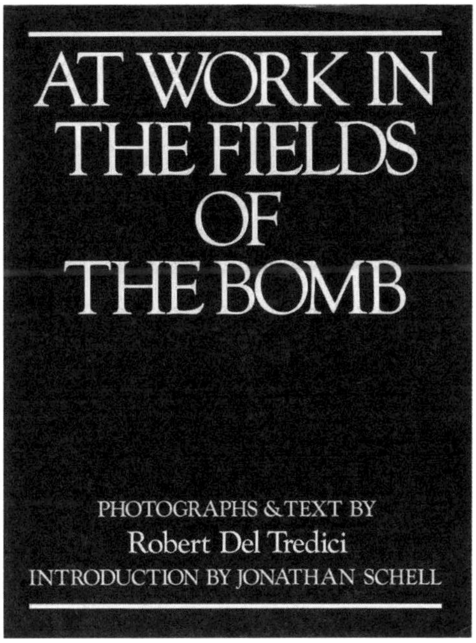

Figure 1.5: Blake Fitzpatrick. *Radioactive Autograph Album: Donna Reed and Greer Garson.* 2002. **Figure 1.6**: Robert Del Tredici. Cover of Robert Del Tredici, *At Work in the Fields of the Bomb.* 1987.

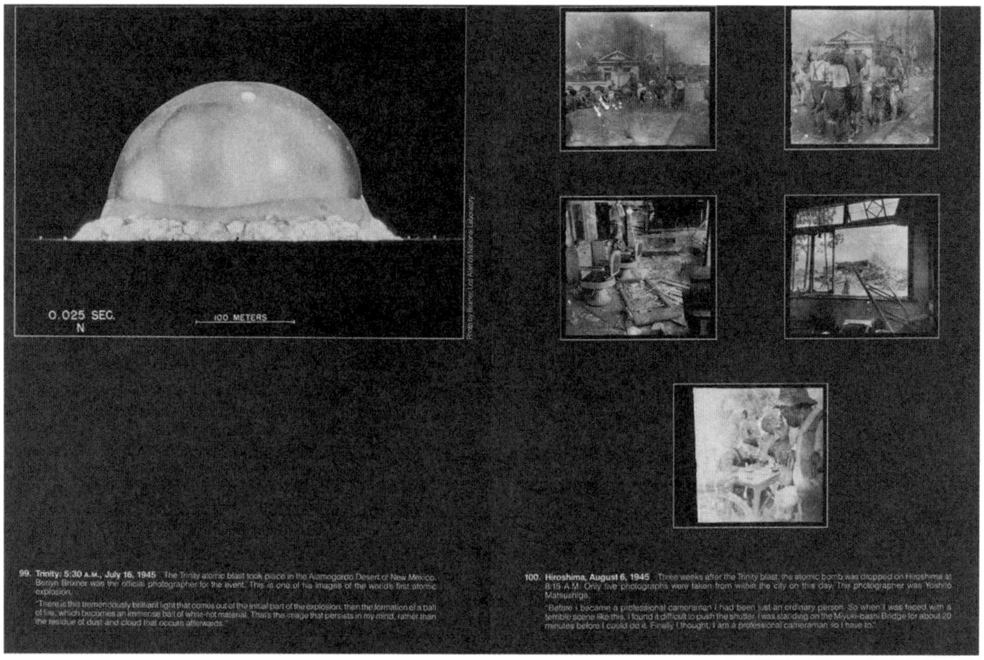

Figure 1.7: Robert Del Tredici. Plates 99 and 100 from Robert Del Tredici, *At Work in the Fields of the Bomb*. 1987. **Figure 1.8**: Yoshito Matsushige. *Aftermath of Bomb Explosion in Hiroshima, August 6, 1945*. 1945.

Close to the end of the sequence of plates is a double-page spread of images on a black ground. [**Figure 1.7**] All the other photographs are set against a white ground. The difference is striking, distinguishing the two pages from the rest of the book. The left page reproduces an image of the world's first atomic explosion, taken by Berlyn Brixner, the official photographer of the Trinity test, and the right page features five shots taken within hours of the Hiroshima explosion by Yoshito Matsushige, an army reporter and photojournalist who survived the attack. Brixner's photograph, *Trinity: 5:30 A.M., July 16, 1945* (Plate 99), depicts the spherical dome generated by the explosion 0.025 seconds after detonation. The superheated gases and debris rising from ground zero produce an organic form that looks like a soap bubble or rotting fruit about to burst.[41] Matsushige's five photographs, grouped under one title, *Hiroshima, August 6, 1945* (Plate 100), show the devastated city and some of its injured inhabitants, many of whom would not survive the extreme heat and radiation from the explosion. Brixner and Matsushige have been called the first atomic photographers. Their cameras bore witness to the first manifestations of the bomb, and their images play a crucial role in Del Tredici's nuclear narrative.

In an article on nuclear criticism and *At Work in the Fields of the Bomb*, Bryan C. Taylor notes the irony of placing Brixner's picture next to those by Matsushige.[42] Brixner had no qualms about the bombing of Hiroshima and Nagasaki, regretting only that he had not received more credit for having photographed the Trinity test. His authorial interest in his photographs contrasts with the reluctant attitude of Matsushige toward the pictures he took. When he was confronted by the dead and dying, he told Del Tredici in a 1984 interview in Hiroshima, "I tried to push the shutter, but I couldn't. It was so terrible. The people were pathetic. I had to wait."[43] (Like all other Japanese photographers during the Second World War, Matsushige was forbidden by the military to depict the dead.) Although he had two rolls of film with twelve exposures each, he could bring himself to take only five shots. Instead of concentrating on the dead bodies he encountered, such as people lying at the bottom of a swimming pool, he focused on the shattered interior of his own home and on a group of high school girls facing away from the camera. In none of his photographs do we see corpses or the faces of the living. We see only the backs and sides of survivors.[44] [**Figure 1.8**] The dead haunt Matsushige's images by forcing us to conjure up what is not seen.[45] The difference between the work of Brixner and Matsushige highlights the difference between photography

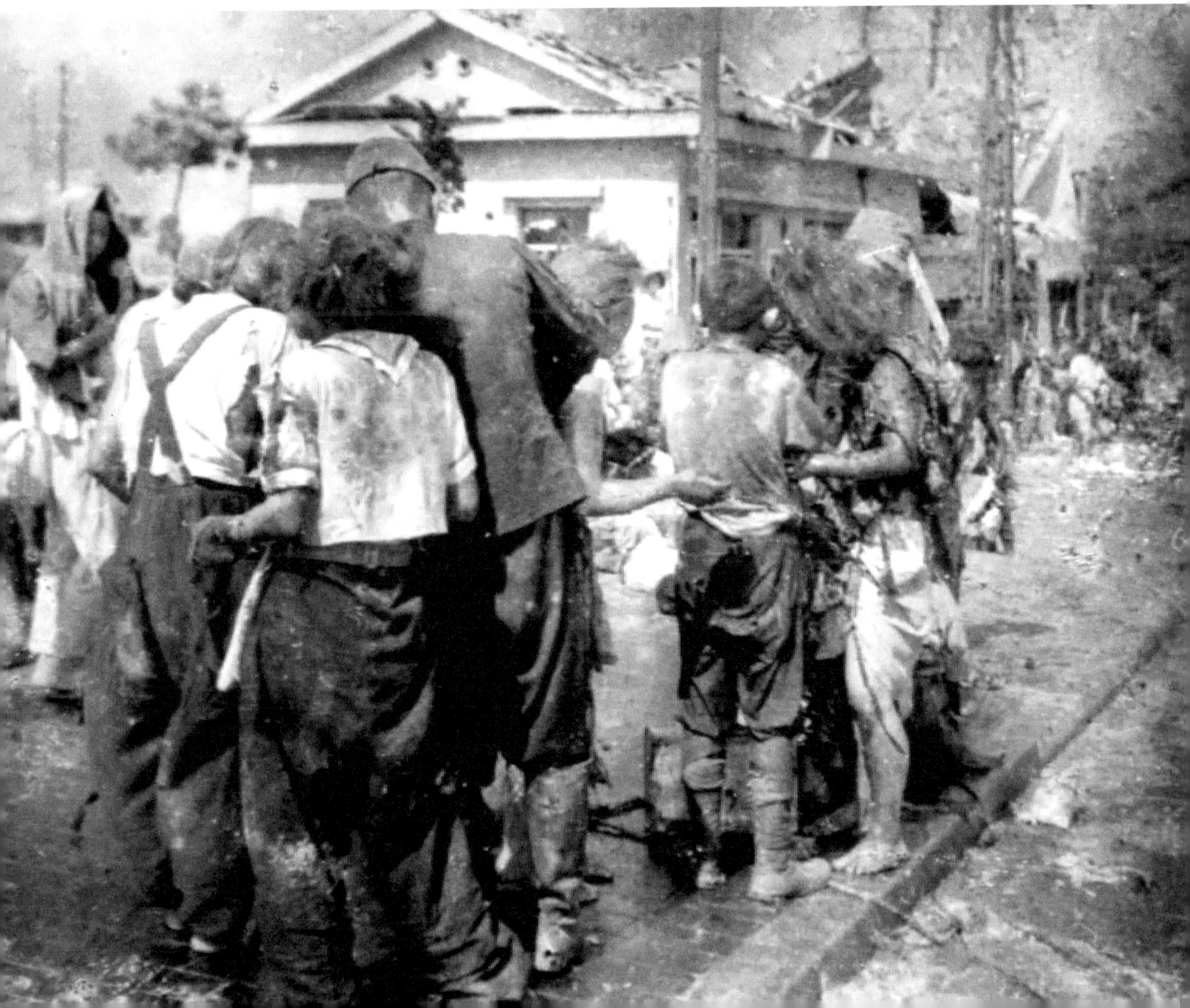

used primarily for technical advancement and photography used critically for documentary purposes.

At Work in the Fields of the Bomb does not make a firm distinction between past and present, between the events of 1945 and those of the 1980s, but instead attempts to bring them together. The book wants to fuse the past with the present. Del Tredici's framing of his project is like Kurt Vonnegut's framing of the novel *Slaughterhouse-Five,* in which Vonnegut refuses to distance the 1945 firebombing of Dresden from later histories of the same city and other sites of destruction.[46] In their separate ways, both the photographer and the novelist are contending that what happened in Hiroshima and Dresden continues to happen. The past cannot be bracketed off from the present. For Fitzpatrick, who has collaborated with Del Tredici and also written on him, *At Work in the Fields of the Bomb* is "a response to nuclear history in which the past comes into legibility in the present."[47]

A favourable review of the book written in the journal *Peace Research* comments on the layers of abstraction that veil the nuclear industry. Del Tredici succeeds in defetishizing the bomb, it concludes, but he fails to bring Canada strongly enough into the narrative.[48] The criticism seems misplaced. Given the different scales of nuclear activity in Canada and the United States, it seems to me that Del Tredici strikes a just balance in his representation of the two countries. His photographs of Canadian uranium mines and tailings ponds at Key Lake, northern Saskatchewan, and Elliot Lake, Ontario, uranium refineries at Port Hope and Blind River, Ontario, a CANDU reactor under construction at the Darlington nuclear complex near Toronto, a deep test shaft for high-level nuclear waste in the Canadian Shield at Lac du Bonnet, and Dr. Rosalie Bertell, a mathematician and health researcher on the effects of radiation, more than adequately embody Canada's involvement in the nuclear cycle.

The images in *At Work in the Field of the Bomb* are ordered sequentially. One early sequence is devoted to uranium mining and refining; it consists of twenty-one shots, including several from Canada.[49] They build a linear narrative of the industrial process, but most importantly they provide juxtapositions that challenge official positions of the nuclear industry, such as the mantra that employees in uranium refining plants should not be overly concerned about health risks. Photographs of workers at the Fernald and Ashtabula facilities in Ohio are juxtaposed against a portrait of Dr. Thomas Mancuso, a government expert on the effects of low-level radiation.[50] The caption beneath the picture of Mancuso states that he studied the radiation records of more than 225,000 nuclear workers over the course of fifteen years and concluded that low-level radiation exposure increased the likelihood that workers would develop cancer. It continues: "When Mancuso published his findings, the government cut his funds, removed him from the study, and confiscated his data."[51] To underscore the significance of Mancuso's findings, Plate 18 of *At Work in the Fields of the Bomb* shows an employee collecting shavings from a "derby," a round block of uranium named for the shape of the hat, for testing in a lab. [**Figure 1.9**] She is touching the derby, which emits alpha, beta, and gamma radiation. The employee works with uranium derbies eight hours a day. She wears overalls, rubber gloves, safety glasses, and earrings, and has a flower in her hair because it is the week before Christmas.

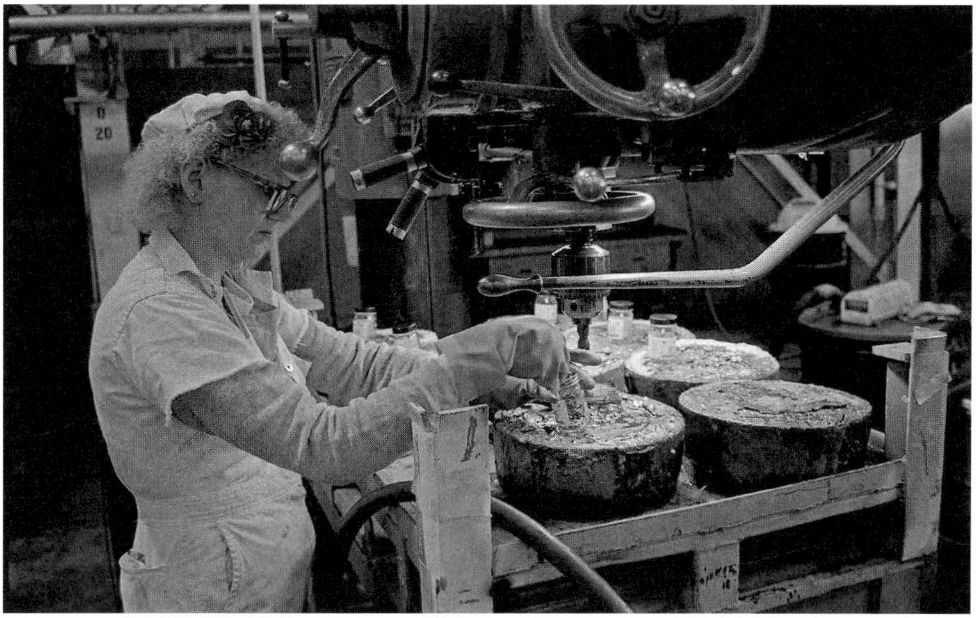

Figure 1.9: Robert Del Tredici. *Sampling the Derby.* 1985.

The caption to Del Tredici's Plate 13, *The Richest Uranium Mine on Earth,* is strictly factual. "The Gaertner pit at the Key Lake Mine [Saskatchewan]," it states, "is the most productive uranium mine in the world. Uranium rock here is richest where the earth is blackest. At a nearby mill, the ore is crushed, refined, and turned into powdered concentrate called yellowcake. Radiation in this pit can be 7,000 times higher than normal background levels." The information in the caption of another Canadian photograph, *Inside the Key Lake Mill* (Plate 66), which depicts a worker fixing a pipe, is also factual. However, the caption for a third Canadian photograph, *Dr. Rosalie Bertell* (Plate 97), expresses a point of view as well as relating facts. [**Figure 1.10**] The first part concentrates on names and affiliations; the second part quotes from an interview with the subject:

> *Dr. Bertell is a doctor of mathematics and a member of the Grey Nuns of the Sacred Heart. Her life's work is the study of radiation on human health. "It finally hit me that when it comes to building nuclear bombs, the military doesn't count the cost in human lives. And from that time on, I really started to look for people in every part of the nuclear weapons cycle, the hidden people, for whom World War III has already started."*[52]

The shift from one voice to another comes as a jolt, forcing the reader to listen in a different register. Del Tredici photographed Bertell speaking in Montreal, at the Unitarian Church of

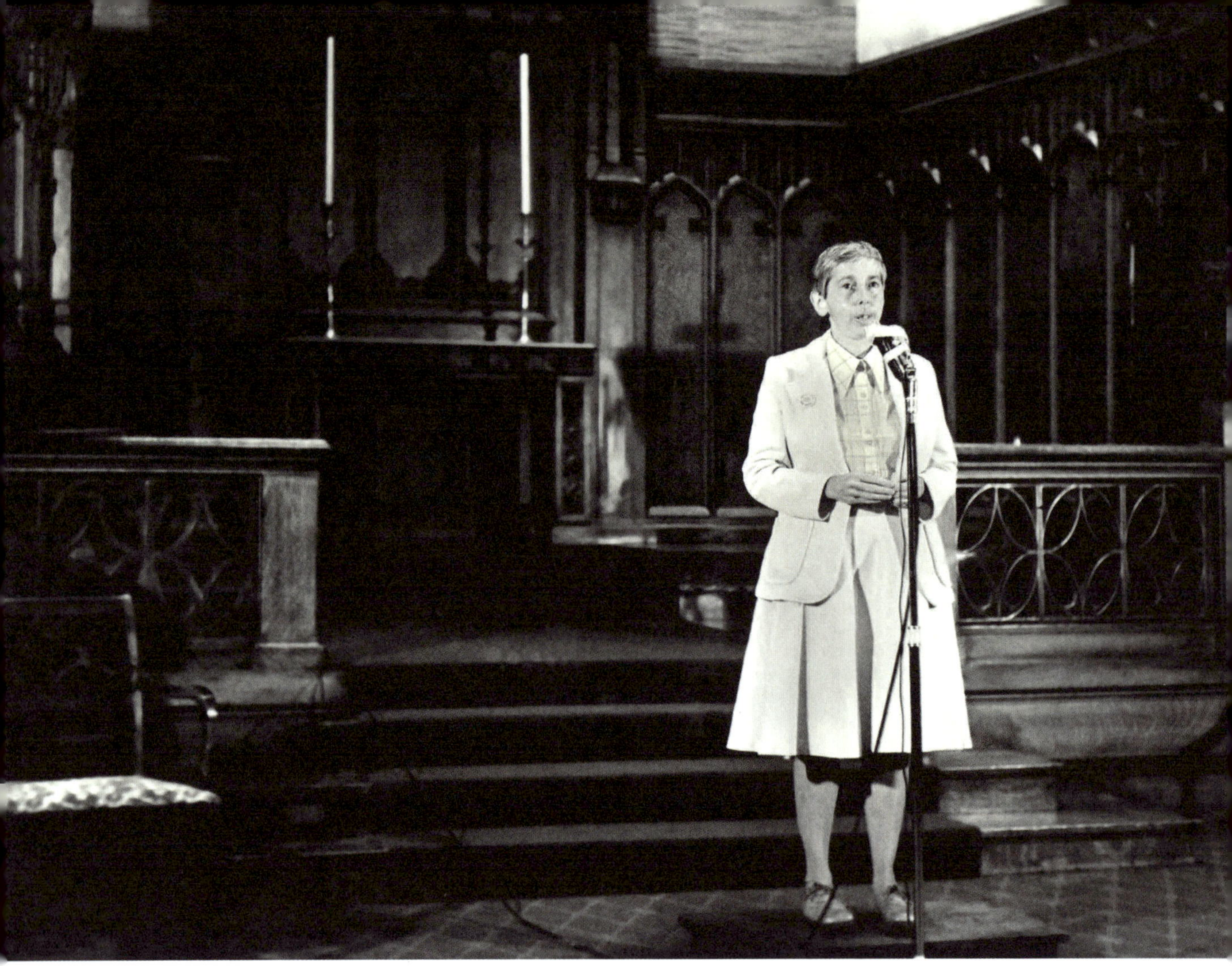

| **Figure 1.10**: Robert Del Tredici. *Dr. Rosalie Bertell.* 1982.

the Messiah, where she is shown standing on a low riser in front of a microphone, with the church sanctuary behind her. Her audience cannot be seen, just like the "hidden people" in the nuclear weapons cycle that Bertell is speaking about. As with Mancuso's findings, Bertell's research demonstrated that radiation can be destructive even at low levels.

The photographs in *At Work in the Fields of the Bomb,* from *Inside the Key Lake Mill* to *Dr. Rosalie Bertell,* avoid the production of spectacle. What makes them compelling is the way they interact with one another while requiring the reader to move back and forth between them. The photo of the mill worker in the yellowcake factory, for example, connects to the shot of *The Richest Uranium Mine on Earth,* which connects to the picture of Bertell. The chain of connectivity is extensive. The image titled *O.K. Liquor, Blind River* (Plate 67) – the slang for yellowcake is "O.K. Liquor" – is related to *Uranium Turned into Gas* (Plate 68), which depicts toxic shipping casks at the Eldorado factory in Port Hope that converts yellowcake into uranium hexafluoride. These casks are the same as those transported by the doomed *Mont Louis.*

A chain of connectivity like that in *At Work in the Fields of the Bomb* characterizes the photobook *Geography Lesson: Canadian Notes* (1997), by Allan Sekula. *Geography Lesson* focuses on the links between the nickel mines of Sudbury and the Bank of Canada in Ottawa; like Del Tredici's project, it refuses to traffic in spectacle. Sekula extends the connections and the rhythms established in his photographs by alternating colour and black-and-white images.[53] Both photobooks address the material geographies of postwar capital. Those of the nuclear complex are represented in *At Work in the Fields of the Bomb* and those of the global circulation of commodities, including uranium, in *Geography Lesson*. The alliance forged between empty landscape painting, commerce, and labour in Canada is represented in Sekula's book by two photographs printed on facing pages. The Ottawa reception room of Eldorado Resources Limited, the uranium mining and processing corporation, is portrayed on one side. [**Figure 1.11**] Hanging on a wall is a landscape painting by A.Y. Jackson. The president's office of the Union of Mine, Mill and Smelter Workers in Sudbury is depicted on the other; it, too, has a landscape painting on a wall. Both photobooks function as social documentary and take a self-consciously historical and critical approach to their material. Each guards against aestheticism. "When photographs are uncritically presented as historical

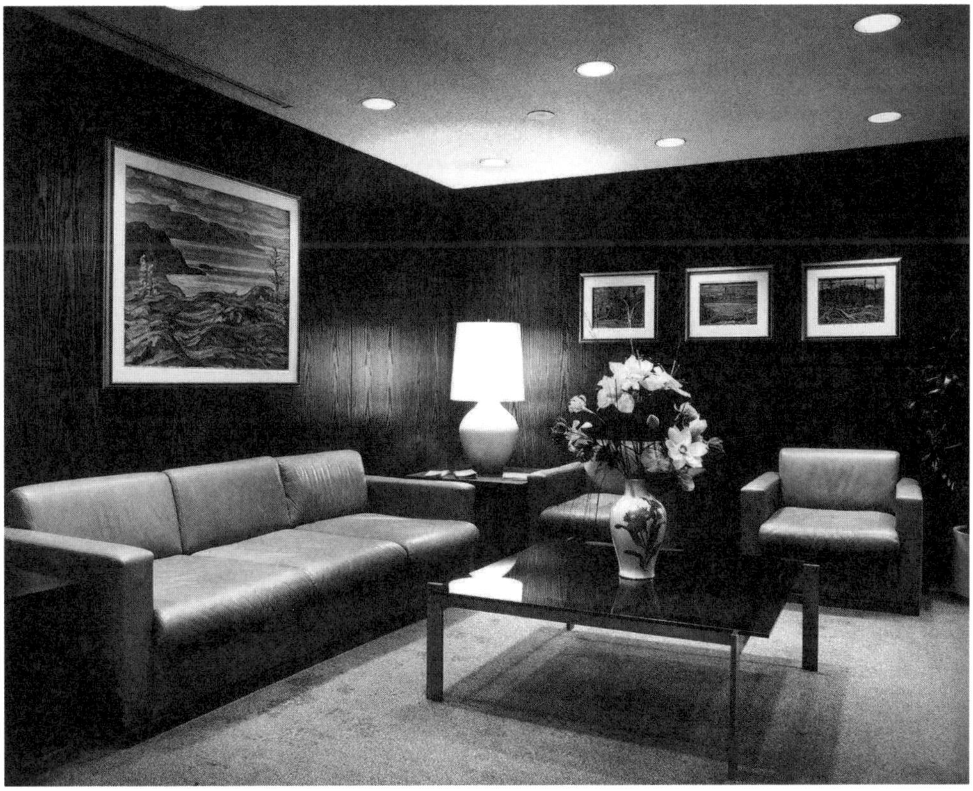

Figure 1.11: Allan Sekula. *Geography Lesson: Canadian Notes* (detail). 1987.

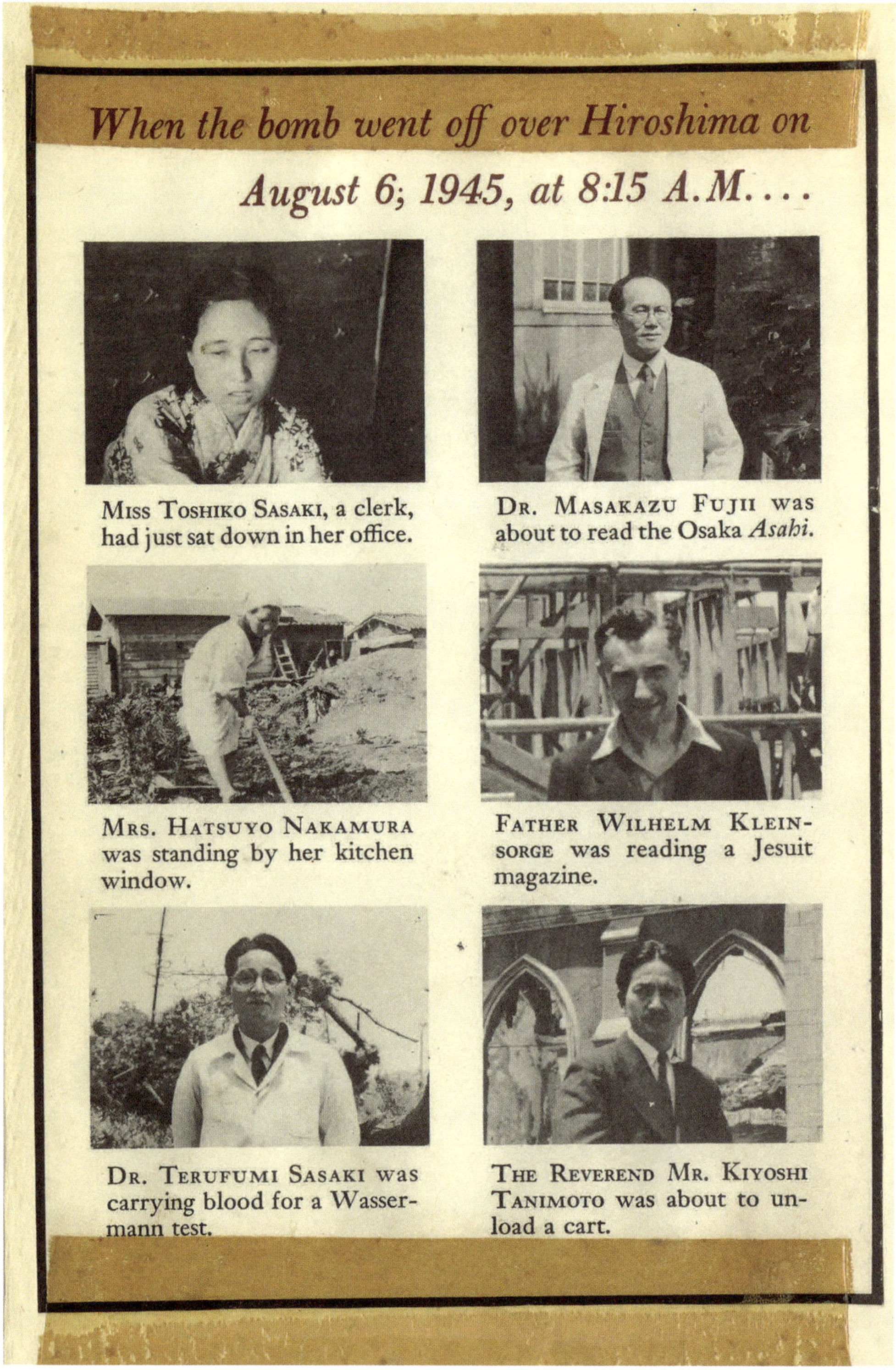

When the bomb went off over Hiroshima on August 6, 1945, at 8:15 A.M. . . .

MISS TOSHIKO SASAKI, a clerk, had just sat down in her office.

DR. MASAKAZU FUJII was about to read the Osaka *Asahi*.

MRS. HATSUYO NAKAMURA was standing by her kitchen window.

FATHER WILHELM KLEINSORGE was reading a Jesuit magazine.

DR. TERUFUMI SASAKI was carrying blood for a Wassermann test.

THE REVEREND MR. KIYOSHI TANIMOTO was about to unload a cart.

documents," Sekula writes, "they are transformed into aesthetic objects."[54] Then critical engagement turns into pretense.

Two American photobooks on the nuclear era appeared in print soon after Del Tredici's publication. In 1992, Richard Misrach published *Violent Legacies: Three Cantos,* and in 1993 Carole Gallagher published *American Ground Zero: The Secret Nuclear War.*[55] Like *At Work in the Fields of the Bomb,* they combine photographs and text to construct critical narratives. They help to make the bomb visible. But unlike Del Tredici's volume, they concentrate on only a single part of the nuclear nexus – the Nevada Test Site and its surroundings.[56]

Following the war, no information on the horrific results of the bombings was disseminated in Japan without first being submitted to American censors. From 1945 until the military occupation ended in 1952, a censorship board policed all aspects of public, and often private, expression in Japan. General Douglas MacArthur, the supreme commander of the Allied Forces overseeing the occupation, functioned more like a shadow emperor than a military commander.[57] The regime of censorship put in place by MacArthur was elaborate. At one point, the Civil Censorship Detachment employed more than six thousand people, the majority of whom were English-speaking Japanese nationals, chosen because they could identify suspicious material and translate it into English.[58] Commentary on the war was subject to intense scrutiny. Writing about the experience of the atomic bomb was not officially proscribed, but it was discouraged and often suppressed.

Although war correspondent John Hersey was permitted to publish his book *Hiroshima* in the United States – it became a sensation when the *New Yorker* magazine devoted an entire issue to it in August 1946 – he was not allowed to publish it in Japanese translation until 1949.[59] The book, which follows the stories of six hibakusha caught in the attack, remains the most enduring journalistic account of the effects of the bomb. Written in the third person, the prose is understated – "deliberately quiet," as Hersey commented later.[60] So are the photographs of the six victims, which were featured on the rear of the dust jacket of the first edition. In my own battered copy of this edition, these photographs have been taped inside the back cover. [**Figure 1.12**] The tape has yellowed with age. I sometimes flip back and forth between text and images when I read the book, just to make sure that the six people in Hersey's story actually survived the bombing.

The first substantial group of photographs of Hiroshima and Nagasaki victims that North American audiences saw was published by *Life* magazine. Its September 29, 1952, issue included the photo-essay "When Atom Bomb Struck – Uncensored," with fourteen pictures of hibakusha.[61] Some shots were by Matsushige, others by Yosuke Yamahata, who during the war was a news photographer making propaganda images for the naval division of the Japanese News and Information Bureau. At dawn on the day after the Nagasaki explosion, he walked into the smoking city and began photographing the ruination and the dead bodies, many

Figure 1.12: *When the Bomb Went Off over Hiroshima on August 6, 1945, at 8:15 A.M.* 1946.

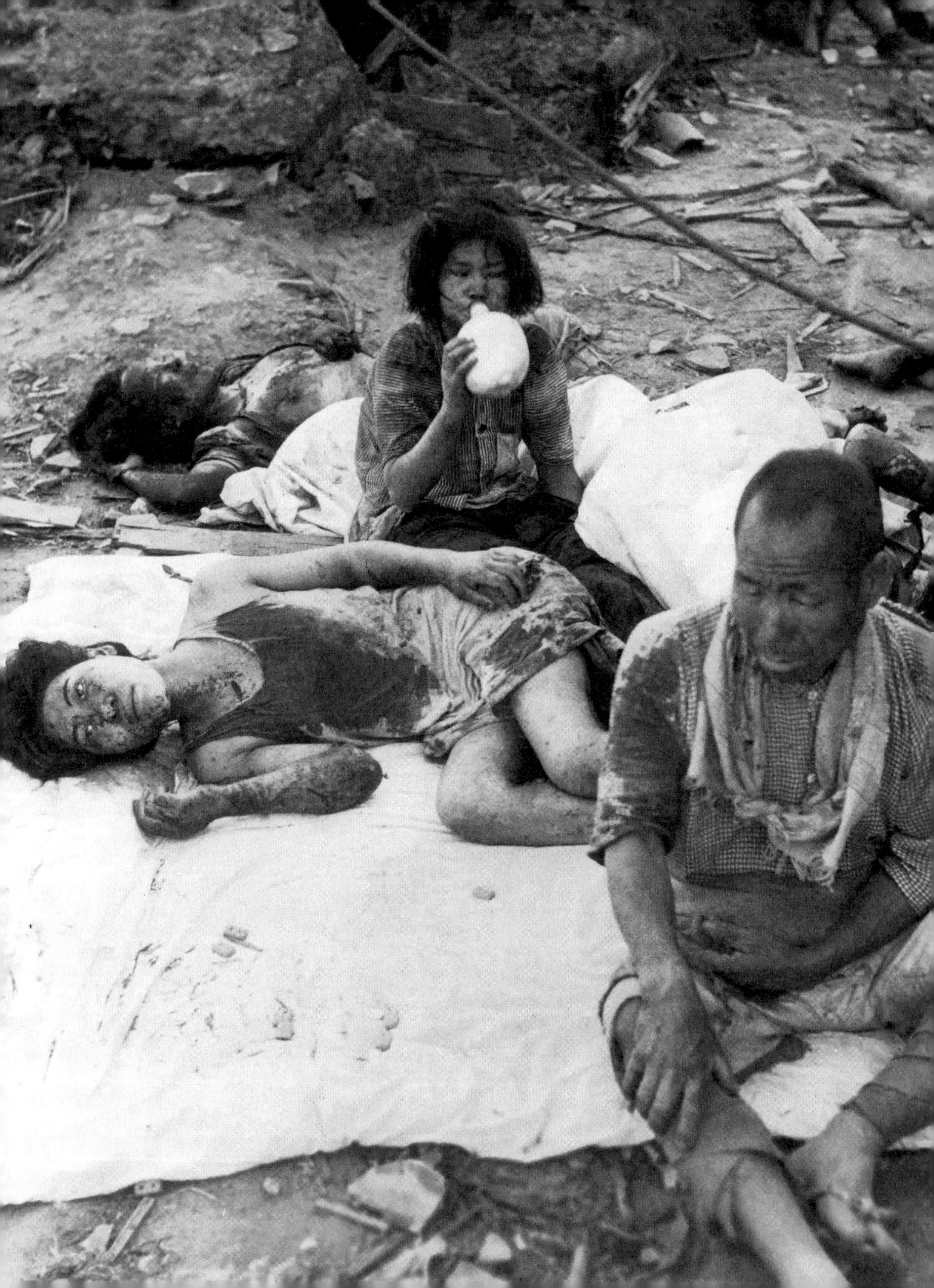

of them charred beyond recognition. [**Figure 1.13**] His photos are the most extensive visual record we have of the immediate aftermath of an atomic bombing. They were published in a book called *Atomized Nagasaki* when the seven-year US press ban was lifted on the circulation of nuclear images and information, and the photographs have been widely published ever since.[62] "Like snapshots from the end of the world, Yamahata's photographs bear witness to the horrors of the past," writes Misrach, himself a witness to nuclear horror.[63] In the opinion of historian Masafumi Suzuki, the images represent "'ground zero' of atomic photography."[64]

Japan has produced the largest number of nuclear photobooks of any country, many of which have received close critical attention.[65] *Atomized Nagasaki* is an indispensable record of the aftereffects of the atomic bomb, but it tells us little about the nuclear cycle and its workings compared to *At Work in the Fields of the Bomb*. Yamahata's photobook is about consequences. That is also the case with many other Japanese photobooks. Ken Domon's *Hiroshima* (1958), Kikuji Kawada's *The Map* (1965), Shomei Tomatsu's *11:02 Nagasaki* (1966), Hiromi Tsuchida's *Hiroshima 1945–1979* (1979), Eikoh Hosoe's *Deadly Ashes* (2007), and Ishiuchi Miyako's *Hiroshima* (2008) all focus primarily on the destructive capacity of the bomb and the death and suffering it caused.[66] The Japanese photobook that comes closest to Del Tredici's in its wide-angle approach is *Photo Document: Japan's Nuclear Power Plants* (1979), by Kenji Higuchi, who is also a founding member of the Atomic Photographers Guild.[67] Concentrating on the lives and factory routines of plant workers, and on their daily exposure to radiation, Higuchi makes visible the banality of the system for which their labour has been solicited. His photographs of nuclear workers at Japanese power plants are the first independent documentation of the industry. The black-and-white images, like the one illustrated of subcontract workers leaving the Tsuruga plant in Fukui Prefecture, are high-contrast prints that bleed to the edge of the page. [**Figure 1.14**] Many are slightly blurred. Stylistically, they are indebted to the example of *Provoke* magazine, published a decade earlier in Japan, but Higuchi undercuts the magazine's exclusive emphasis on the visual by including captions and texts, turning *Japan's Nuclear Power Plants* into a work of social documentary. *Provoke* declared that it wished to "free photography from subservience to the language of words"; Higuchi brings words and photography together.[68]

The global footprint of the nuclear body can be measured by its vast infrastructural networks. These networks are part of what is revealed in *At Work in the Fields of the Bomb*. They are likewise revealed in *The Journey* (1986), Peter Watkins's fourteen-hour film that maps the normalization of the bomb and nuclear catastrophe (and includes a series of images by Del Tredici), and in Blaine Campbell's large-format photographs taken at TRIUMF, Canada's national laboratory for particle and nuclear physics at the University of British Columbia.[69] Interested in finding ways to visualize applied mathematics and quantum theory, Campbell photographed TRIUMF infrastructure during an artist's residency there.[70] His

Figure 1.13: Yosuke Yamahata. *Last Drink*. 1945.

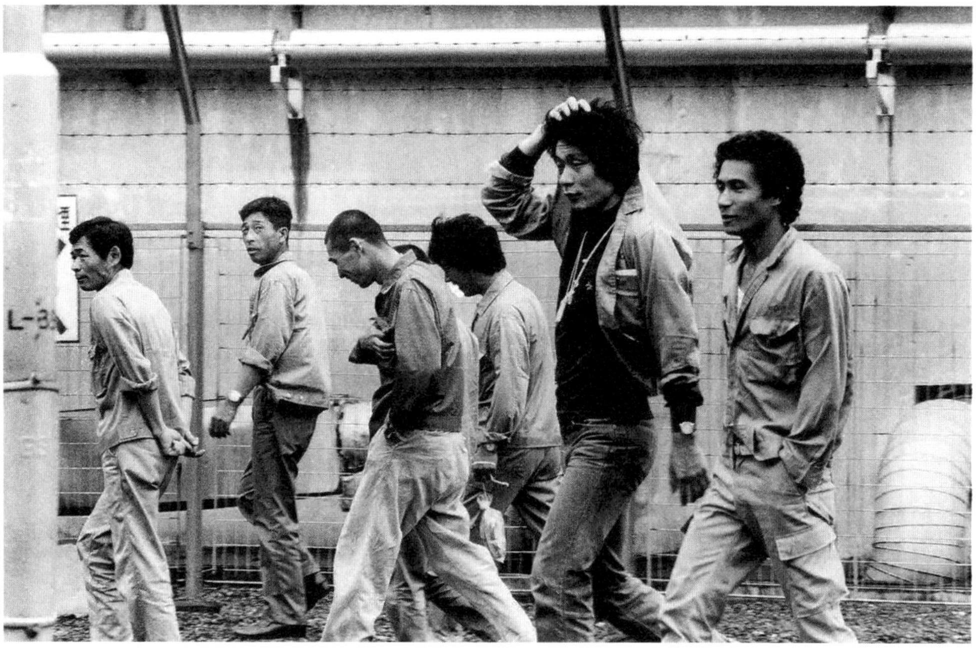

Figure 1.14: Kenji Higuchi. *Subcontract Workers, Tsuruga Nuclear Power Plant, Fukui Prefecture.* 1977. **Figure 1.15**: Blaine Campbell. *Omnis Experientiae Miraculum (Tigress).* 2016.

Omnis Experientiae Miraculum (Tigress) (2016) is a large, detailed shot of a gamma-ray spectrometer, a tangle of coloured wires and tubing. [**Figure 1.15**] The eye has trouble piecing the parts together, reminding us that nuclear networks are not easily revealed by photography. "If I learned anything through the fields of the bomb," Del Tredici notes, "it is that the closer you get to nuclear weapons the harder it is to see them."[71] The infrastructural networks addressed in his book include research centres, weapons laboratories, test sites, uranium mines, tailings pond walls, enrichment facilities, waste disposal tank farms, missile-launching platforms, command centres, and even nuclear museums.[72] In addition to threatening the human body, the networks put the human psyche at risk.

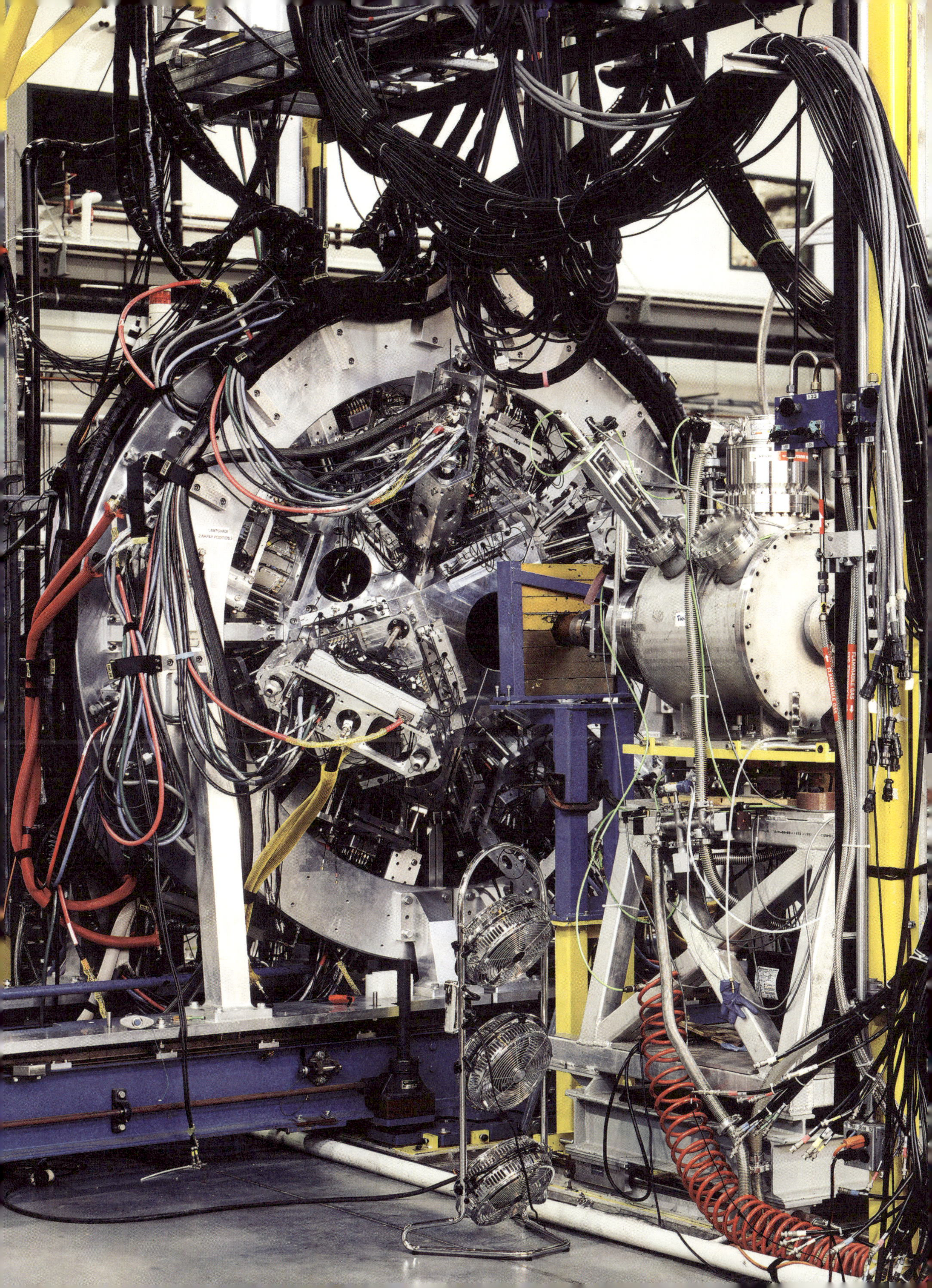

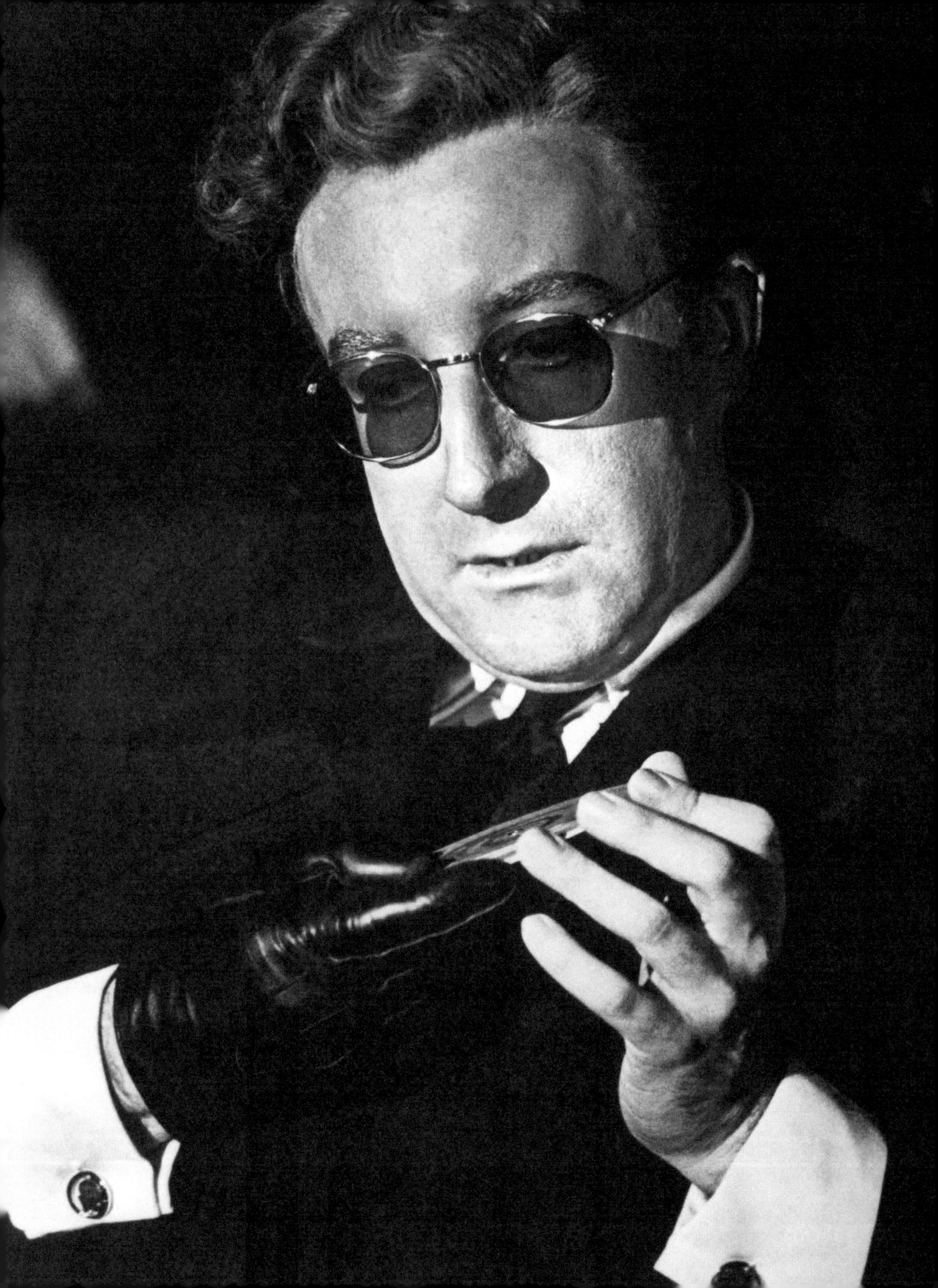

2
Chalk River

THE IMAGES WE CARRY IN OUR MINDS CONDITION HOW WE PERCEIVE NUCLEAR risk.[1] "Some fool in a laboratory might blow up the universe," the nuclear physicist Ernest Rutherford wrote to a friend in 1903 while undertaking radiation research at McGill University, producing an image that entered into the consciousness of a public already anxious about technological change.[2] The stereotype of a white-coated boffin playing with test tubes in a lab contributed to the invention of Professor Radium, the comic-book character who became a "human radium ray" as the result of an accident. Professor Radium is one of countless fictionalized mad scientists. Dr. Strangelove, played by Peter Sellers in Stanley Kubrick's 1964 film of the same name, is the most iconic of crazed nuclear physicists. He has no qualms about calculating with a slide rule the megadeaths that would result from all-out nuclear war with the Soviet Union. [**Figure 2.1**] In a publicity still for the film, he holds a slide rule between his good hand and his bad hand. The bad hand is sheathed in a black glove. Sigmund Freud has something to say about scientific risk. More than a decade before Enrico Fermi split the atom and discovered nuclear fission in his squash court at the University of Chicago in 1942, Freud declared that science had become so powerful it was in danger of eradicating the human species down "to the last man."[3] This chapter extends the concerns of the previous chapter on

Figure 2.1: Publicity still from *Dr. Strangelove or: How I Learned to Stop Worrying and Love the Bomb*. 1964.

the nuclear industry by examining public perceptions of radiological risk, particularly at the Chalk River complex.

Another figure that has entered the collective consciousness of a public anxious about the dangers of nuclear technology is Godzilla. The monster's signature weapon is his radioactive "atomic breath." Godzilla is a product of the atomic age, roused from deep sleep in primordial ooze by nuclear detonations. His skin resembles the keloid scars that developed on the bodies of survivors of Hiroshima and Nagasaki. Since the creature's invention in 1954, the same year that the United States detonated Castle Bravo, a thermonuclear weapon with the force of twelve thousand Hiroshimas, Godzilla has become a metaphor for how nuclear technologies have threatened civilization by outrunning the ability of experts to control them. In the original Japanese film, Godzilla's breath was generated by an atomic furnace inside his belly, which erupted through his mouth as deadly rays.[4] Physics tells us that radiation is invisible, odourless, and tasteless, but Godzilla shows us that radioactive breath is blue, hot, and smelly.[5] There are more than two dozen Godzilla movies. In many of them, the monster stands in the middle of a burning city, shooting blue rays through long-pointed teeth at a terrified population below.

Godzilla functions as a paradigm for the trauma incited by the bombings of Hiroshima and Nagasaki. It is a response to psychic shattering. Images of nuclear violence have been in wide circulation since 1945 – sometimes as fiction, sometimes as fact – like a palimpsest of unconscious recollection that has shaped public perceptions of nuclear catastrophe and radiological hazard. What role have photography and what I am inclined to call the "Godzilla effect" – a dispersed phenomenon larger than the monster itself – played in shaping perceptions of radiation risk? How, for example, have they altered the way we might read a newspaper photograph of an aging NRU (National Research Universal) reactor that has sprung a leak and been shut down for repairs? [**Figure 2.2**] A sign in the foreground of the shot, occupying the lower left side of the image and bearing the black-and-yellow trefoil symbol for ionizing radiation, warns against approaching the machine.

Chalk River Laboratories is owned by a private consortium. Until 2014, it functioned under the umbrella of a Crown corporation, Atomic Energy of Canada Limited (AECL Chalk River). The facility began operations during the Second World War as part of the Manhattan Project. The nuclear complex is on the Ottawa River, 180 kilometres northwest of the Canadian capital, away from major population centres.[6] The relocation by the British government of its Cavendish Laboratory from Cambridge to Montreal, to allow for closer collaboration on the Manhattan Project, contributed to its development. Chalk River Laboratories remains a major Canadian nuclear research facility today, specializing in CANDU reactor technology. Looking to the future, it hopes to participate in the development of a new generation of reactors, small modular reactors (SMRs). The government supports the initiative, arguing that it will help to advance resource extraction in the North.[7]

In a 2009 email message sent to me from AECL Chalk River, researchers seemed to concur with historian and physicist Spencer R. Weart's observation that public perceptions of

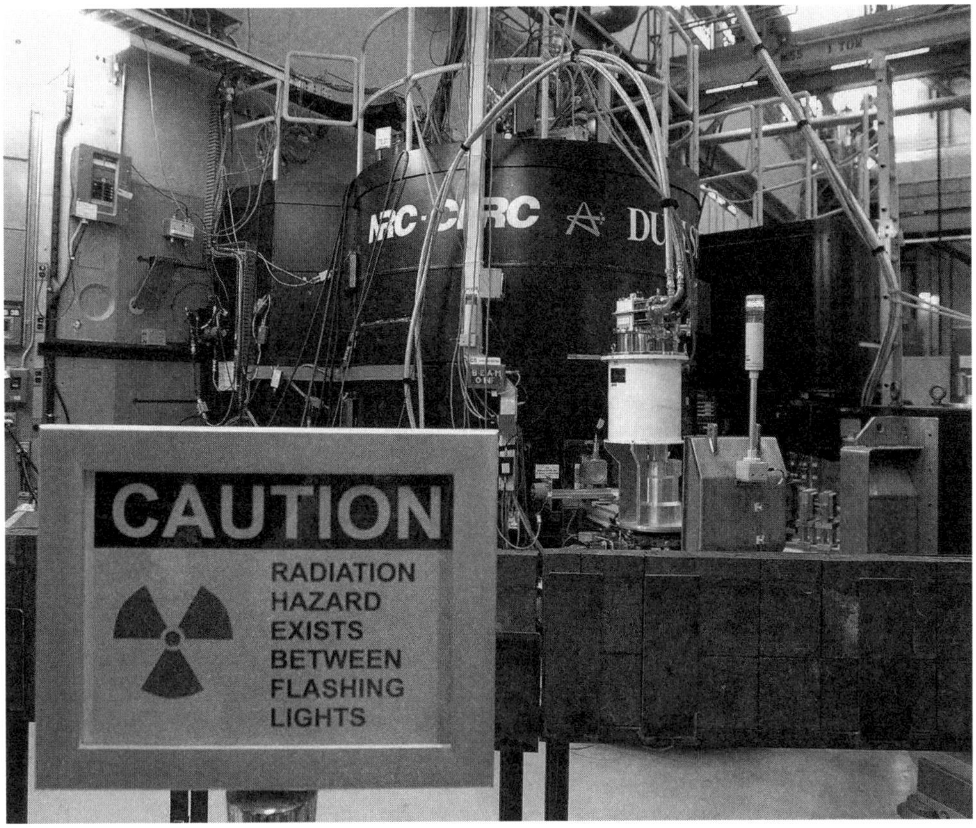

Figure 2.2: Fred Chartrand. *The 53-Year-Old Chalk River Reactor Has Been Out of Operation for Repair of a Radioactive Water Leak.* 2007.

nuclear risk are conditioned by the images we carry with us. The message arrived shortly after the NRU began leaking heavy water for the second time in eighteen months, requiring it to be shut down for extensive repairs. It also arrived after I had been refused entry, for security reasons, to Chalk River's unclassified photo archives. Would I be willing to participate in a survey, the message asked, on public attitudes toward radiation? "The public has an irrational attitude to the radiation hazards presented by nuclear, medical and environmental factors," the email stated, which is "distorted by information filters that change the balance between perceived benefit and detriment." The survey intended to "measure the deviation between public and expert views" in order to better understand the imbalance.[8] When its organizers stated that public attitudes were skewed negatively by "information filters," they were referring to broadcast and print media, social media, film, video, television, photography, and other forms of visual and written representation through which the public receives information on nuclear issues.[9] They were presumably thinking of an image like *Atomic Energy of Canada Limited*, by Chris Lund, in which a figure behind glass, dressed in rad-safe gear, seems to be trapped at his work station. [**Figure 2.3**] In the opinion of the organizers, public attitudes were skewed by irrational fear. They had been turned upside down, as it were, by the Godzilla effect.

Along with suggesting that the balance between public and expert views on radiation is unstable, the AECL Chalk River survey recognized that in radiological risk-benefit equations, one side of the equation is inseparable from the other. Both observations are correct. Public attitudes toward radiation vary widely according to news of fresh disasters, and positives and negatives are not easily disentangled. As discussed in the Introduction, this instability can be explained by the notion of the pharmakon, which refers to a remedy that can both injure and cure.[10] The first atomic bomb, for example, was a spherical device choked by wires and known as "the gadget," which was detonated in 1945 at Alamogordo, New Mexico. [**Figure 2.4**] Uncannily similar to that bomb, the machine in the government-issued photograph *Detecting*

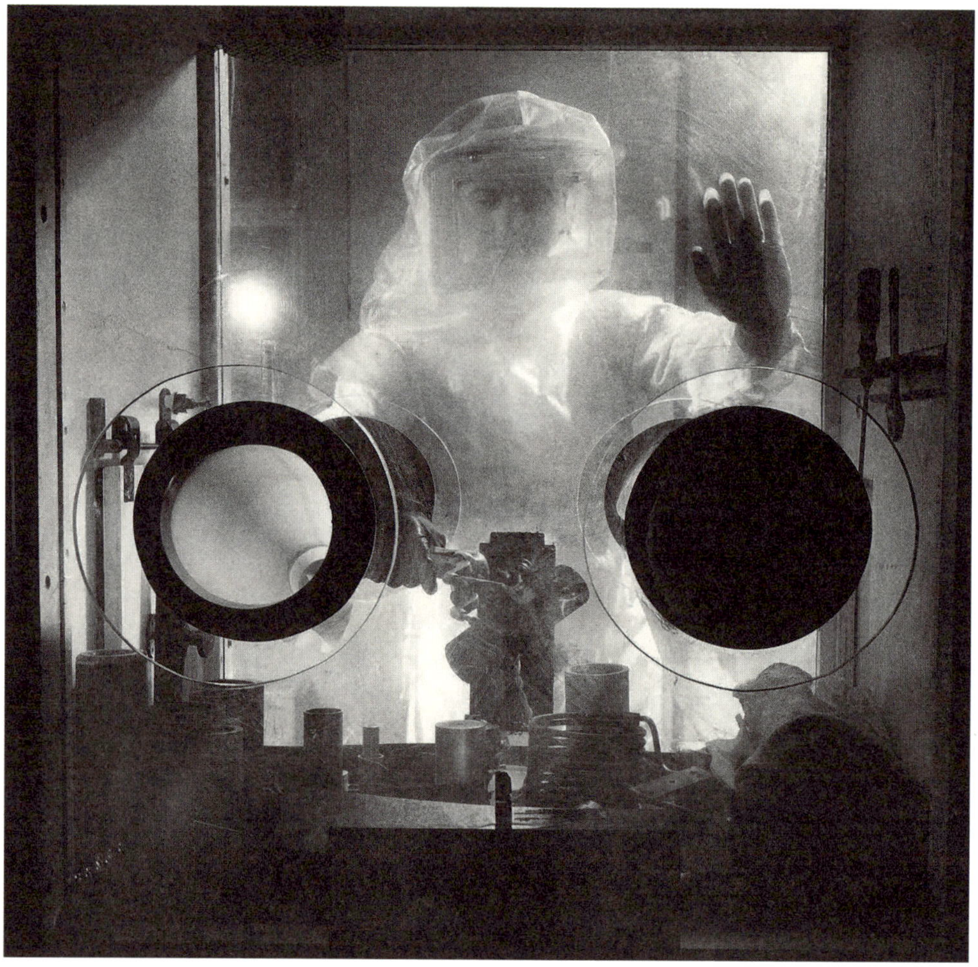

Figure 2.3: Chris Lund. *Atomic Energy of Canada Limited, Chalk River.* 1955.

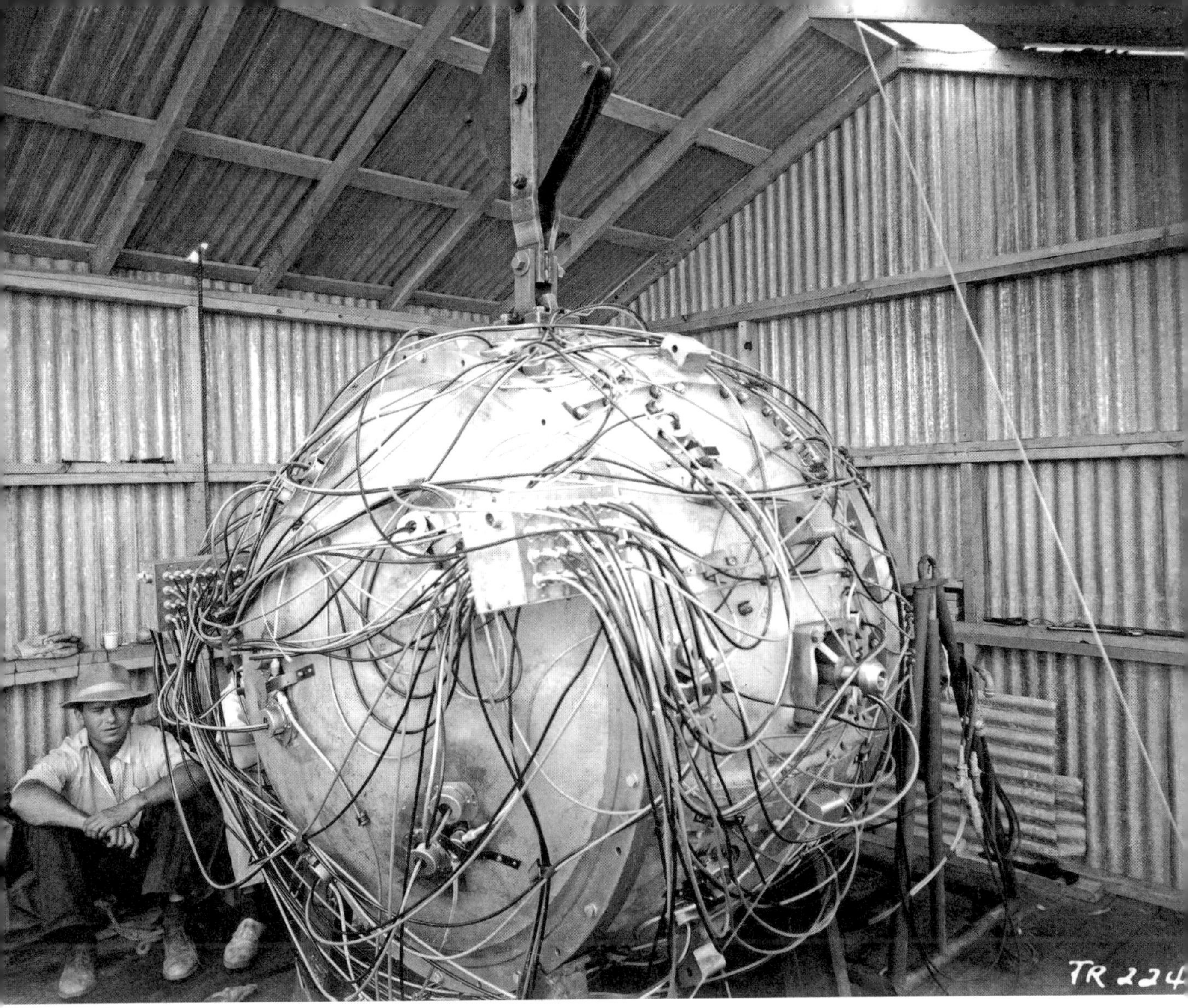

Figure 2.4: US Department of Defense. *Nuclear Device Detonated at the Trinity Test Site,* Alamogordo, New Mexico. 1945.

Disease with Atoms looks like a cyborg Medusa sprouting wires instead of snakes from its head. [**Figure 2.5**] The machines dwarf the human figures around them. In the shot of the disease-detection apparatus, the top half of the woman's head is cut off, and in the photograph of the gadget the man wearing a fedora seated on the floor blends into his surroundings. (The arm and the foot of the individuals to his left and right are even less obvious.) The disease-detection machine appears to be more life threatening than life enhancing, a visual contradiction that exemplifies the frustration expressed by the survey organizers about filters that distort perception.

Another example of the pharmakon is an instrument called the gamma-ray atomic camera, a lightweight, portable device that requires no electric power and can reach into places

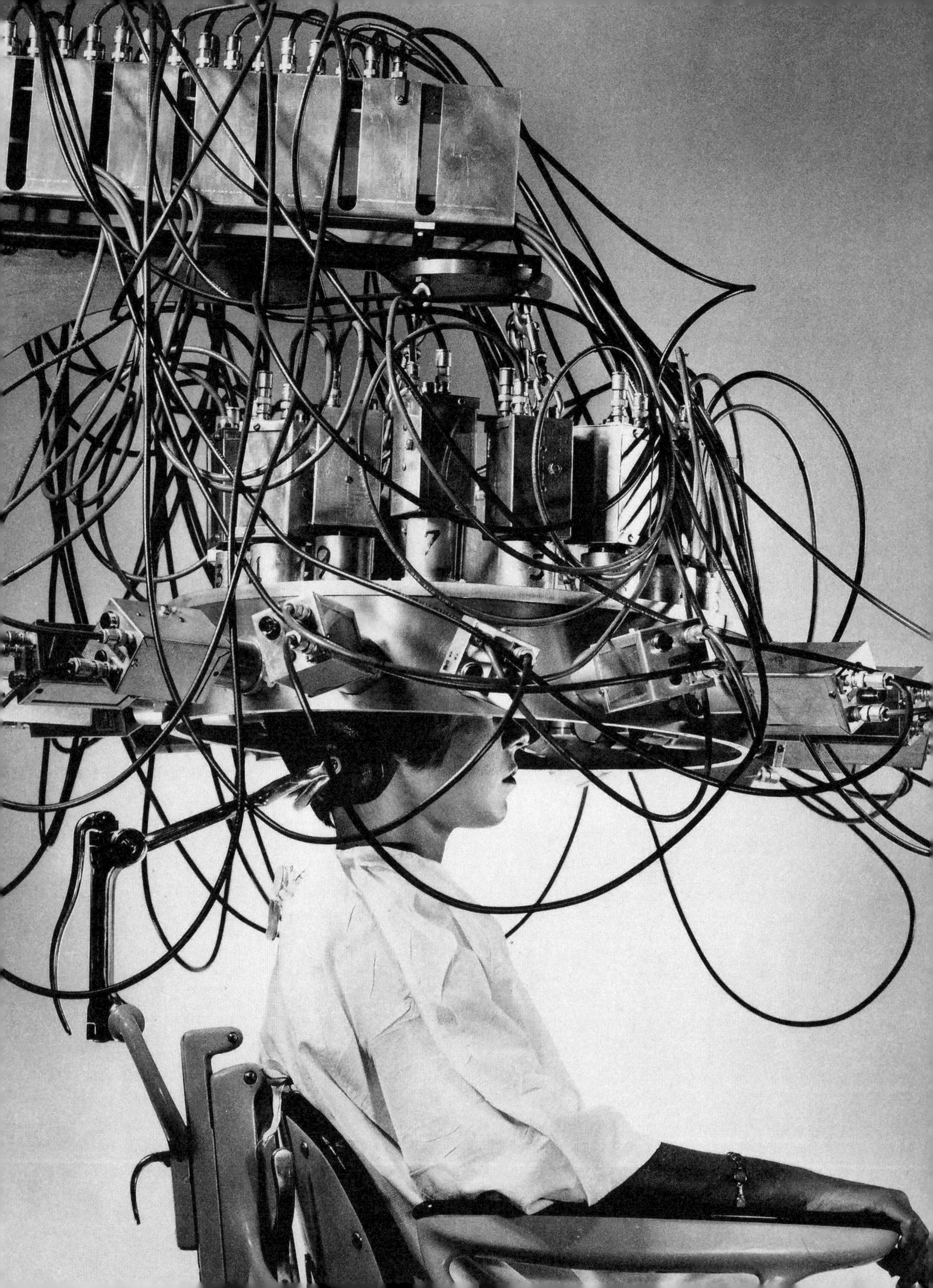

inaccessible to its predecessor, the industrial X-ray machine. The first atomic cameras were loaded with capsules of radioactive iridium that emitted gamma rays to record on film structural flaws in metal. They produced a specialized form of knowledge that extended beyond previous perceptual limits.[11] At AECL Chalk River in 1957, damage to an atomic camera contaminated nearly five hundred workers. In an article titled "Who Will Police Our Peacetime Atoms?" *Star Weekly* magazine recounted the incident:

> *Innocent and harmless though these [iridium] capsules look, they are deadly. Atomic Energy of Canada Ltd., the only supplier in this country, ships them in lead containers weighing up to 500 pounds ... The technician [who handled the camera] undoubtedly knew this, as he must also have known why he was not permitted to remove the capsule from its shielded compartment in the camera. Unshielded and held one yard from his body, it would give off more radiation in one hour than an atomic worker is permitted in six months. Nevertheless ... he removed the capsule and tried to change its shape by cutting the edge with a hacksaw.*[12]

By hacking through the iridium capsule, the technician contaminated hundreds of workers. Human error in the nuclear industry is no more frequent than in any other industrial endeavour, but it has less predictable consequences. The sociologist Charles Perrow, who studied the partial meltdown of the Three Mile Island reactor in Pennsylvania in 1979, concluded that "as technological systems become more complex, disasters that appear to result from a confluence of bad consequences become, in fact, unavoidable, as a failure in one part causes a failure in another and another in ways that no designer could predict."[13] Extraordinary disasters, he determined, are inevitable. Technoscientific society is unable to control nuclear risk, and this has produced a sequence of accidents leading from Chalk River to Three Mile Island to Chernobyl to Fukushima Daiichi. In Ted Grant's photograph of a boy who developed cancer as a result of the Chernobyl disaster, the subject stares down the camera. [**Figure 2.6**] Light strikes his bald head in the same way that it strikes the hood of the automobile pictured behind him. The juxtaposition of the boy's head and the car's hood is unsettling, as the boy may never have the chance to own or drive a car, never mind a luxury car like the one in the image.

At Chernobyl in 1986, a failed turbine test in Reactor No. 4 blew a hole in the reactor's roof and spewed radiation across Europe and the world, producing one of the two worst nuclear peacetime disasters in history. The other is the triple meltdown at Fukushima Daiichi. The Chernobyl explosion could potentially have been much worse than it was. It could have

Figure 2.5: US Information Service. *Detecting Disease with Atoms: A Patient Sits Underneath a Complicated Contrivance with Dangling Wires Called a Multidetector, Which Can Pinpoint a Brain Tumor*. c. 1950s.

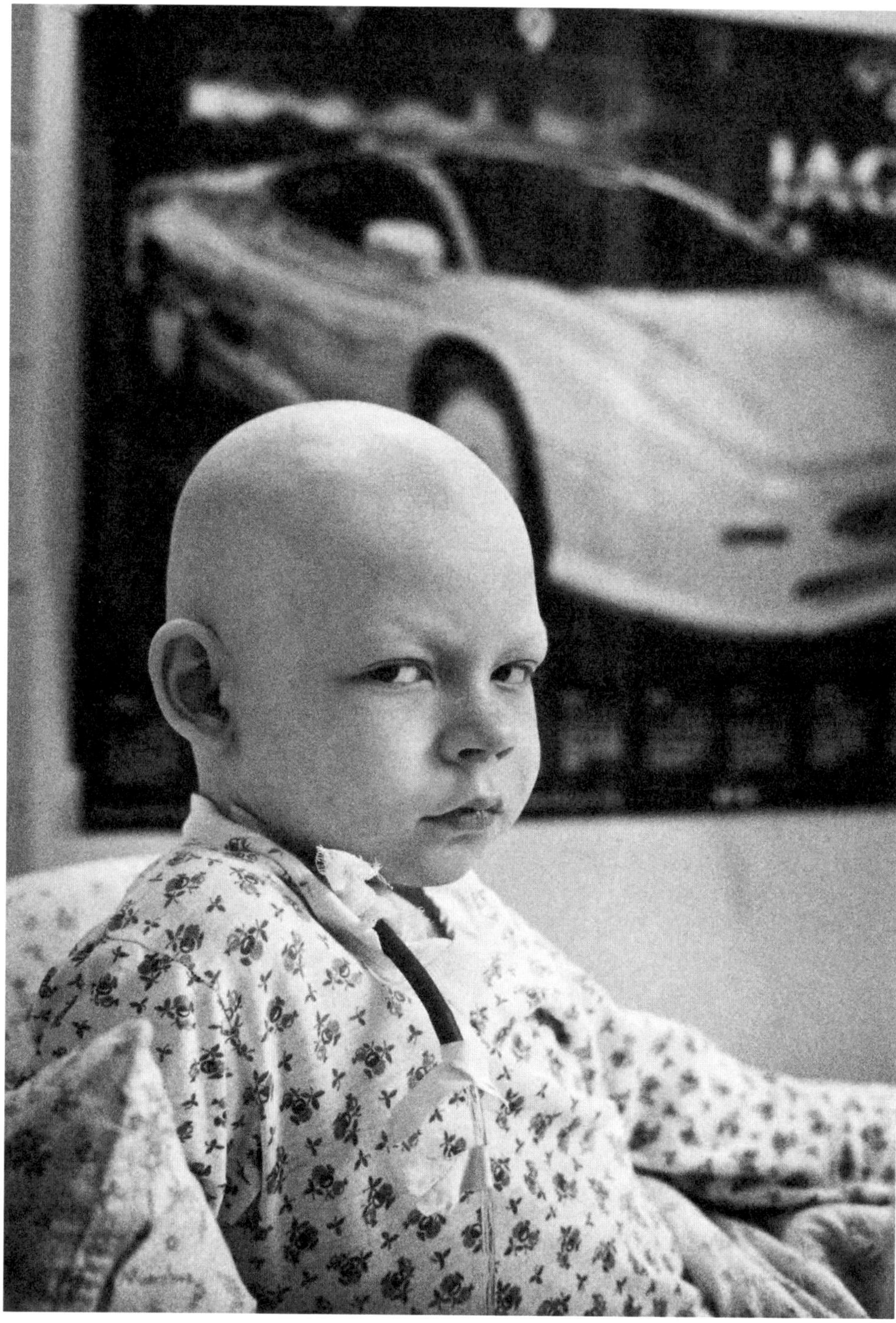

Figure 2.6: Ted Grant. *Untitled (The Children of Chernobyl)*. 1992.

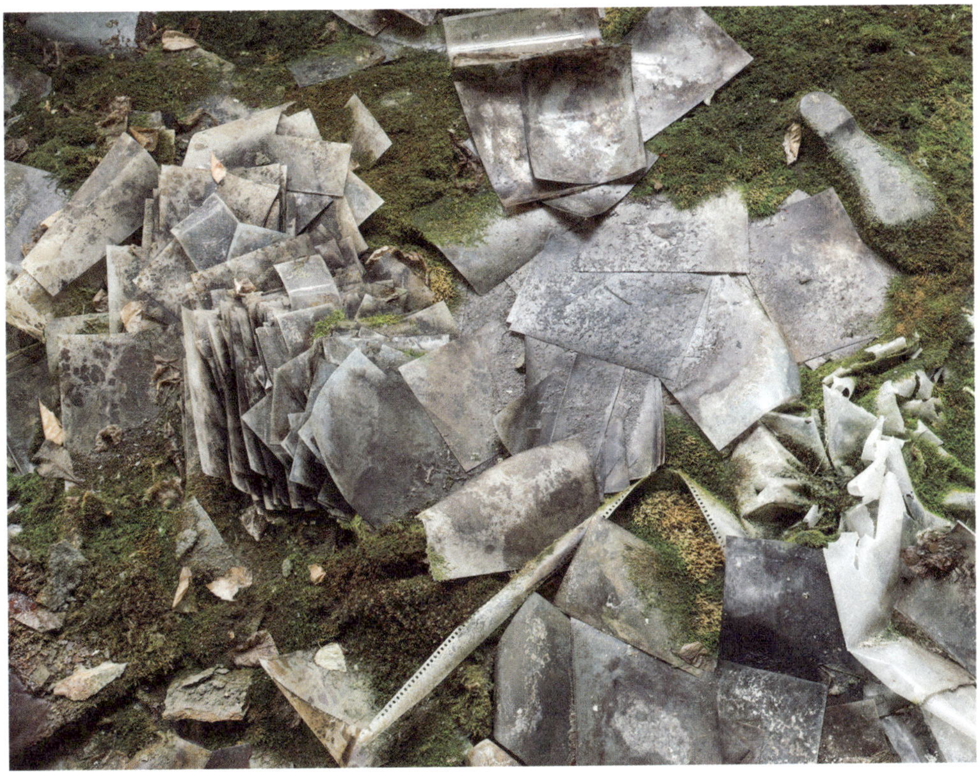

| **Figure 2.7**: David McMillan. *Negatives and Shoe, Photographer's Studio, Prypiat.* 2016.

generated a second, larger explosion that would have made much of Europe uninhabitable.[14] The exclusion zone around Chernobyl, including the adjacent city of Prypiat, remains uninhabitable. Historian Daniel Beer refers to Prypiat as "a latter-day Pompeii," a description borne out by David McMillan's photographs of the city, which the artist began taking in 1994.[15] [**Figure 2.7**] *Negatives and Shoe, Photographer's Studio, Prypiat,* is one of thousands of shots in McMillan's visual archive of nuclear abandonment. The discoloured, decaying negatives in the studio are scattered on moss across the plane of the photograph, and the sole of the overturned shoe is discernible at the upper right.[16] For a photographer, the ruined studio of another photographer functions as a *memento mori.* It addresses not only the transience of life, but also the transience of images. Compared to the artist's own visual archive, the abandoned Prypiat negatives are spectral ruins.

Until the meltdown at Three Mile Island, hourglass-shaped cooling towers were generally viewed as signs of progress. That changed after 1979. In public consciousness, the nuclear industry became linked with drastic failure, death, and the ubiquity of toxic waste – and so did images associated with it. Why else would Robert Del Tredici's photograph of a woman walking her children and a dog on a suburban road in the direction of two cooling towers seem menacing? [**Figure 2.8**] There is a reason why images of the dead at Chernobyl and

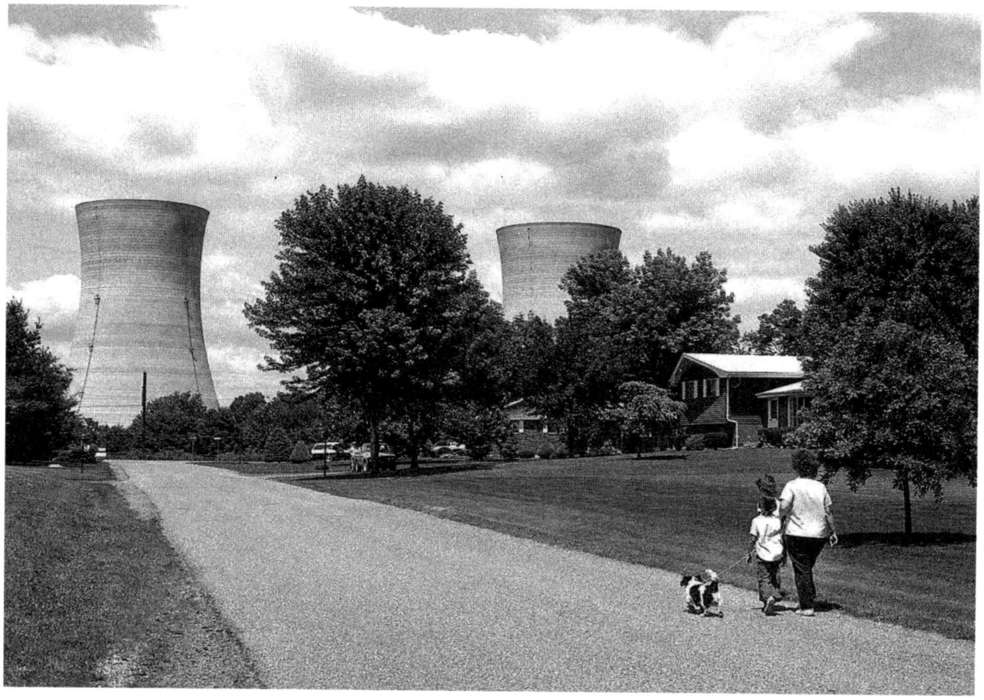

Figure 2.8: Robert Del Tredici. *Peck's Road, Londonderry Township*. 1979. **Figure 2.9**: *Plastic Suit Protects against Atomic Radiation*. 1960.

Fukushima, buried four metres underground in zinc- or lead-lined coffins, are difficult to find. Authorities have kept them out of the public eye; photography is forbidden at nuclear funerals.

The accident with the atomic camera at AECL Chalk River is a "paradox of necessity," to borrow a phrase from the poet Octavio Paz.[17] If nuclear systems were to function properly, radiological cameras were required to search for flaws and leaks in reactors, bomb casings, and other essential metal components. But if a camera failed, it put nuclear workers at further risk. A camera was also used to illustrate the anti-radiation clothing worn by workers who were assigned to clean up after the accident. [**Figure 2.9**] A photograph shows a "jumper," also known as a "gamma sponge" or a "glow boy," dressed up like a giant insect with truncated feet – a bloated antecedent to Darth Vader, sheathed in white instead of black.[18] But the worker in rad-safe gear is not a fictional character. If a puncture were to occur in his plastic suit, which is inflated with compressed air funnelled from an unseen generator, he would become radioactively contaminated. Another slang term for nuclear workers sent into toxic areas after accidents is "dose fodder." In Japan, cleanup workers at Fukushima Daiichi were called "nuclear gypsies."

Since the hacksaw accident occurred at Chalk River, radiation safety gear has become more reliable. Canadian scientists at the Radiological Analysis and Defence group have designed improved equipment for military personnel. Some of the new headgear is represented by Mary Kavanagh, who participated in the Canadian Forces Artists Program in 2012–13, in

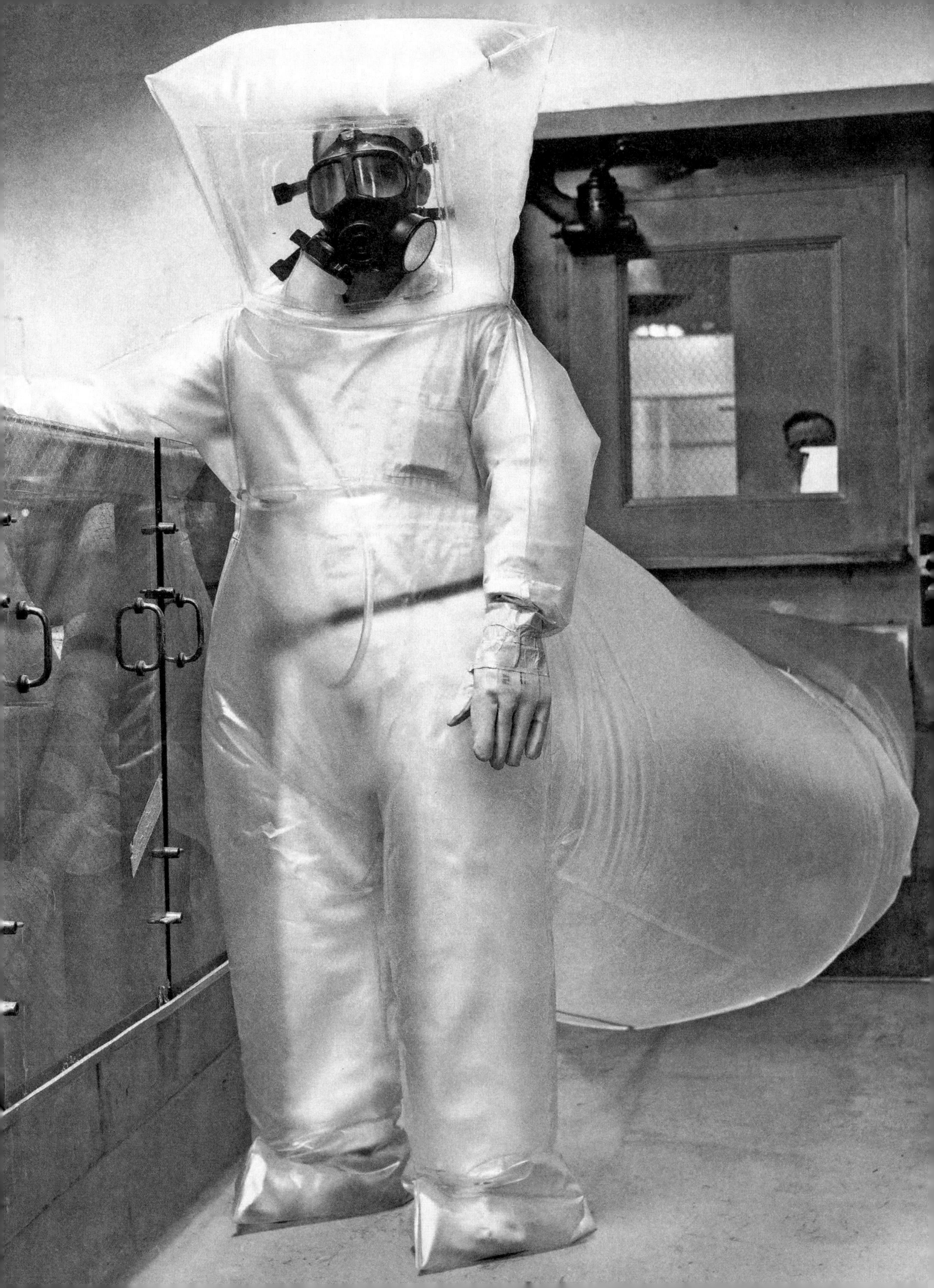

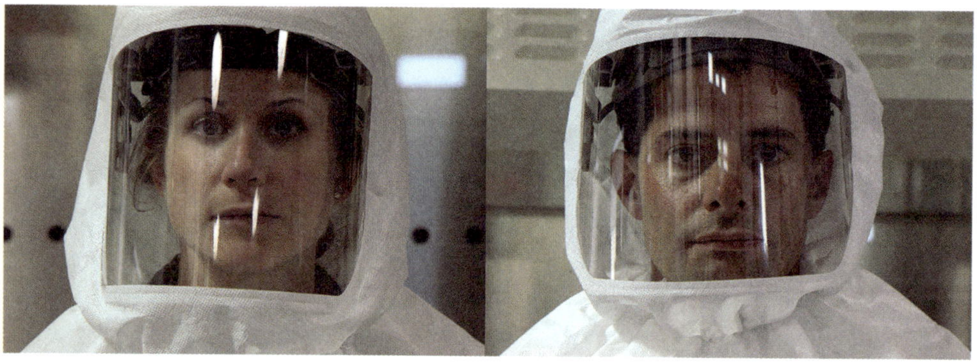

Figure 2.10: Mary Kavanagh. *The Expulsion (in White)*. 2012–13.

The Expulsion (in White). [**Figure 2.10**] The photograph is a double portrait of a woman and a man looking deadpan at the camera through the convex Plexiglas of their white helmets.[19] Their shiny regalia promises a brave new world of immunity from chemical and nuclear threat – except that no protection against nuclear threat can ever be foolproof. There is no returning to Eden, least of all for the military Adam and Eve in the picture.

The shutdown of the leaking NRU reactor at Chalk River began in May 2009 and continued until August 2010. The reactor went "critical" in 1957 – that is, it started self-sustaining operations – the same year as the atomic camera incident. It needed to be drained of heavy water and then repaired by remote means to avoid endangering workers. At the time of the shutdown, it supplied approximately half the world's medical radioisotopes. Through Nordion, a subsidiary first of Eldorado Mining and Refining and then of AECL Chalk River, Canada had pioneered the development of cancer treatments using cobalt-60 and other radioisotopes.[20] The NRU reactor's closure caused a severe shortage in supplies and led to delays in cancer diagnoses, heart monitoring, and other medical procedures. Although safer ways are now being developed to manufacture medical isotopes, those at Chalk River are produced by bombarding highly enriched uranium – the kind of uranium in nuclear weapons – with neutrons. The procedure produces highly toxic waste and the problem of how to dispose of it.

In the eyes of the AECL Chalk River survey organizers, the benefits of making radioisotopes for medicine outweighed the risks. The public's perception of risk was too high, its perception of benefit too low. They may have been correct; I am no expert in nuclear risk analysis as it relates to medical concerns. In *America's Nuclear Wastelands,* Max S. Power writes that because "the hazards produced by nuclear contamination are unseen and often latent, people ... have no sensory clues to help protect themselves," and this generates anxiety.[21] Polls demonstrate that the greatest concern of the public since the end of the Cold War has been the handling of nuclear waste.[22] Power also writes that "people are more likely to dread things nuclear" because of their primary association of the nuclear with war.[23] This is the Godzilla effect at work. The dread arises especially at moments of nuclear danger, when imagery from the past is recognized by the present as its own.[24] Del Tredici's photograph of reindeer carcasses in the Harads Same-produktor meat packing plant in Swedish Lapland, which shows meat unfit for consumption due to radioactive caesium-137 fallout from Chernobyl, suddenly acquired new resonance in 2011 following the triple meltdown at Fukushima Daiichi. [**Figure 2.11**] Reindeer occupy a central place in the lives of Laplanders. The contaminated reindeer, filling a room-sized freezer to capacity, their legs pointing toward the ceiling like otherworldly stalagmites inside a primordial futuristic cave, were eventually sent to mink farms as feed. Del Tredici was so stunned by this information that he forgot to ask the plant manager what happened to the meat and excrement of the mink.[25]

Nuclear images, whether documentary or fictional, instruct us about nuclear realities at the same time as they conjure up hallucinations. The space between reality and hallucination is where the Godzilla effect is most strongly felt. Consider a newspaper photograph of Conservative minister of natural resources Lisa Raitt and her press secretary arriving at a news conference, less than two months after the NRU reactor at AECL Chalk River began leaking heavy water. [**Figure 2.12**] It was published in the *Globe and Mail* on June 4, 2009. The two women look grim, like many officials who are forced to face the media after a disaster.

Figure 2.11: Robert Del Tredici. *The Becquerel Reindeer, Harads Same-produktor, Harads, Lapland, Sweden.* 1986.

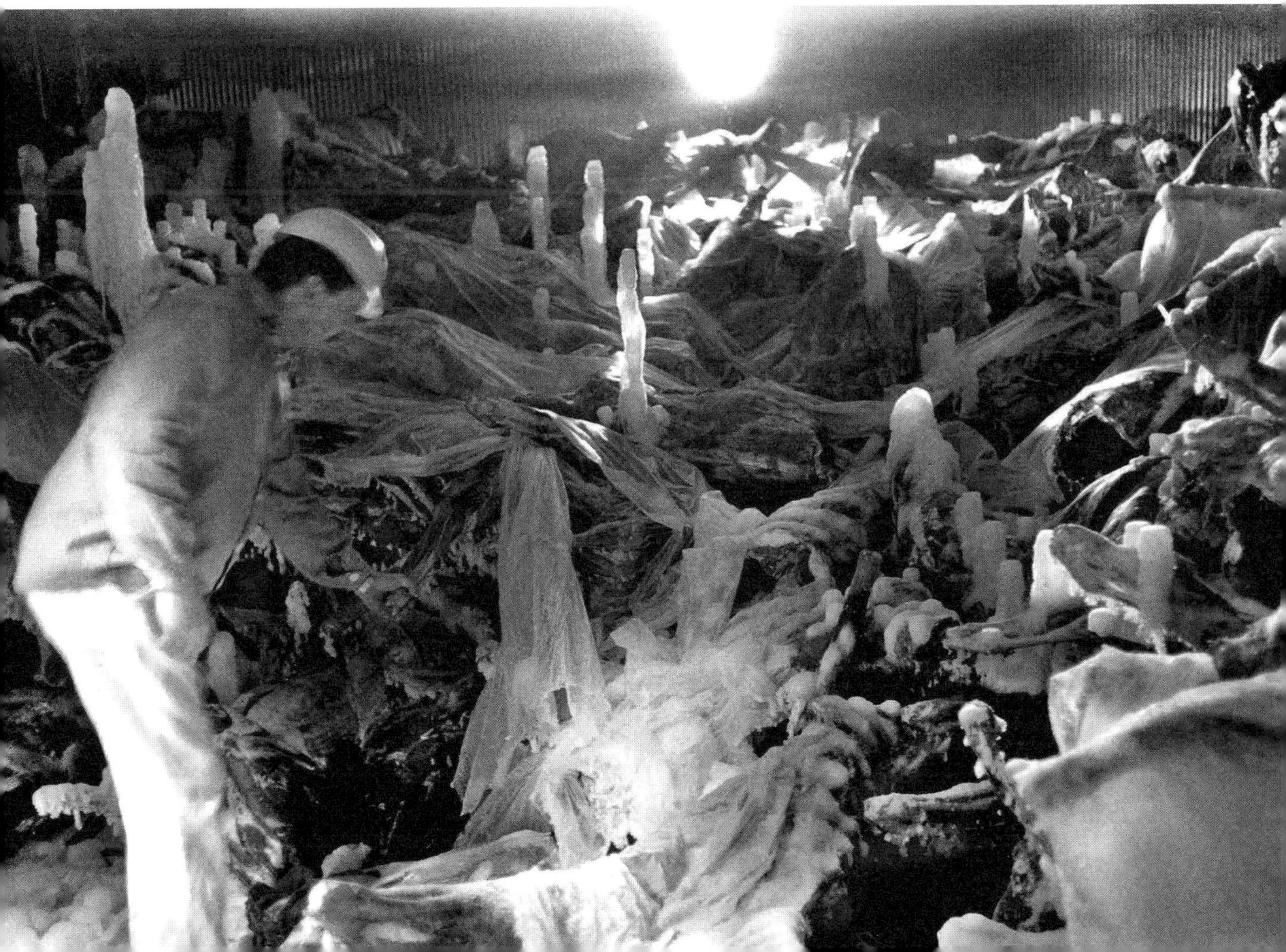

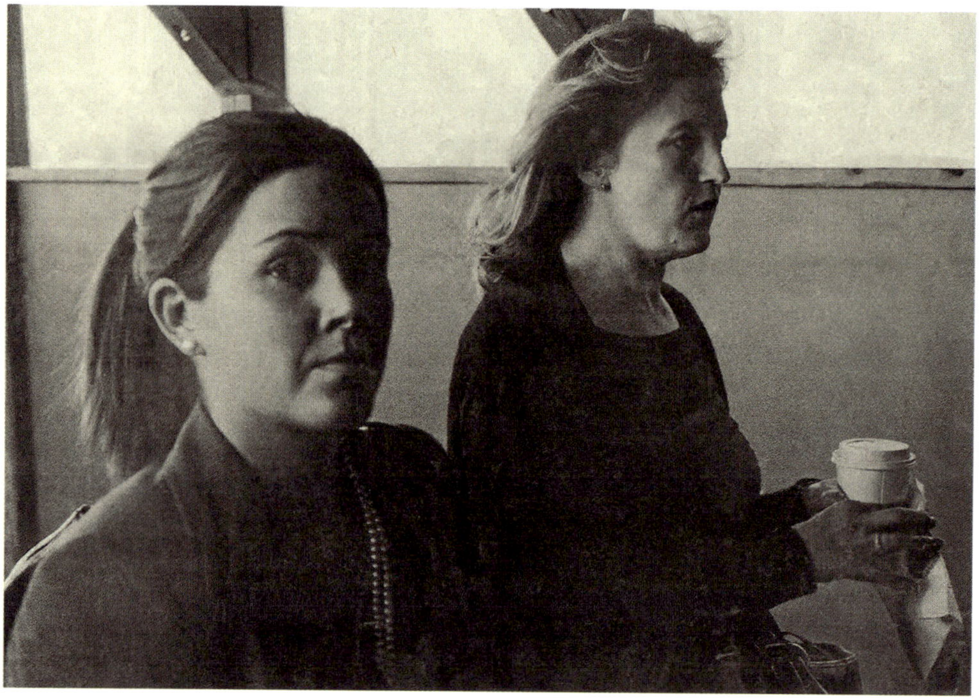

Figure 2.12: Chris Wattie. *Natural Resources Minister Lisa Raitt (Right) Enters Parliament Yesterday with Press Secretary Jasmine MacDonnell.* 2009. **Figure 2.13**: Associated Press. *Checks for Radioactive Contamination, NRX Reactor, Chalk River, Ontario.* 1947.

In this instance, the disaster has been compounded by an indiscretion. Following an interview at the offices of CTV News, Raitt, or her secretary, mistakenly left behind secret documents about the nuclear industry, an error that became public at an already sensitive moment. Raitt revealed nothing about the contents of the documents, leaving the public to imagine the worst.

The first AECL Chalk River reactor, the ZEEP, would have enabled Canada to build its own atomic bomb. C.D. Howe, who was "Minister of Everything" in the King government during the Second World War, including Canada's participation in the Manhattan Project, announced the successful construction of the reactor immediately following the bombing of Hiroshima.[26] A second reactor, the NRX, was finished two years later. In December 1947, the Associated Press was given permission to take a series of photographs for distribution as press images. One of these, *Checks for Radioactive Contamination,* shows how AECL Chalk River protected itself against radiation accidents and spills. [**Figure 2.13**] A health inspector in a white lab coat squats in a hallway beside an array of cleaning equipment. A caption accompanying the image, which was staged for the camera, explains that he is using a portable Geiger counter to check for radioactive contamination. The cleaning supplies in front of him are standard issue, what might be found in any school or business office at the time: brooms, mops, rags, work shoes, dustpan, vacuum cleaner, floor polisher, aluminum bucket, O-Cedar

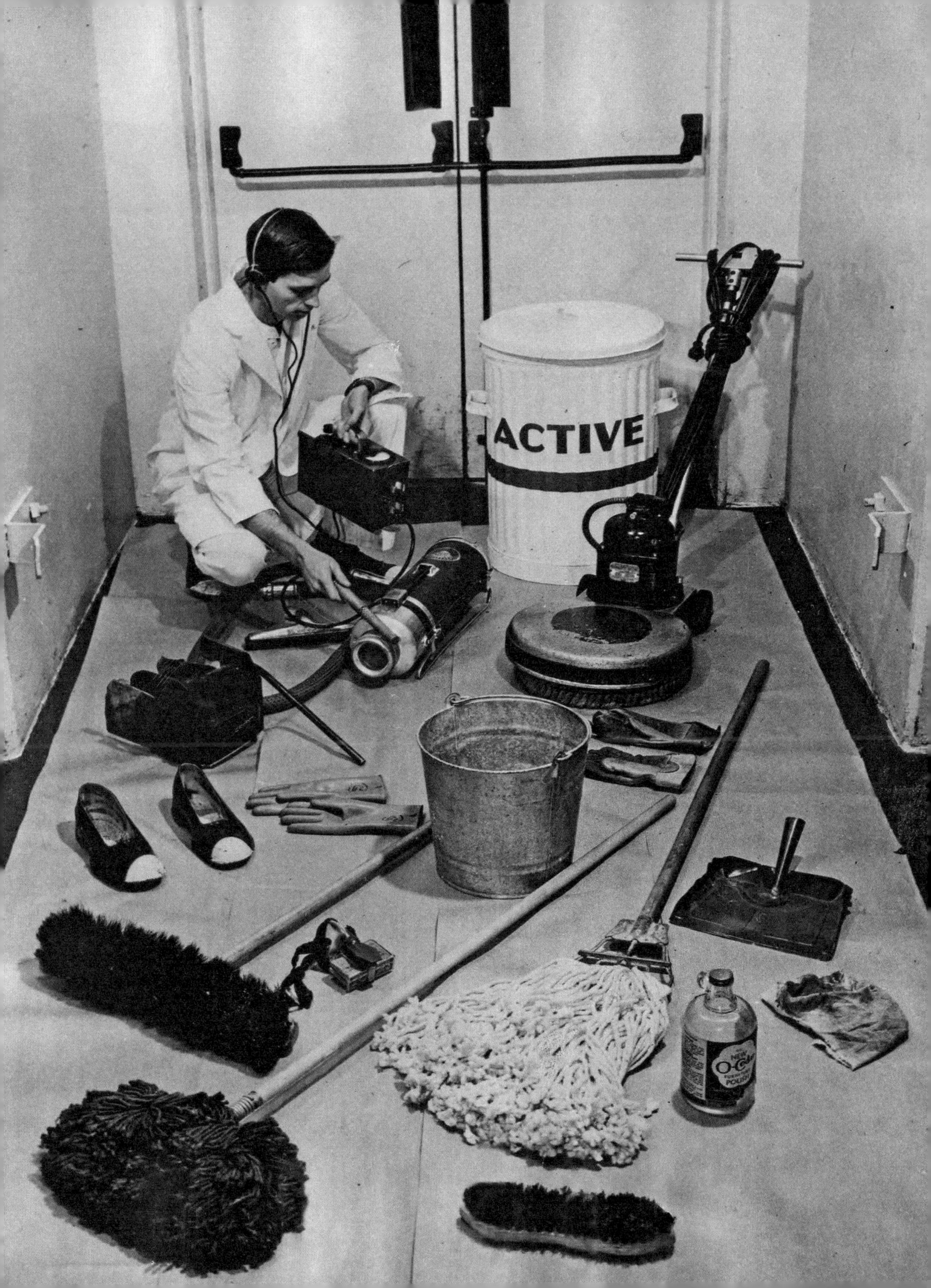
ACTIVE

cleaning fluid, and two pairs of gloves. If an item is found to be contaminated, the caption states, it is placed in the garbage can marked "RADIOACTIVE" to the inspector's left (in the shot the word is half cut and reads "ACTIVE").

Technically, there is nothing wrong with the picture. Remove the human figure, and it looks like a competent advertisement for cleaning products. The problem lies with the content. At a micro level, the man should be wearing a haz-mat suit and gloves (it is also difficult to imagine how he will fit the floor polisher into the garbage can should it turn out to be hot). At a macro level, the mop-and-pail equipment is incommensurate – absurdly incommensurate – with the technological sophistication of a nuclear-generating plant. The radiological risks posed by the reactor, should a problem occur, will not be solved by O-Cedar cleaning fluid.

When this picture was taken, Chalk River scientists and administrators had enough evidence about the dangers of radiation to know that dealing with an accident would require more than mops and pails. Information about how the Hiroshima and Nagasaki bombs had affected the human body – about radiation damage and long-term poisoning as well as thermal burns and keloid scars – had been thoroughly studied by those involved in the Manhattan Project. The US Strategic Bombing Survey had published a report in 1946, titled *The Effects of Atomic Bombs on Hiroshima and Nagasaki,* that was detailed and graphic.[27] Nuclear scientists at AECL Chalk River had also been informed about the circumstances of Louis Slotin's death from a lethal dose of radiation in New Mexico. The Canadian physicist was employed at the Los Alamos National Laboratory to undertake criticality testing, first with uranium, then with plutonium cores. This involved bringing fissile materials to a point of near criticality, a process that insiders call "tickling the tail of the dragon," which carried the risk of setting off a chain reaction. "I am one of the few people left here who are experienced bomb putter-togetherers," Slotin boasted.[28] On May 21, 1946, as Slotin worked on a plutonium core, his screwdriver slipped, causing a burst of radiation equivalent to that of an atomic bomb detonated 1,500 metres away.[29] He died nine days later and was buried by his parents in his hometown of Winnipeg. Since then, Slotin has been the subject of poems, novels, and a play. First and worst among these offerings is a poem published in the *Los Alamos Times* immediately after his death. "May God receive you, great-souled scientist!" it begins.[30] Last and best is a book of poetry called *Bloom* by Michael Lista, published in 2010.[31] The title evokes not only the term given to a sudden release of radioactivity but also the character of Leopold Bloom in James Joyce's *Ulysses,* whose position as an outsider parallels Slotin's, and brings the two figures together around Slotin and his death.

From the early 1940s, many of the scientists and workers at AECL Chalk River lived a short distance away in the town of Deep River. It was a similar but smaller version of the American atomic cities of Oak Ridge in Tennessee, Richland (Hanford) in Washington, and Los Alamos in New Mexico, which were built to service the Manhattan Project.[32] The media sometimes referred to them as "little Edens" because of their extensive recreational facilities.[33] Following the war, as the new atomic age was beginning to develop a visual vocabulary for its activities, views of the towns and the nuclear sites to which they were attached started to

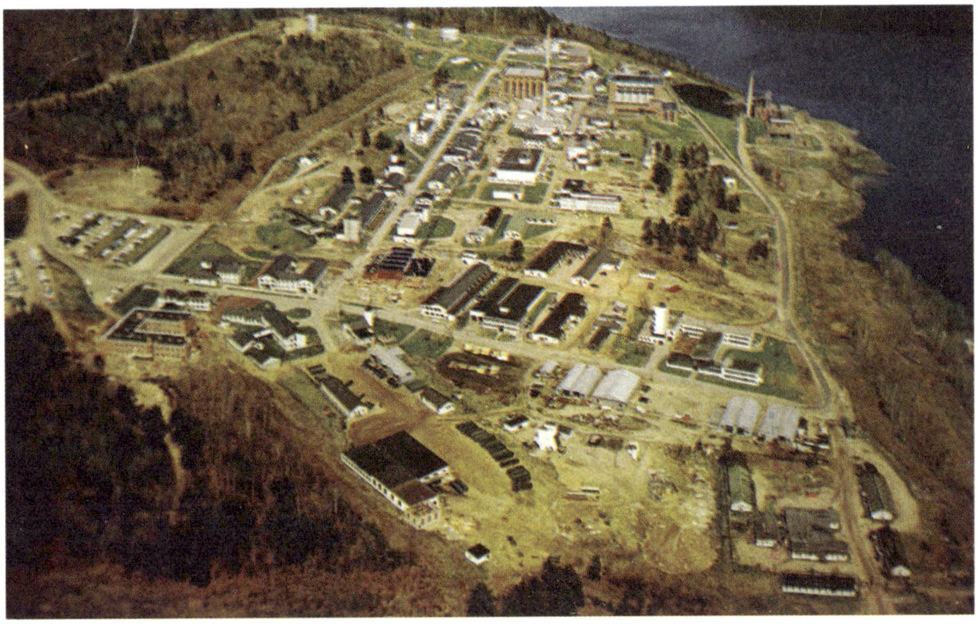

Figure 2.14: Postcard. *The Chalk River Research Centre of Atomic Energy of Canada Limited, on the Ottawa River near Chalk River, Ontario.* 1959.

show up on postcards. Some were taken from the air, such as *The Chalk River Research Centre of Atomic Energy of Canada Limited, on the Ottawa River,* a card published in "Spectrome Color" in Dryden, Ontario. [**Figure 2.14**] Although I am wary of atomic postcards, in the same way I am wary of instrumental photography in general, they have something to say about hopes and anxieties at the time of their making. As a product of tourism, the modes of address and reception of atomic postcards are peculiar to the medium. Handwritten messages on the backs of such cards are almost invariably positive, like the photographic images on the front.[34] "Hope both of you are keeping well in the heat, it's much cooler this way, just fine for the holidays," correspondent J. Owen Hughes writes in 1959 on the back of the Chalk River card, which pictures trees and water surrounding the complex. Rectos and versos, fronts and backs, tend to reinforce one another's reassuring tone of optimism, and even postcards mailed from the edge of danger rarely stray from a lingua franca of cheerfulness.

Canadian boom-and-bust uranium mining towns are not referred to as "little Edens."[35] A 1955 *Life* magazine photo-essay on Blind River, Ontario, remarks that "the town throbs with a Klondike Fever, and the primeval bush around it is being trampled out of recognition."[36] This is hardly a description of paradise. The region, including Elliot Lake, had recently been identified as the world's richest uranium field, and the town was being turned upside down by miners, drillers, and stock promoters, who are discussed and pictured in the *Life* photo-essay. We learn that the railway station agent and his wife, who appear in a picture taken by the African American photographer Gordon Parks, have made their fortune in uranium stocks.

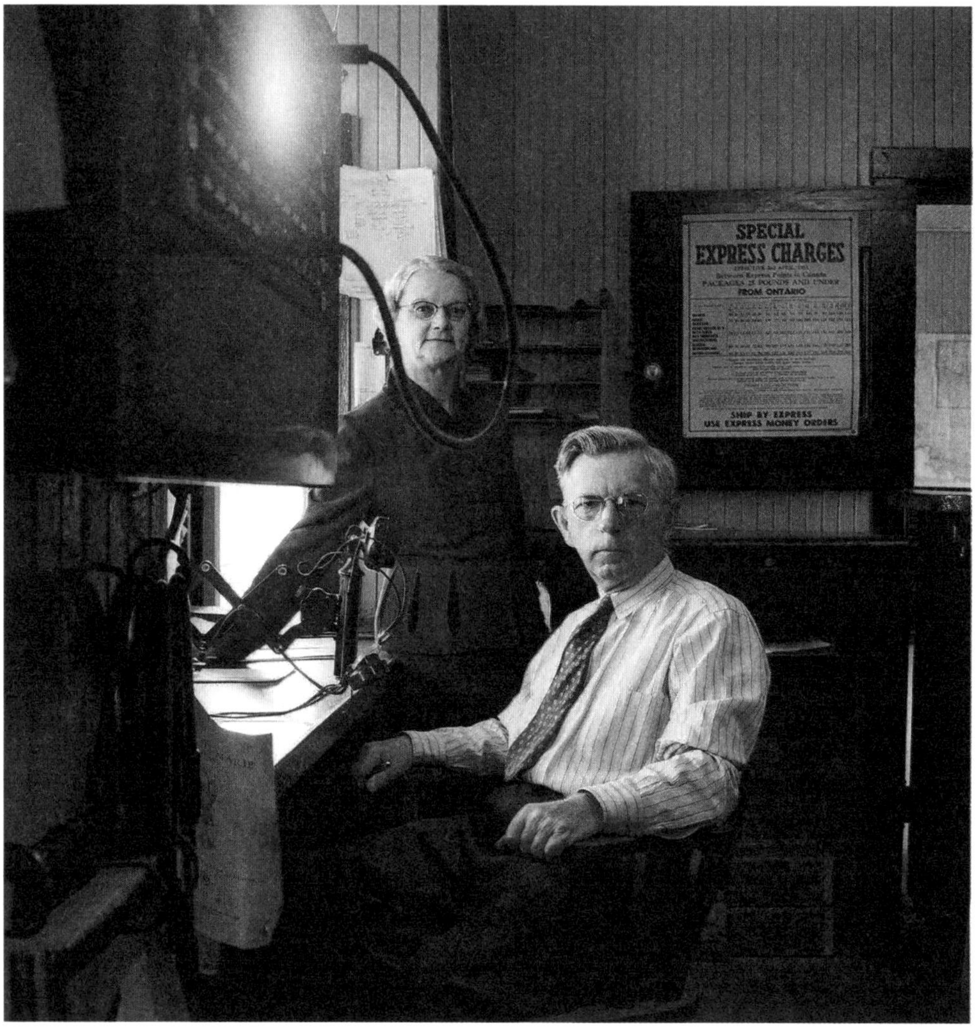

Figure 2.15: Gordon Parks. *Beneficiaries of Boom Are Station Agent and Mrs. Charles Egan.* 1955.

[**Figure 2.15**] The shot shows them in the station agent's office, their mouths drawn tight; the agent's hands are clasped hard to his chair. One of the stock promoters was the infamous entrepreneur and financier Joe Hirshhorn, who sold shares of mining companies and talked grandly about constructing an Italian-style piazza in the town, decorated with Henry Moore and Jacob Epstein sculptures. Hirshhorn did not build the piazza. His unrealized vision for Blind River did not progress beyond nuts-and-bolts construction of the sort represented on a postcard from the time, which shows a blasted rock face towering over the newly built Trans-Canada Highway near the town. [**Figure 2.16**] Hirshhorn donated his Moores and Epsteins to the museum on the mall in Washington, DC, that bears his name. While he was

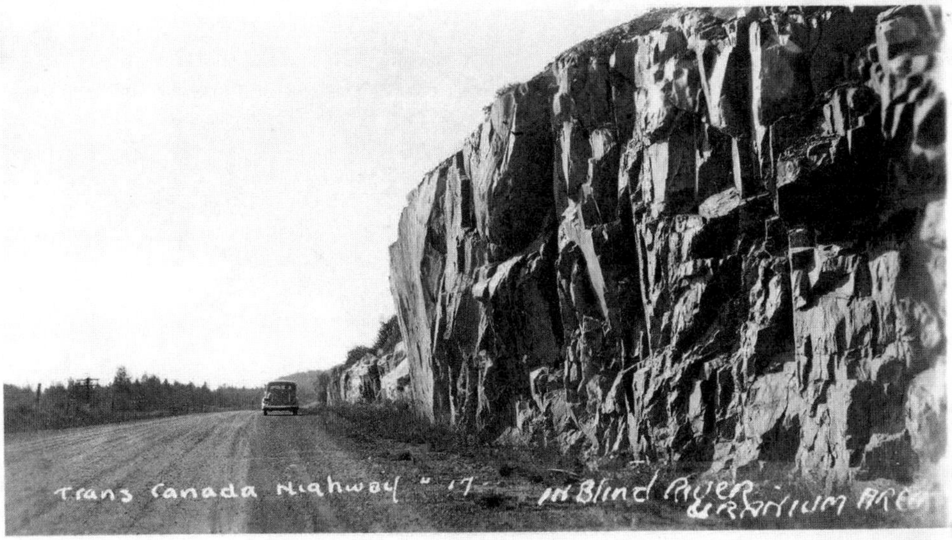

Figure 2.16: Postcard. *Trans Canada Highway "17" in Blind River Uranium Area.* 1955.

being feted in the US capital, the region responsible for his fortune was being contaminated with radioactive waste. *Stanrock Tailings Wall, Elliot Lake, Ontario,* by Del Tredici looks, at first glance, like a photographic homage to a painting of a flooded landscape by a member of the Group of Seven. [**Figure 2.17**] On closer inspection, it reveals a man-made retaining wall, a ten-metre-high embankment of earth behind the sludge in the foreground and the dead trees in the middle ground. Edward Burtynsky took a photograph of the same scene, *Uranium Tailings #12, Elliot Lake,* a decade later. In both works, we are looking at a defeated landscape. The wall holds back a toxic lake of uranium tailings trucked in from nearby mines supplying uranium ore to Chalk River and to the United States for nuclear weapons and nuclear power.

Runoff from uranium tailings ponds in the area contaminated the entire Serpent River system, which flows into the north shore of Georgian Bay on Lake Huron, affecting the health of the Serpent River First Nation. In her book *This Is My Homeland,* Lorraine Rekmans remarks that the lakes in the region "were used as a dumping ground for radioactive waste ... The ground-waters under the tailings basins are virtual rivers of poison."[37] In addition, land on the reserve was used to build a sulphuric acid plant to supply the uranium mines. Bonnie Devine, artist and member of the Serpent River First Nation, undertook extensive research on uranium mining to produce her 2004 exhibition *Stories from the Shield.*[38] A mixed-media drawing, *Set for "Rooster Rock," the Story of Serpent River,* was bound into a large book for the

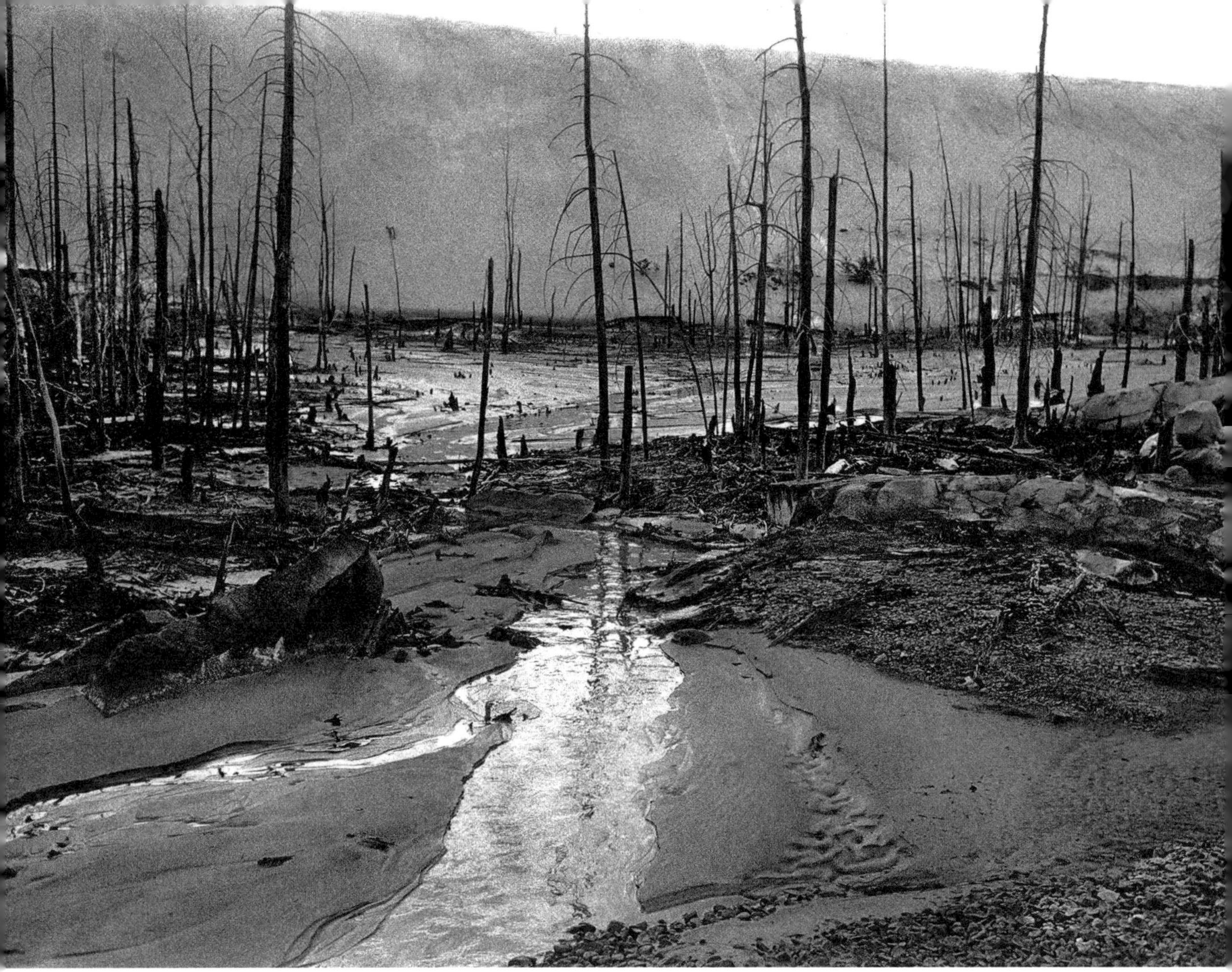

| **Figure 2.17**: Robert Del Tredici. *Stanrock Tailings Wall, Elliot Lake, Ontario.* 1985.

show. [**Figure 2.18**] The dazzling yellow of the sulphur pile contrasts with the greys and blacks of the poisoned land around it. The Serpent River First Nation has suffered the consequences of the mining, and it continues to suffer from the consequences. In her article "The Radioactive Colonization of Indigenous Bodies," Robynne Mellor comments on the Serpent River First Nation and Indigenous populations elsewhere, including the Sahtu Dene people of Great Bear Lake, who are also affected by uranium mining.[39] She observes that white mineworkers are epidemiologically at less risk of developing stomach cancer than Indigenous populations living near the mines, because they do not depend on the contaminated land for sustenance. Their food is shipped in from the outside, she explains, and they move on when a mine closes. The local First Nations stay on the land and consume what it offers.

Ara Mooradian, who obtained his doctorate in physical chemistry from the University of Saskatchewan, arrived at AECL Chalk River during the 1950s. In a 1995 interview, he recalled his life there:

> *I moved to Deep River to work in the Chemical Extraction Division of AECL. My first job at AECL was separating plutonium from irradiated fuel. I also separated isotopes that we were selling all over the world; things like strontium, cesium and phosphorous.*
> *In the early days, people did many things for recreation: yachting, hockey, tennis, art and pottery. We even had a theatre, the Strand Theatre, where Bookworld is now located. Deep River was, and still is, a pretty active community! There are darn few communities with a population of 5,000 that have their own symphony orchestra.*[40]

Mooradian does not pause to catch his breath between describing his professional work with plutonium, strontium, and caesium and recounting recreational pursuits that were available to the inhabitants of Deep River. In other postcards from the Chalk River and Deep River area, *Yacht & Tennis Club, Deep River, Ontario, Canada,* and *A.E.C.L. Staff Hotel, Deep River, Ontario, Canada,* recreation and pleasure are also primary.[41] [**Figure 2.19**] The promise of yachting on the Ottawa River, playing tennis at the club, and staying in the hotel operated by Atomic Energy of Canada overrides the uncertain environment in which the activities take place. Poet and critic Susan Stewart observes that souvenirs and postcards speak in the nostalgic language of longing.[42] They provide a miniaturized, hold-in-the-hand spectacle that

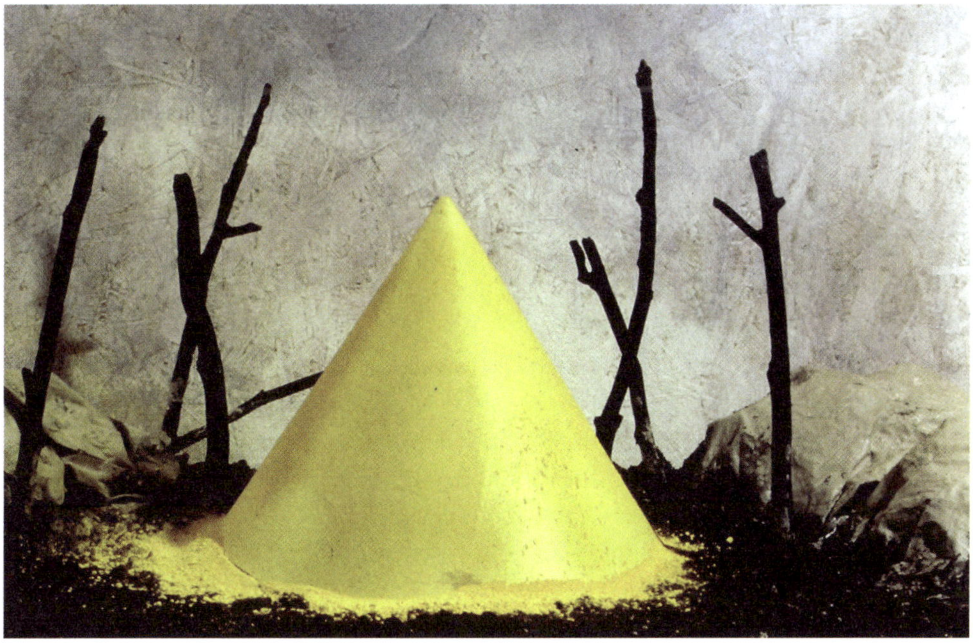

Figure 2.18: Bonnie Devine. *Set for "Rooster Rock, the Story of Serpent River."* 2001.

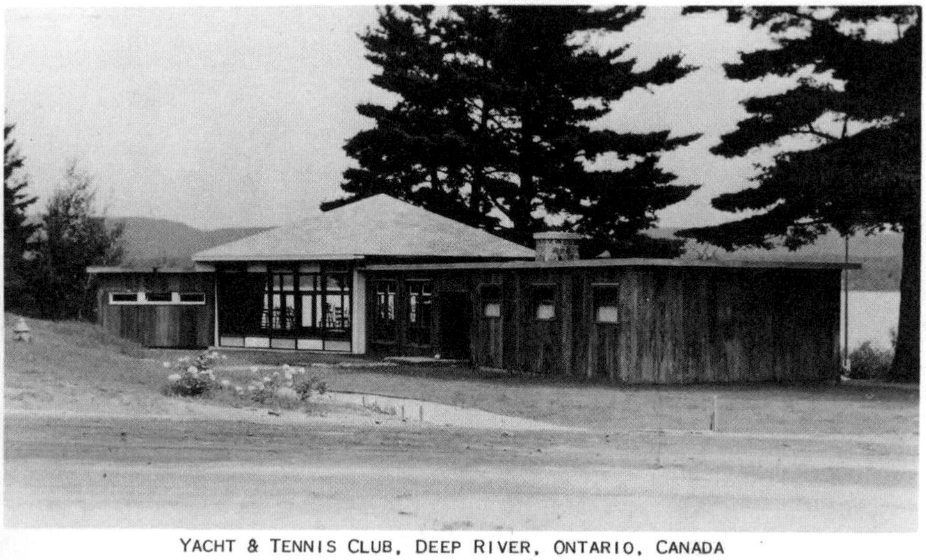

Figure 2.19: Postcard. *Yacht & Tennis Club, Deep River, Ontario, Canada.* c. 1950.

domesticates what is pictured, even if what is pictured, like Deep River and AECL Chalk River, is implicated in traumatic historical events. If the authority of the state and organizations such as AECL Chalk River are identified with bureaucratic vastness, the citizen subject is associated with miniaturization. The recreational postcards, printed in Toronto by Canadian Lost Card Company, were most probably commissioned by AECL Chalk River.

Mooradian presents his life in Deep River as if the near meltdown of the NRX reactor on December 12, 1952, and other serious Chalk River accidents in 1957 and 1958, had never occurred. In 1952, the NRX was being restarted after routine alterations and maintenance, and through a series of human and design errors, part of the cooling system was brought to a white heat, setting off explosions that destroyed the core of the reactor and released a gaseous bloom of radioactive particles into the atmosphere.[43] More than a million gallons of radioactively contaminated water flooded the reactor basement in what was the first major reactor accident in the world.

The Canadian military and the US Navy were brought in to clean up the physical damage and pump the contaminated water into shallow trenches near the Ottawa River. One of those enlisted for the job was future president Jimmy Carter, then a nuclear engineer with the US nuclear submarine program. Interviewed shortly after the 2011 meltdown at Fukushima Daiichi, Carter recalled that "the same thing happened in Canada." As a young naval officer, he reported, he had been dispatched to help dismantle the "very highly radioactive core" of the Chalk River reactor after its partial meltdown. Wearing protective gear and armed with wrenches and vise grips, he and his team had ninety seconds to dash into the reactor and "take off as many nuts and bolts as we could." "I had radioactive urine for six months," Carter remembered, "and I thought I would never have another child but Amy came later."[44] President Carter was dose fodder. He was, as the newspaper headline of the interview put it, "Jimmy Neutron." In 1952, when reporters questioned C.J. Mackenzie, president of Atomic Energy of Canada, about the NRX accident, he assured them that everything was under control. A "pinhole leak," he declared, had occurred in the reactor.[45] His disavowal of what had happened was an attempt to manage public perceptions of radiation risk. He feared that if he were to admit to the gravity of the accident, the public would respond irrationally. People would begin to feel Godzilla's breath on their necks.

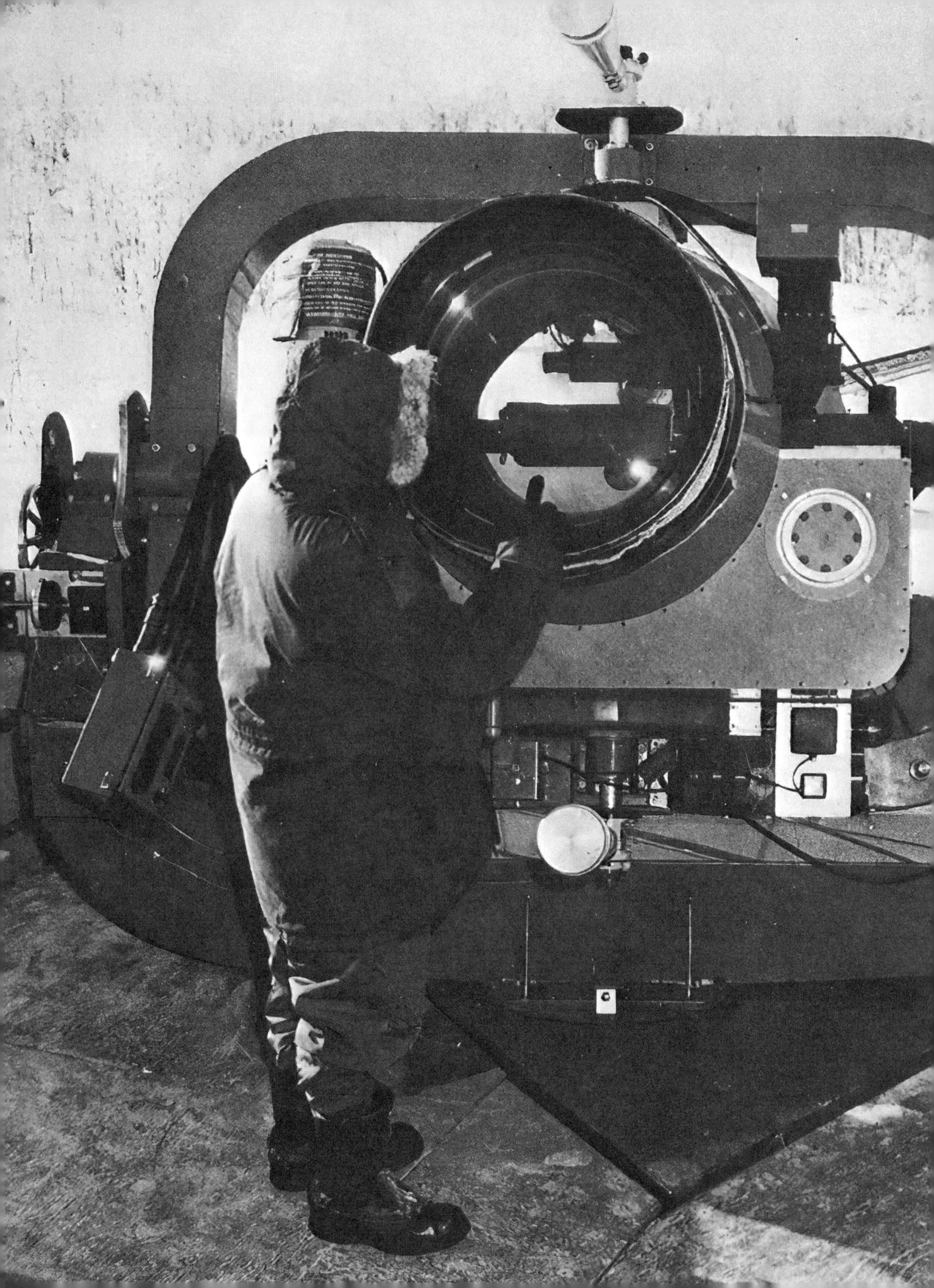

3

Cold War Defence

WHEN THE SOVIET UNION ACQUIRED THE BOMB IN 1949 AND DEVELOPED LONG-range military aircraft capable of delivering it to American cities, Canada became the zone over which bombers would have flown in the event of war. As in the children's game "pig in the middle," Canada was stuck between the superpowers. Preceding chapters have concentrated on the representation of nuclear energy and the bomb; this chapter focuses on defence systems designed to protect against nuclear attack. In the 1950s, three radar systems were constructed across the North – the Pinetree Line, the Mid-Canada Line, and the Distant Early Warning (DEW) Line – to warn against incoming Soviet bombers. They were jointly funded by the United States and Canada, though Canada was worried about the amount of control held by its partner.[1] Canada also joined the States in funding the installation of a satellite detection system following the 1957 launch of Sputnik by the Soviet Union. It employed specialized instruments to identify the orbiting space capsules. Weighing more than three tonnes apiece, Baker-Nunn cameras with multiple-axis mounts were set up at military bases in Prince Albert, Saskatchewan, and Cold Lake, Alberta, to track the satellites. [**Figure 3.1**]

A fourth radar system called the Northern Warning System, which consists mostly of unmanned radar stations, was built to replace the DEW Line at the end of the Cold War. Data from the stations are streamed south to Canadian and US military command centres to identify incoming intercontinental ballistic missiles.

Figure 3.1: NORAD. *Baker-Nunn Camera for Satellite Detection and Tracking, Cold Lake, Alberta.* 1961.

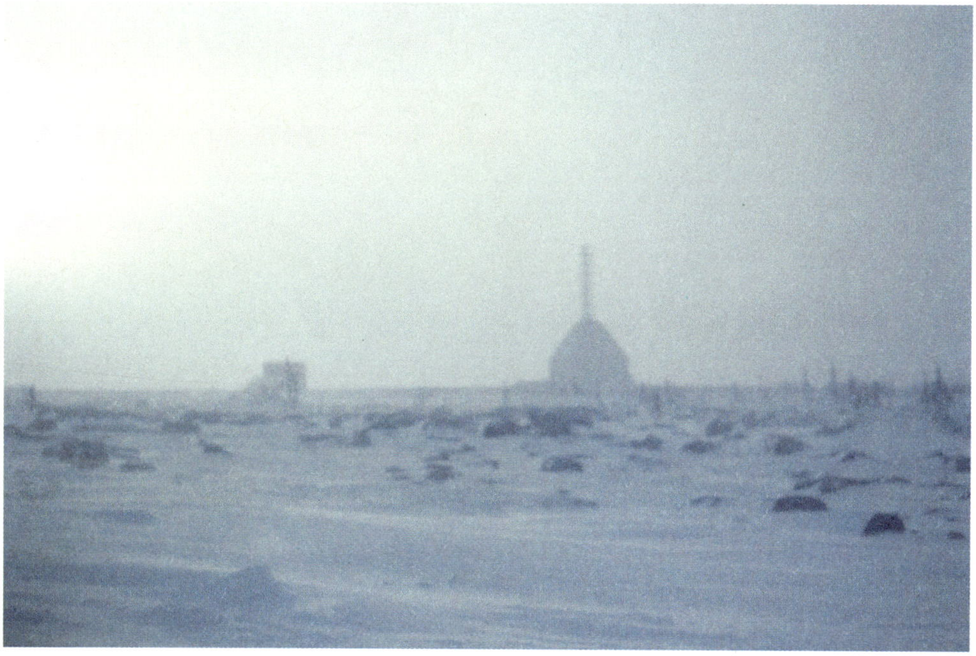

Figure 3.2: Warren Langford. *Fort Churchill, Manitoba, January 1963*. 1963.

In the Cold War game of pig in the middle, Canada occupied an inferior position. It represented the pig over which the superpowers tossed a ball back and forth. For the game to be effective, the United States had to show restraint in dealing with Canada, at least during public exchanges. When US secretary of defense Robert McNamara told Congress that supplying nuclear-tipped Bomarc-B missiles to Canada would draw Soviet fire away from America in the event of an attack, President Kennedy chastised him for being too blunt.[2] Kennedy was furious, as were many Canadians. However, Warren Langford, a career public servant in the Canadian government, favoured arming the country with nuclear weapons regardless of the risks outlined by McNamara. "Canada's refusal, on very weak and illogical grounds, to allow her NORAD forces to be armed with nuclear warheads," he wrote in a group-authored paper, "creates a serious hole in the defense of North America which could be disastrous in an all-out war."[3]

Langford's paper was written in November 1962, just after the Cuban Missile Crisis. Langford had recently joined a special training program for cold warriors at the National Defence College in Kingston, Ontario. His superiors had recommended him for the program because they thought he exhibited the necessary leadership skills to assume a senior position in government. He had the right stuff. The program involved visiting Cold War sites in Canada and other parts of the world. During his travels, Langford took Anscochrome and Kodachrome slides with his Aires Viscount 35mm single-lens reflex camera. He was "an apprentice snapshooter," to borrow a phrase from a book on his Cold War tourism, whose "slides reflect both the dream of world travel and the nightmare of nuclear confrontation."[4] A trip to Fort Churchill, Manitoba, took him to the site of a joint Canada-US cold-weather testing facility for ballistic missile guidance systems. (Fort Churchill was also supplying equipment to the DEW Line at the time.) Among Langford's images is a slide of the Aerobee rocket launch tower, taken when the January sun was frozen to the horizon. [**Figure 3.2**] The slide shows the tower – fifty-three metres high to ensure accurate launches and surrounded by corrugated steel cladding to keep the missiles warm – as an alien presence in the Arctic environment. It is a fantasy outpost of ice and snow, the tower silhouetted against the sky. The shot strikes me as more nuclear nightmare than travel dream. It demonstrates that Langford, as an amateur photographer as well as a career public servant, knew what he was doing.

Fifteen years after Langford visited Fort Churchill, a malfunctioning Soviet spy satellite – Cosmos 954 – broke up when it re-entered the earth's atmosphere, scattering debris across northern Canada. Some of its pieces were highly radioactive. Codenamed Operation Morning Light, a search got under way to locate the fragments.[5] At first, the Soviet Union disclaimed any responsibility for the cleanup, but after Canada demonstrated that debris had fallen along a six-hundred-kilometre path from Great Bear Lake to Baker Lake, it provided compensation. The television comedy show *Saturday Night Live* was onto the story immediately. Its January 28, 1978, episode reported that giant mutant lobsters were coming out of the Canadian North to invade the United States. Huge papier mâché crustaceans swarmed the set at the end of the show.[6]

Writer and artist Douglas Coupland knows better than most Canadians what it means to play nuclear pig in the middle. He is the child of an air force officer (as I am), and he spent some of his early years in Germany growing up on an airbase (as I did). During the Cuban Missile Crisis, he was living on the West German base of Baden–Söllingen, on a primary flight path between the Soviet bloc and Western Europe. During the Cold War, Germany was the other country in addition to Canada over which a Third World War was likely to begin.[7] Looking up and seeing a nuclear flash, Coupland has written, was a recurring image in his dream life and everyday thoughts. In one variation of the imagined image, he is in a Vancouver apartment gazing out to sea when someone points through the window: "I look and see that the sun is growing too large too quickly, like a Jiffy pop popcorn foil dome, glowing orange, like an electric stove element."[8] Jiffy Pop foil domes can also explode. In another variation, he closes his eyes and burns the image of a landscape onto his retina by blinking, a procedure he was taught in art school to spur the creative process, and is reminded that nuclear flashes invert our vision of the world in a similar way. At the bottom of a manipulated photograph Coupland made to accompany a republication of his essay "The Wrong Sun," he has written in capital letters, "WINNER: NONE."[9] [**Figure 3.3**] If the text were not there, you might imagine the image to be a schematic representation of a fireworks display rather than a depiction of missile flight paths.

"After the blast, you might feel slightly disoriented," advised *11 Steps to Survival,* a Canadian civil defence manual published in 1961.[10] Released by the Emergency Measures

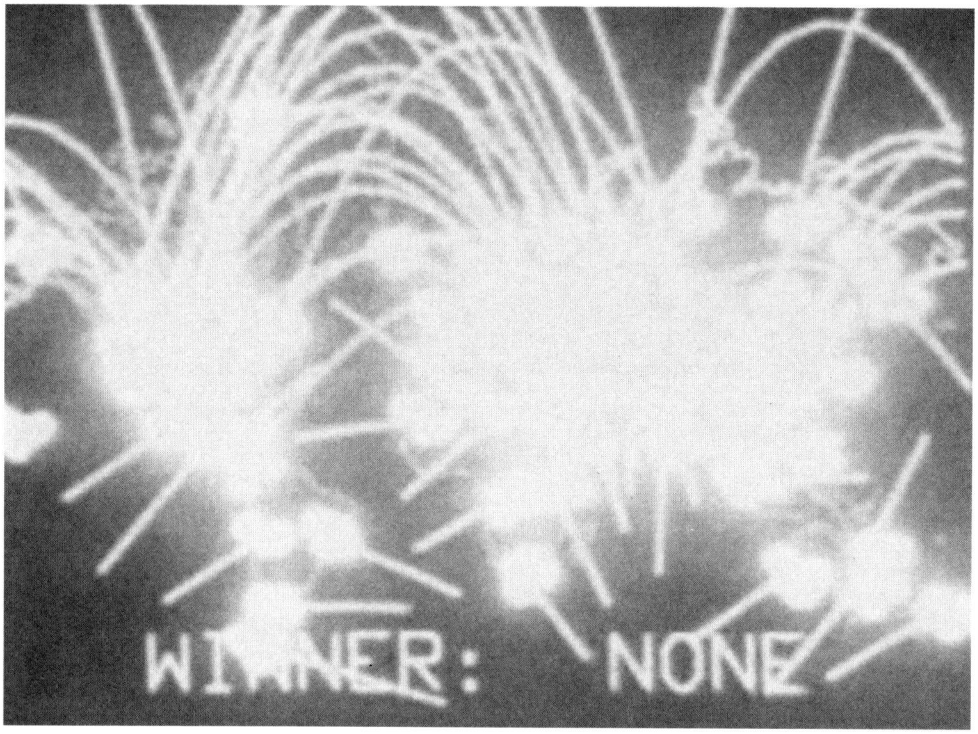

| **Figure 3.3**: Douglas Coupland. *WINNER: NONE.* 2015.

Organization, Ottawa, it was one in a series of defence manuals sponsored by the government. In the early decades of the Cold War, government leaders regularly informed citizens that nuclear conflict could be defended against, even as they themselves prepared for the worst. In the foreword to one of the manuals, Prime Minister John Diefenbaker wrote that "protection against radioactive fallout [can] be achieved by family shelters."[11] During the Berlin Crisis of 1961, which began with nuclear sabre rattling by the United States and the Soviet Union and ended with the erection of the Berlin Wall, President Kennedy gave a speech to the nation, assuring the American people that a nuclear attack could be survived. "The lives of families which are not hit in a nuclear blast and fire can still be saved," he declared, "if they can be warned to take shelter and if that shelter is available."[12] Diefenbaker and Kennedy were trafficking in false promises. With the development of thermonuclear weapons in the early 1950s, the possibility of surviving an atomic attack had become improbable, regardless of whether shelters were available or not. Wind also spreads radioactive fallout all over the map, poisoning air, water, and food supplies.

In *Give Me Shelter: The Failure of Canada's Cold War Civil Defence,* Andrew Burtch identifies three stages in Canadian civil defence from the late 1940s to the early 1960s.[13] From 1948 to 1954, planners used Second World War models, assuming that the atomic bomb was just another weapon that could be successfully absorbed. With the deployment of thermonuclear weapons in 1952, that fiction could no longer be sustained, and a new stage of planning began. It emphasized evacuation preparedness in so-called target cities. The publication of civil defence pamphlets became intensive between 1954 and 1959, and it remained intensive during the third phase from 1959 to 1962, when a "national survival" strategy called for warning systems and fallout shelters. Even then, however, civil pamphlets in Canada and the United States claimed that a nuclear attack would not differ greatly from a natural disaster, and that those who took proper precautions would pull through. All you needed was a safe place to go in an emergency, if possible a fallout shelter with stores of food and water.

Following Kennedy's speech, there was a bomb shelter construction boom, and the rush to build spilled across the border into Canada.[14] The first prefabricated underground shelter in Canada, so the manufacturer claimed, went on display in Toronto ten days after Kennedy's speech. [**Figure 3.4**] "Opens for Viewing Thursday August 4th," a sign in front of it announced, as if suburban real estate were being advertised rather than a bomb shelter. In the photograph, the couple inspecting the structure seem less sure. They stand well to the left of the entrance, while a uniformed guard oversees access. For those who preferred to build

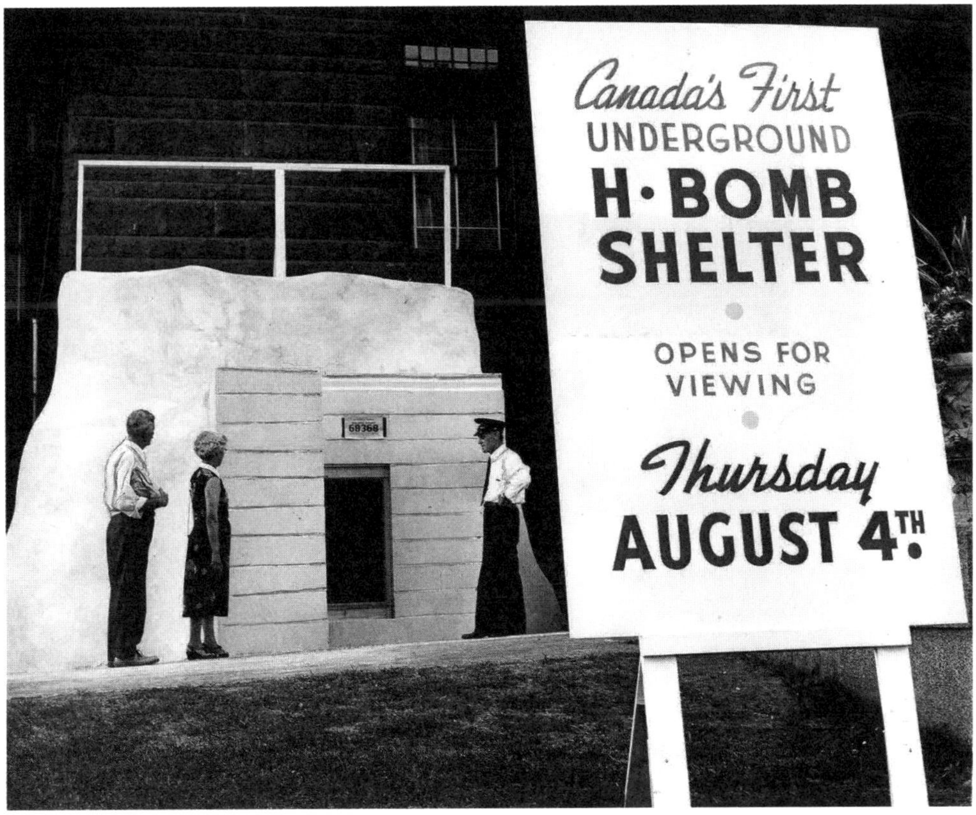

Figure 3.4: Harold Whyte. *Canada's First Underground H-Bomb Shelter.* 1961.

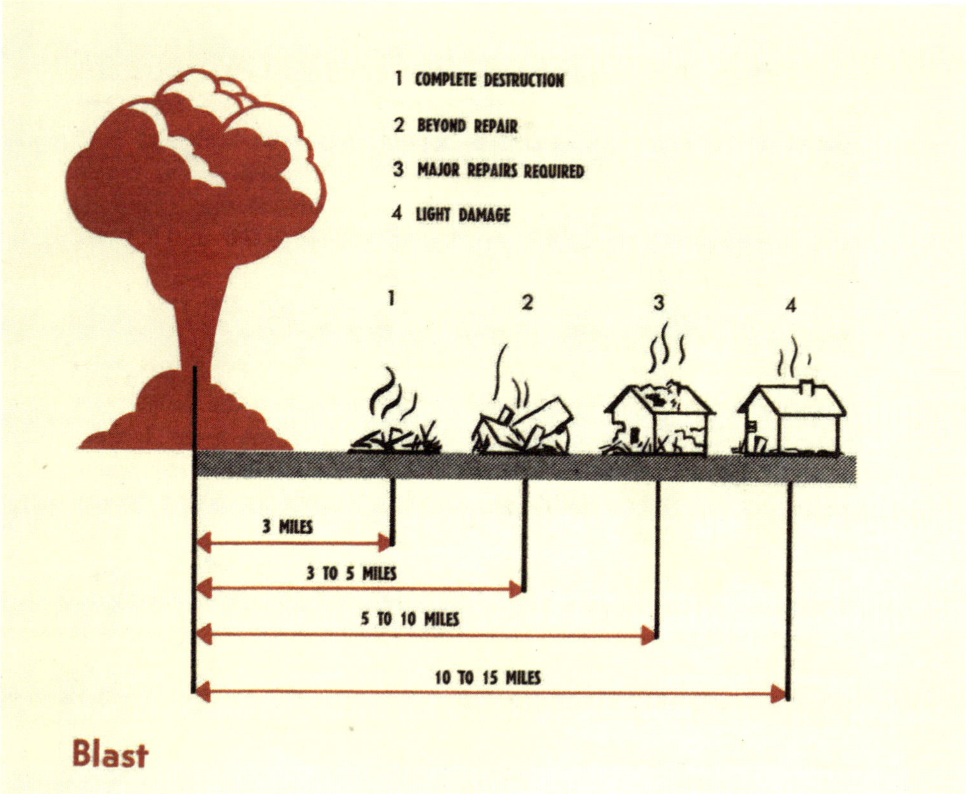

Figure 3.5: *Blast.* 1961.

family shelters themselves, the government circulated pamphlets with titles such as *Your Basement Fallout Shelter* and *Blast Shelter* that demonstrated how it was done.[15] In addition, a nationwide network of early warning civil defence sirens was built to sound the alarm.[16]

Most civil defence pamphlets were mass produced. For reasons of cost and graphic clarity, they were generally illustrated with line drawings rather than photographs. The drawing *Blast,* from *11 Steps to Survival,* provides the reader with a set of calculations for measuring the distance between a mushroom cloud, coloured red for emphasis, and a series of houses damaged by the explosion. [**Figure 3.5**] But not all pamphlets were mass produced, and two exceptions are worth examining for their use of photography. Both had nicknames: "The Hockley Incident" (1952) and "The Bloody Book" (1955). Prepared as training manuals for professional use by firemen, police, and soldiers, they were not distributed to the general public, in part because some illustrations were too gruesome or too technical for mass consumption. "The Hockley Incident," named for the location of an imagined attack in England, was published by the Home Office Fire Service Department in Britain as *Technical and Tactical Study Courses Manual.*[17] In the witching-hour scenario presented by the manual, "It is nearly midnight. Somewhere an enemy aircraft is carrying toward Hockley an atomic bomb. At midnight

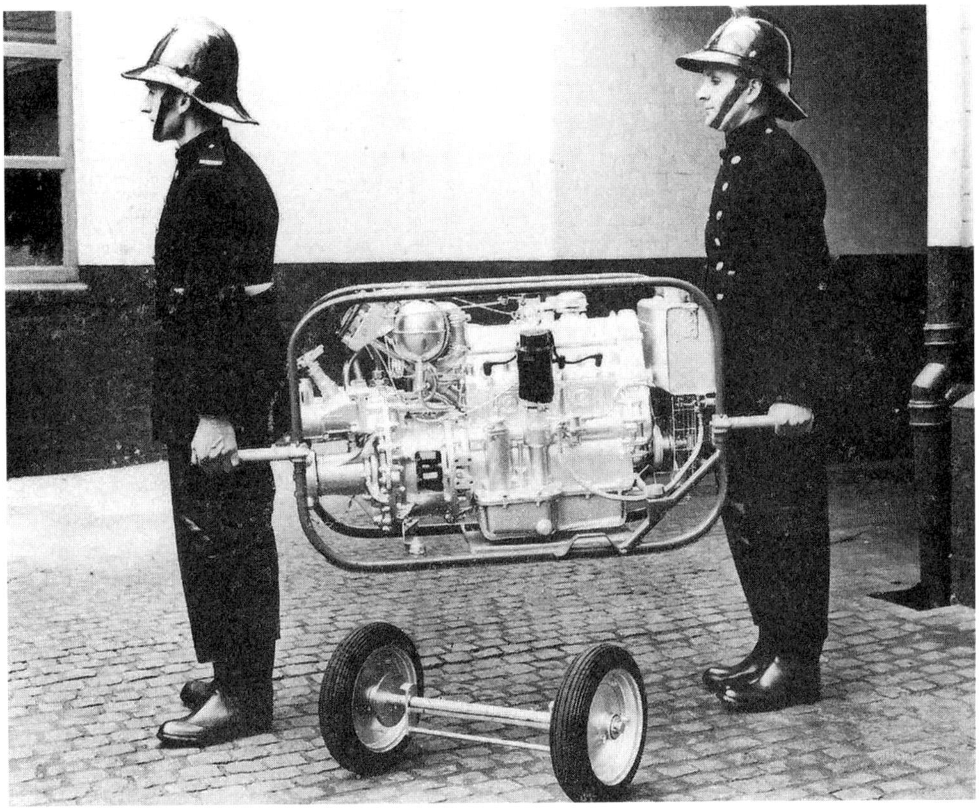

| **Figure 3.6**: *Light Portable Pump – Carried by Two Men.* 1952.

precisely, it falls over Hockley and explodes at 600 feet."[18] The manual's primary purpose was to instruct fire departments in Britain and the Commonwealth on the impact of an atomic attack on urban centres. Treated as secret and confidential – only a limited number of copies were printed – the mimeographed manual was illustrated with photographs and lithographs. The fire marshal of Ontario, William J. Scott, travelled to London to participate in the courses, returning to Canada with the manual and lessons learned about nuclear conflagration. Some of the information may have been useful, but not much of it. One photograph, *Light Portable Pump – Carried by Two Men,* features a water-pumping machine thought to be capable of fighting thermal fires triggered by an atomic detonation. [**Figure 3.6**] To protect themselves, the firemen are wearing helmets modelled on those worn by French *pompiers* in the nineteenth century, and their pumping equipment is arcane. Like the mop-and-pail gear supplied at Chalk River for dealing with radiation accidents, it is wildly incommensurate with the task at hand.

The purpose of the second publication was to teach civil defence volunteers about war injuries, including those caused by nuclear war, by simulating them as powerfully as possible. In 1955, a Canadian government bulletin announced a forthcoming publication, a "super-deluxe" civil defence book sponsored by the Department of National Health and Welfare.

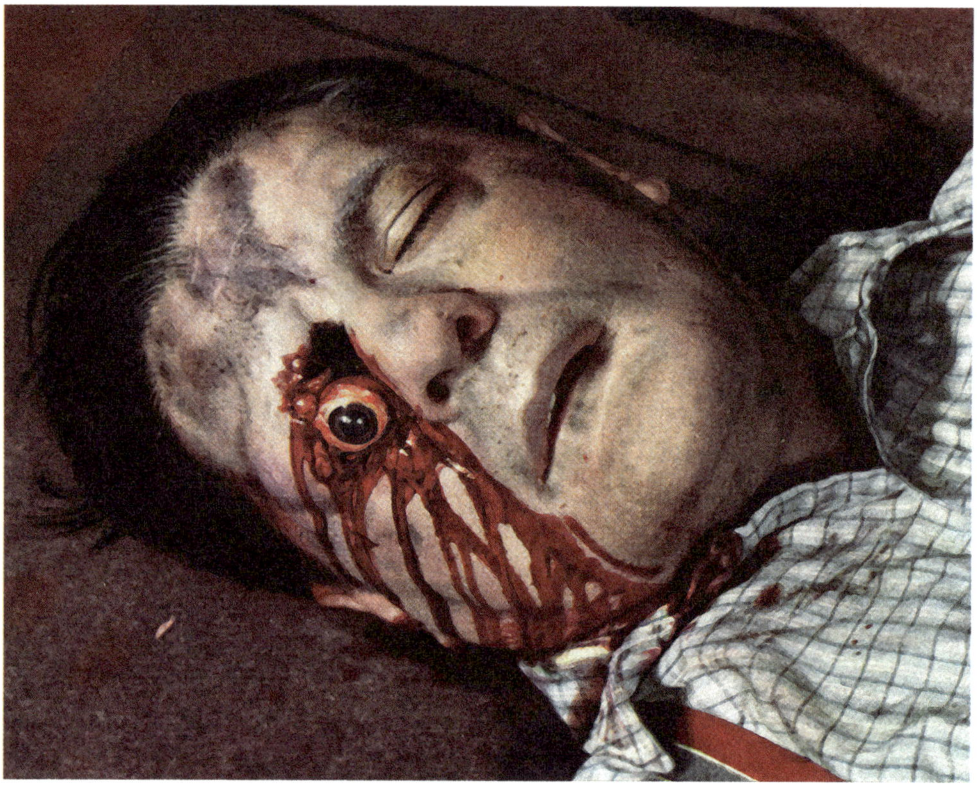

Figure 3.7: C. Ellis Kerr. *Eye Enucleation.* 1955.

"The Bloody Book," as it was known, would feature forty Kodachrome illustrations and be one of the most highly produced presentations on civil defence anywhere in the world.[19] If volunteers were to learn about first aid in emergency situations, realism was crucial, the bulletin stated.[20]

When the Bloody Book appeared in print, it was formally called *Casualty Simulation.*[21] Although the title accurately describes its contents, the nickname refers to its most striking characteristic: the vivid colour images of wounded and burned bodies. A lot of artificial blood was spilled to make the book.[22] Four kinds of red fluid were manufactured, which the editors described in the same way as they might have referred to the contents of a box of chocolates. The varieties were dark coagulated, light coagulated, thick liquid, and thin liquid. The editors' explanation for the large amount of fake blood was its ability to simulate the real.[23] One close-up photograph, identified as *Eye Enucleation,* shows the face of an injured man, with his right eye dislodged from its socket and rivulets of blood streaming down his cheek. [**Figure 3.7**] The image simulates the injury by using a pig's eye, the lids of which have been removed, fixed into place with string. The kind of realism portrayed in the Bloody Book also resembles the fantasy realism of nuclear horror films. Both genres traffic in artifice and mutation. The film *Godzilla,* conceived in the aftermath of the devastation of Hiroshima and Nagasaki, appeared in Japanese theatres a year before the Bloody Book came out in Canada. Both relied on fantasy and fabrication in responding to nuclear catastrophe.

The editors' explanation for the use of Kodachrome photographs was the same as for the utilization of fake blood – an ability to simulate the real. Kodachrome 35mm film, invented in 1935 by Eastman Kodak, was widely valued for its "nice bright colours," as Paul Simon phrases it in the song "Kodachrome" ("Makes you think all the world's a sunny day").[24] The editors asserted that realism was the future of civil defence instruction because it provided heightened authenticity. Their claims for the project correspond closely to those made for documentary film and photography by the filmmaker and theorist John Grierson between the wars.[25] Grierson coined the term "documentary" in 1926, in a review of the purportedly anthropological film *Moana* by Robert Flaherty, who had earlier made *Nanook of the North.* Documentary realism, according to Grierson, referred to films and photographs that treated non-fiction material with creative licence. He favoured re-enactment strategies to stage real-life scenes, not unlike the photographic re-enactments used to produce the Bloody Book.[26] The echo of Grierson's ideas in *Casualty Simulation,* I would argue, derives from the impact he had on the National Film Board of Canada (NFB), which he directed from its formation in 1939 until the end of the war, and which in turn had an impact on all media produced in the country around that time. At the NFB, Grierson put his ideas

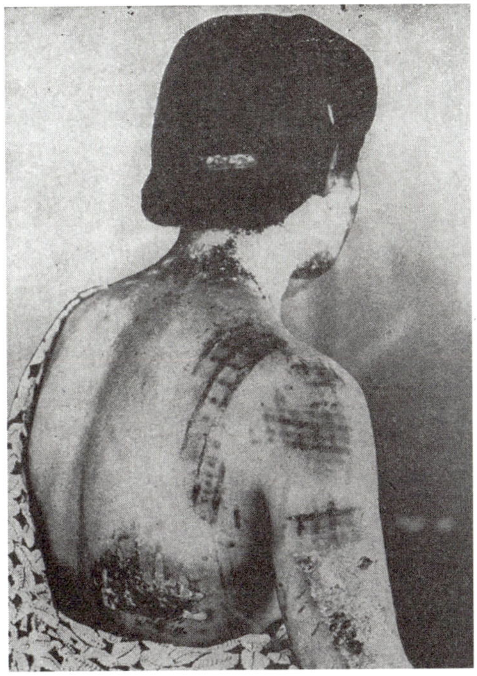

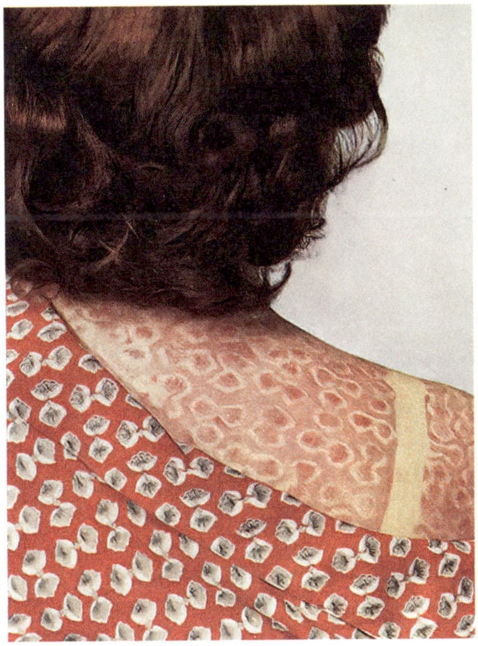

Figure 3.8: *Flash Burn through Patterned Clothing.* 1945. **Figure 3.9**: C. Ellis Kerr. *Flash Burn through Patterned Clothing.* 1955.

into effect across a wide canvas.[27] He thought that propagandistic film and photography could help to solve social problems and build national consensus.

Some illustrations of the traumatized body in the Bloody Book are modelled on pictures of injured survivors from Hiroshima and Nagasaki. The original photographs and the simulated illustrations of bodily harm function rhetorically by manipulating the body to convey the effects of extreme trauma. The Bloody Book illustration *Flash Burn through Patterned Clothing* is based on a frequently published photo of a Hiroshima victim, a woman suffering skin burns caused by thermal radiation [**Figs. 3.8 and 3.9**]. In the original black-and-white shot, which appeared in several restricted US scientific reports, including *Radiological Defense: A Series of Indoctrination Lectures on Atomic Explosion, with Medical Aspects,* from 1949, the woman's back and right shoulder are prominently pictured.[28] She is posed facing away from the camera. A caption states that at the time of the attack on Hiroshima, she was wearing a geometrically patterned kimono and had her back toward the epicentre of the blast. In the name of verisimilitude, her body is positioned in the same relationship to the camera as it was to ground zero at the time of the explosion. The light-coloured parts of her kimono reflected the heat rays from the blast, *Radiological Defense* observes, leaving her skin undamaged in those areas, whereas the dark-coloured parts absorbed the heat and burned her skin.[29] Her back and arm are covered in grids, as if a game of X's and O's were being played there. A similar pattern appears in the simulated image of the Bloody Book. *Flash Burn through Patterned Clothing* represents burn patterns on a woman's right shoulder and upper back, but to distinguish it from the original photograph, the patterning on the blouse is organic rather than geometric. The forms look like miniature dog bones.

At least four organizations were responsible for photographing the hibakusha, survivors of the attacks on Hiroshima and Nagasaki. They were the Atomic Bomb Casualty Commission, which was a joint enterprise of the United States and Japan, the US Atomic Energy Commission, the US Strategic Bombing Survey, and the US Department of Defense.[30] All wanted to gather information about the physiological effects of the fires, radiation, and trauma caused by the bombings. "The burns that occurred may be classified as 'flash burns,'" *Radiological Defense* states, "which are the result of the direct action of radiant energy across the entire width of the spectrum, which resembles that of the sun."[31] Another photograph of a hibakusha shows a soldier who was standing 6,500 feet away from ground zero and who suffered similar flash burns. He wears a cloth cap in one shot and is bare-headed in another, to illustrate how the cap had protected part of his head. His condition is mimicked in the Bloody Book by a picture of a man whose flash burn covers half his face. [**Figure 3.10**]

The US censorship board suppressed scientific publications about radiation effects, especially in Japan.[32] When a Tokyo news service reported that "many of those who received burns cannot survive the wounds because of the uncanny effects which the atom bomb produces on the body" and that "even those who received minor burns, and looked quite healthy at first, weakened after a few days," US authorities clamped down hard.[33] The Pentagon accused Tokyo of promulgating propaganda. The Japanese, a government spokesman declared, wished "to

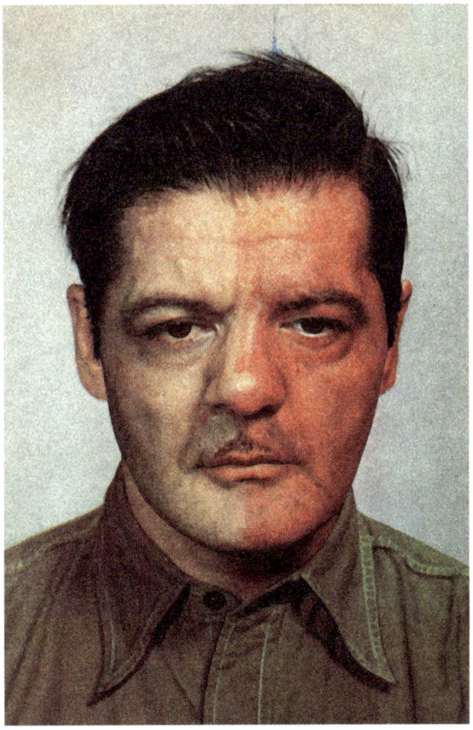

Figure 3.10: C. Ellis Kerr. *Man with Flash Burn on Face*. 1955.

capitalize on the horror of the atomic bombing in an effort to win sympathy from their conquerors."[34]

The scientific photographs from Hiroshima operate through a logic of surfaces, as do the simulated illustrations in the Bloody Book. There are three varieties: the surface of the body and of skin; the surface positioned behind the body, the background that if empty makes intelligible the traumatized skin; and the surface of the photograph. The Hiroshima survivor in the patterned kimono, for example, is presented against an amorphous grey ground from which most details have been expunged by photographic manipulation; the soldier with the on-and-off hat stands against a neutral drop cloth; and the shoulder of the model wearing the red blouse is offset by an empty pale-blue background. The positioning of the victims against blank surfaces allows the wounds to become more visible. The surfaces heighten the details yielded by the skin; because the surfaces are empty, the body becomes legible. For a medicalized photograph to function successfully, the background must be neutral or blank so that the injured body can offer up its data in a calculated way. The Hiroshima and Nagasaki photographs were taken in first aid stations and hospital environments – the spaces where the hibakusha lived because of their injuries from the bomb – but these backgrounds were covered over so that the wounds would be shown clearly. In the Bloody Book, the background surfaces of the studio were also covered over.

The material surface of a photograph is light-sensitive paper on which an image is printed. Photographic surfaces are no less rhetorical than those of skin and background, and can conceal as much as they reveal. The bright-white edges of photographic paper indicate that an image has been cut from the world and reordered into a rectangular form. The cutting is what allows the shot to work as an image in medicine and civil defence illustration, as in any other illustrated subject matter. As a small, intimate object that can be held in the hand, a photograph and the data it contains can be manipulated and multiplied, which is to say, the information it contains about the body is malleable. The artificial blood produced for the Bloody Book represents yet another layer of mediation between actual victims and actor victims. All three varieties of surface – the surface of bodies, the surface of backgrounds, and

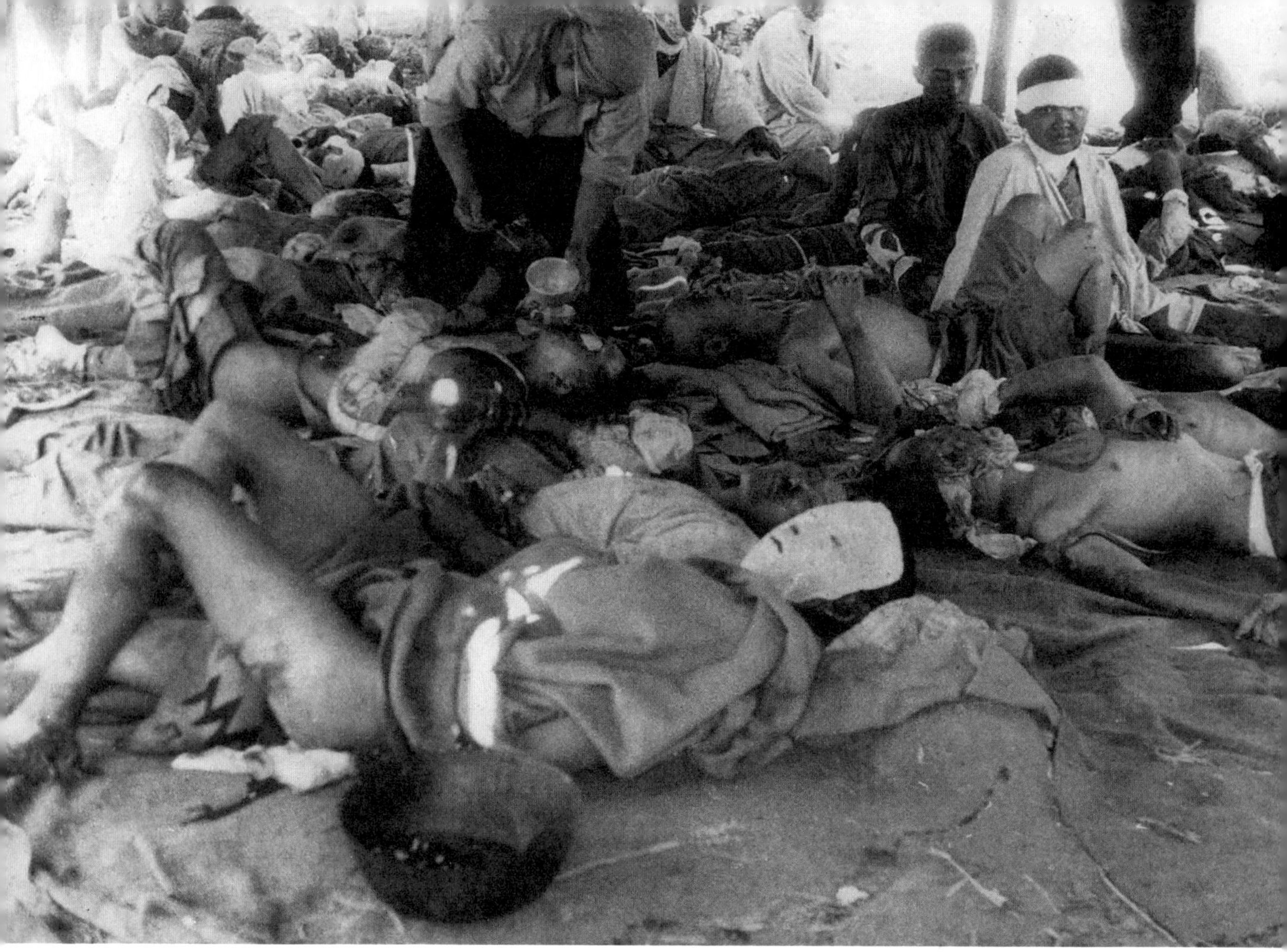

| **Figure 3.11**: Yotsugi Kawahara. *A Scene in a Provisional First-Aid Station.* 1945.

the surface of photographs – instrumentalize the body and point to the instability of the representations. The photographs provide information that can be used to cure and to heal, as was the intention in the Bloody Book, but they can also hide the environment in which the damage occurred.

In the days and weeks after the bombings, Hiroshima and Nagasaki victims were crowded into makeshift spaces.[35] Yotsugi Kawahara photographed them lying on the floor in a first aid tent erected by the Second Hiroshima Army. [**Figure 3.11**] In his picture, it is impossible to tell where one body begins and another leaves off. The logic of surfaces is obscured. The background reads as an extension of the foreground, making the overlapping chaos of skin and bodies unreadable. Apart from the figure in the foreground with the bandaged face, and a second figure in the middle who leans over a bowl, the photograph struggles to cohere. Its incoherence reflects the incoherence of the event and its aftermath. Because of the visceral impact of the photographic and filmic record of atomic destruction, US censors suppressed visual media in Japan more than the written record. In late 1945, a team of Japanese cameramen shot footage of the ruined cities for a documentary. Their film was confiscated before release, with explicit instructions that not a single copy was to remain in Japan.[36] Photographs of the bombings were accorded similar treatment. With few exceptions, those taken in the days and weeks after the attacks were forbidden from being shown or published. It was not until after the occupation, on the seventh anniversary of the attacks, that Yosuke Yamahata's photographs of Nagasaki could be published.[37] He had hidden more than one hundred negatives from censors until it was possible to publish the images in 1952.[38]

In 1955, the American curator and photographer Edward J. Steichen organized *The Family of Man* exhibition for the Museum of Modern Art, New York, and included a Yamahata photograph of an injured boy or girl – it is difficult to tell which – eating a rice ball.[39] Travelling the world under the auspices of the US Information Agency, *The Family of Man* became the most popular photography show ever mounted. The agency's purpose was to soft sell capitalism and to resist communism with whispers rather than shouts. When the exhibition closed in 1962, it had been seen by more than 9 million people in sixty-one countries, including large audiences in Canada.[40] Its brand of humanism has been the subject of critique for more than half a century, written about by Roland Barthes, Allan Sekula, and many other commentators on photography.[41] At their crudest, Steichen and the exhibition suggested that Indigenous peoples living in the Northwest Territories were no different from wealthy investment bankers living on the top stories of skyscrapers in New York. This theme of commonality meshed seamlessly with the global ambitions of American Cold War liberalism. Steichen's show did not include Yamahata's photographs of charred bodies, some of which had been published in *Life* magazine in 1952 (see Figure 1.13). They undercut its proposition that people the world over, regardless of differences in geography and culture, all share in the same experiences. Only those in Hiroshima and Nagasaki had been subjected to the full force of the atomic bomb. But within the exhibition's ideology of universality was lodged an anti-nuclear subtheme. A wall text in the show, reproduced in the exhibition catalogue opposite the shot of the child with the rice ball, quoted philosopher and anti-nuclear activist Bertrand Russell: "The best authorities are unanimous in saying that a war with hydrogen bombs is quite likely to put an end to the human race."[42]

Russell's message was at odds with the assurances delivered by Diefenbaker and Kennedy. If you had access to a fallout shelter, they advised, the prospects of making it through were excellent. But the fallout shelters that were reserved for heads of state were of a different order from those on offer to the public. They were immense bunkers designed to house senior members of government in the event of war. Kennedy had several to choose from, including Project Greek Island, which was built beneath the Greenbrier luxury hotel in West Virginia. Diefenbaker had a network of fifty "emergency government headquarters" scattered across the country, nicknamed Diefenbunkers, to protect him against nuclear holocaust, including one in the rural farming community of Carp, west of Ottawa.[43] The Carp shelter was finally decommissioned in 1994 and designated a national historic site. Several years later, it reopened as Diefenbunker: Canada's Cold War Museum. Built to accommodate 565 people for up to a month, the four-story underground shelter has massive blast doors as well as an elaborate

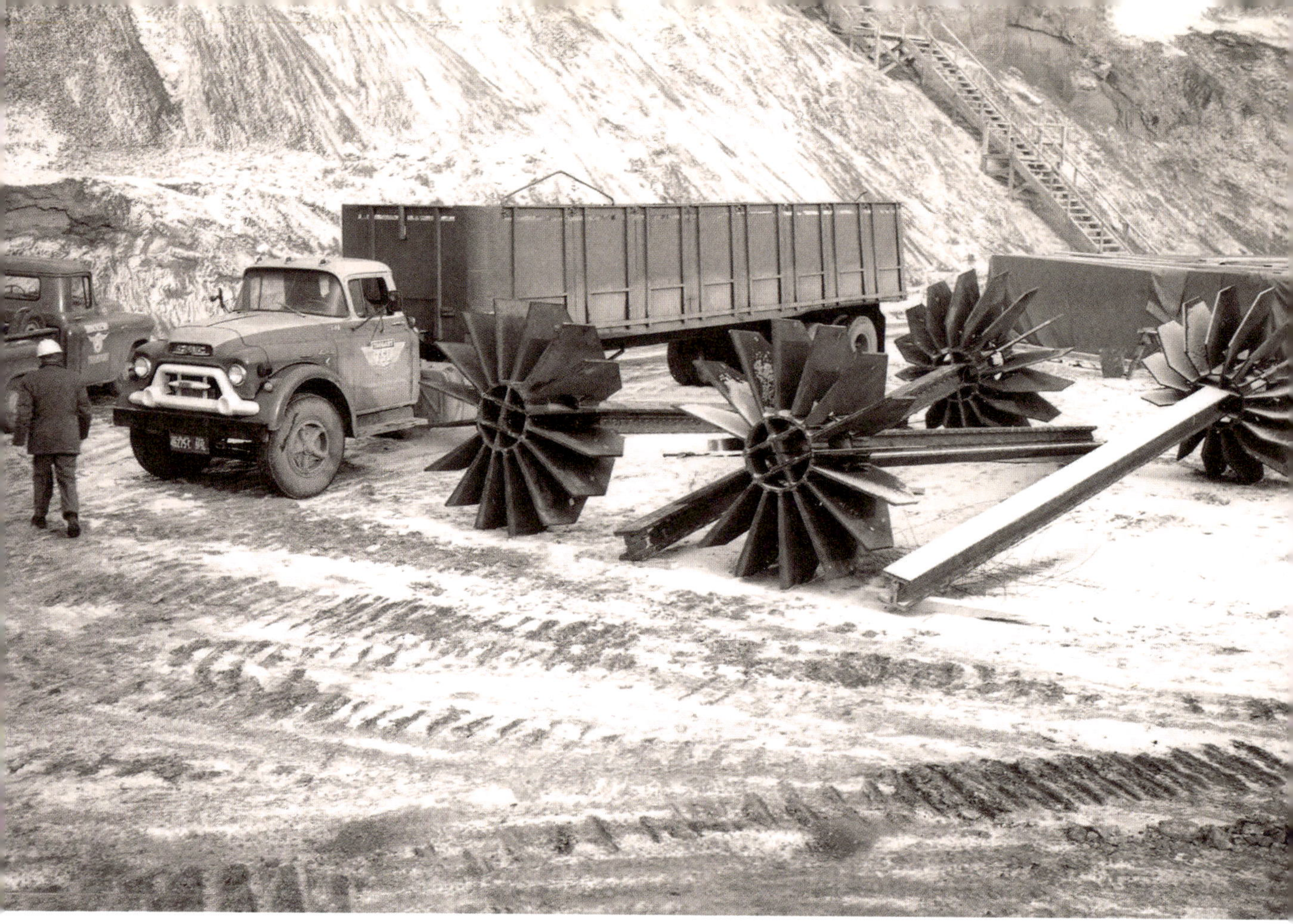

Figure 3.12: Unknown. *Diefenbunker: Reinforced Steel Used to Support Columns during Construction.* 1960.

ventilation system to prevent radioactive infiltration.[44] It even had its own broadcast studio. Photographs exist of the construction site and of the original interior spaces, ranging from kitchen to cafeteria to sleeping quarters to sinks in washrooms. [**Figure 3.12**] However, decorum prevented pictures from being taken of the men's room urinals and the morgue until the photographer Mark Ruwedel did so in 2015. Stenciled across the walls of the men's room, in both official languages, are instructions to "PLEASE FLUSH URINALS/TIRER LA CHASSE D'EAU S.V.P." [**Figure 3.13**] The decorative effect is that of linguistic wallpaper.

In Ruwedel's shot of the morgue, racks for corpses can be glimpsed through the cold-storage doors. [**Figure 3.14**] "Glimpsed" is the operative word. The Diefenbunker and the Greenbrier were secret undertakings, and the general populace knew nothing of their existence. Britain also had its own cache of secret bunkers from which to govern the country in the event of nuclear war. Among the few who saw them were Spies for Peace, a group of activists who broke into a shelter in Reading, Berkshire, and copied a map of the country's nationwide network of secret shelters.[45] Disclosure of the network damaged claims by British, Canadian, and other governments that a nuclear war was defensible. It was not.

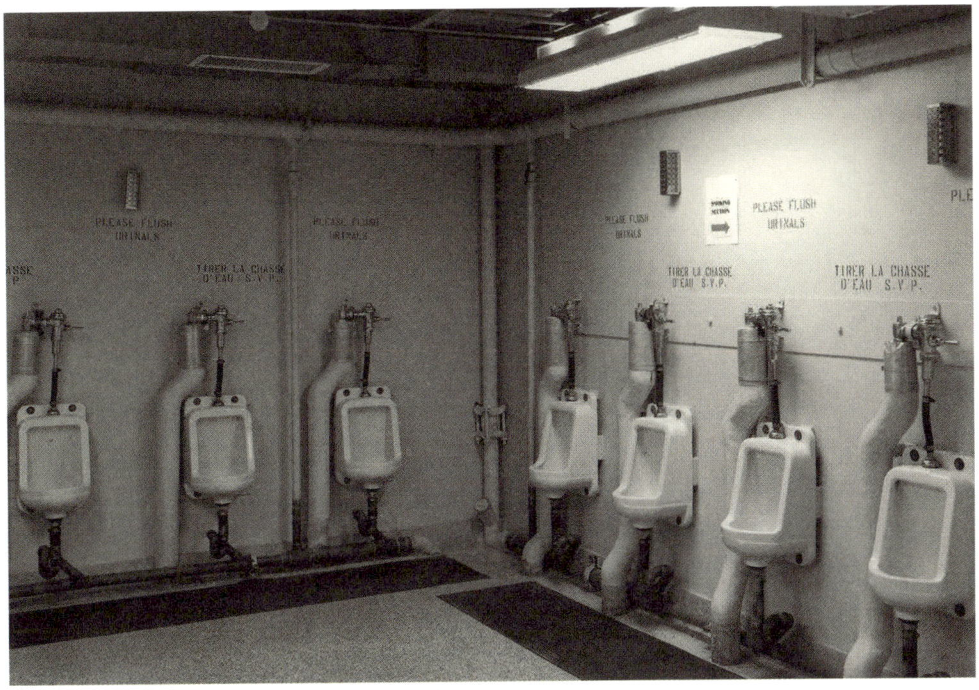

Figure 3.13: Mark Ruwedel. *A Cold Winter #7: Men's Room Toilets.* 2015. **Figure 3.14**: Mark Ruwedel. *A Cold Winter #8: Morgue.* 2015.

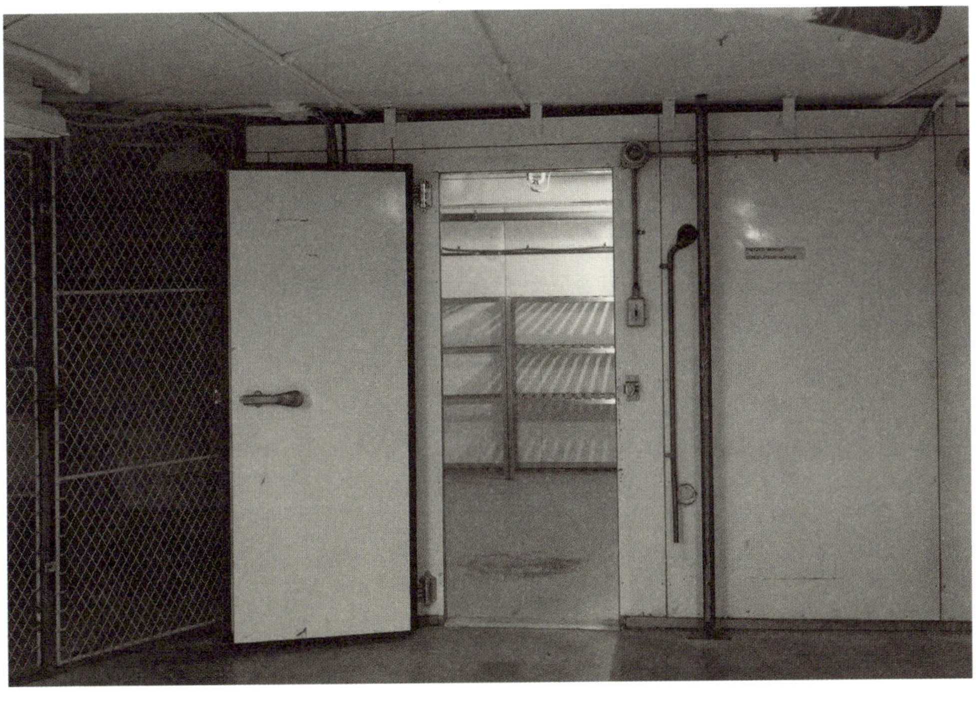

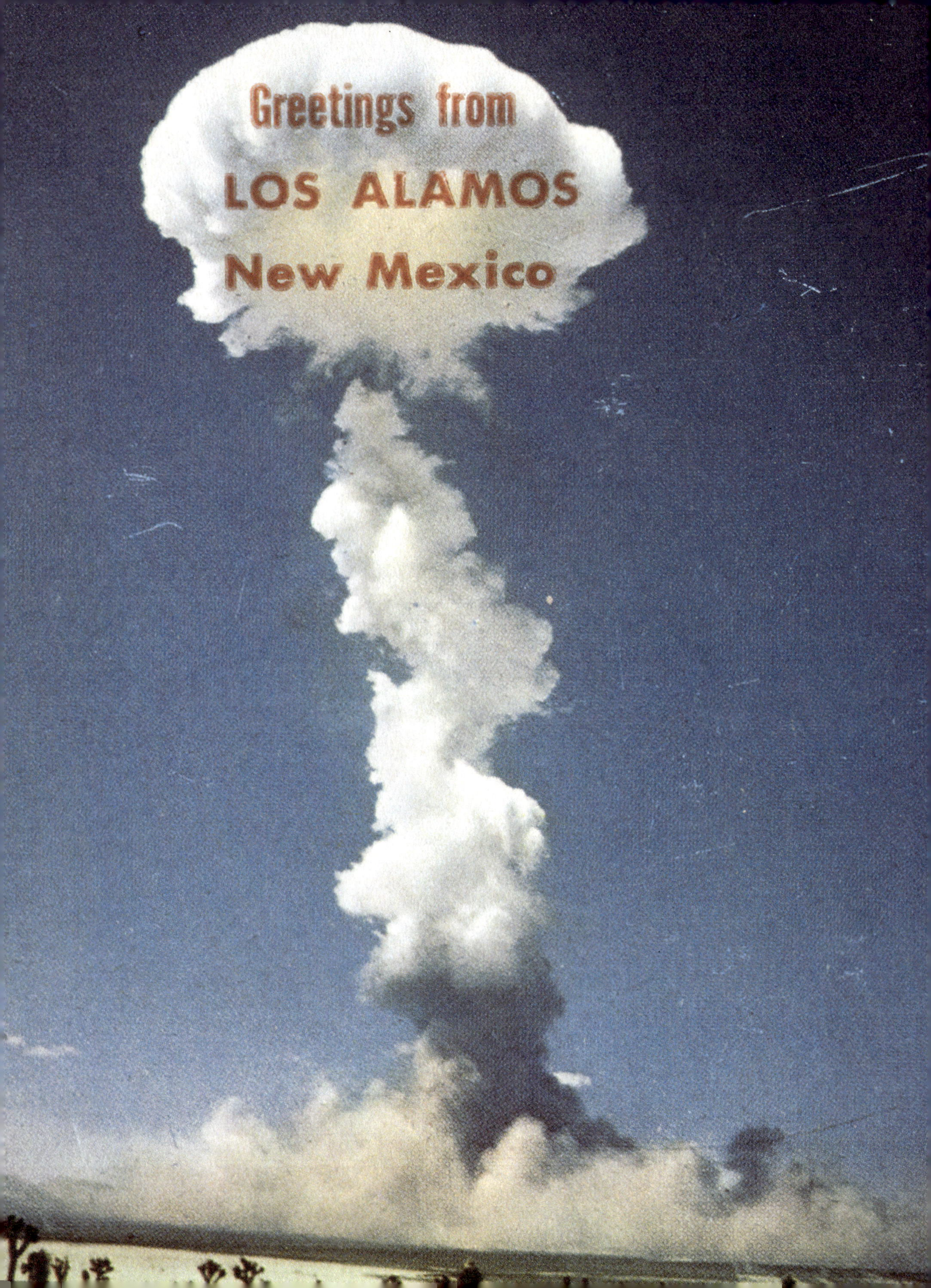
Greetings from
LOS ALAMOS
New Mexico

4
Cloud Shadows

IN 1945, THE WORLD WAS BENT INTO A DIFFERENT SHAPE BY THE ATOMIC BOMB and would never be the same again. The unprecedented force of the bomb was symbolized by the image of the mushroom cloud, which was endlessly reproduced as a spectacle of mass destruction, largely through photographs. The spectacle attracted not only the producers of popular culture, who often traded on it for profit, but also artists who frequently endeavoured to make strange again what had become familiar. Both figure in this chapter. The producers of magazines, record album covers, advertisements, and souvenirs helped to domesticate the bomb, as did American, Canadian, and other governments that wanted to create a "comfort level" around it.[1] Artists tried to unsettle this process of domestication.

A colour postcard from the early 1950s, trademark Mirro-Krome, represents an atomic explosion at the Nevada Test Site. [**Figure 4.1**] Emblazoned in red letters across the mushroom cap at the top of the card is this message: "Greetings from LOS ALAMOS New Mexico." The letters might just as well say, "Greetings from Hiroshima and Nagasaki." Like other postcard images of disastrous events, such as the Chernobyl meltdown, the card naturalizes the catastrophic. The bright blues and reds serve to feminize the explosion. Remarks by nuclear officialdom operate in a similar vein. An employee at the Pantex Nuclear Weapons Final Assembly plant in Amarillo, Texas, stated in the early 1980s that handling warheads was "just

| **Figure 4.1**: Postcard. *Greetings from Los Alamos, New Mexico.* Before 1955.

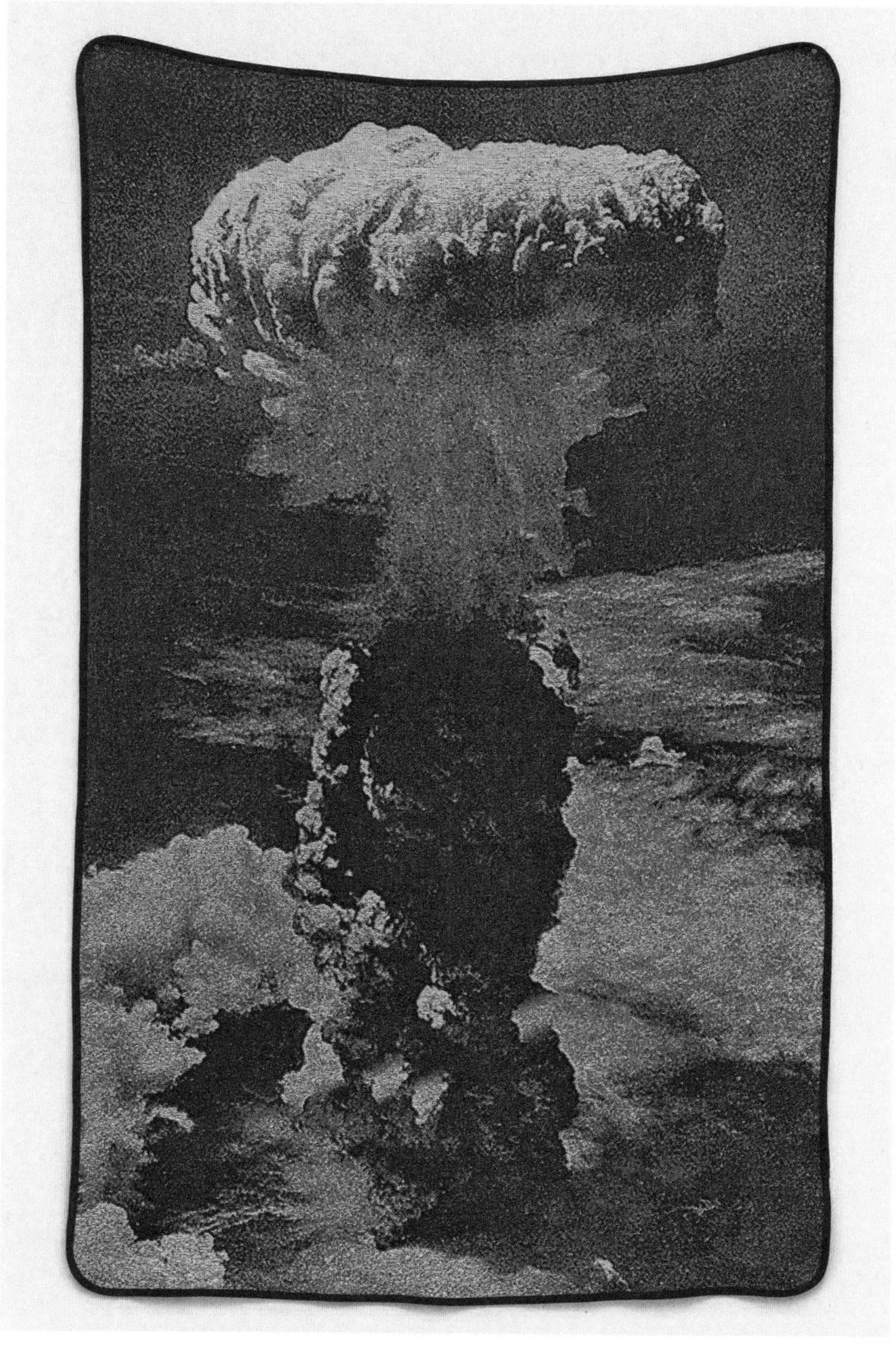

Figure 4.2: Douglas Coupland. *Nagasaki Blanket*. 2011.

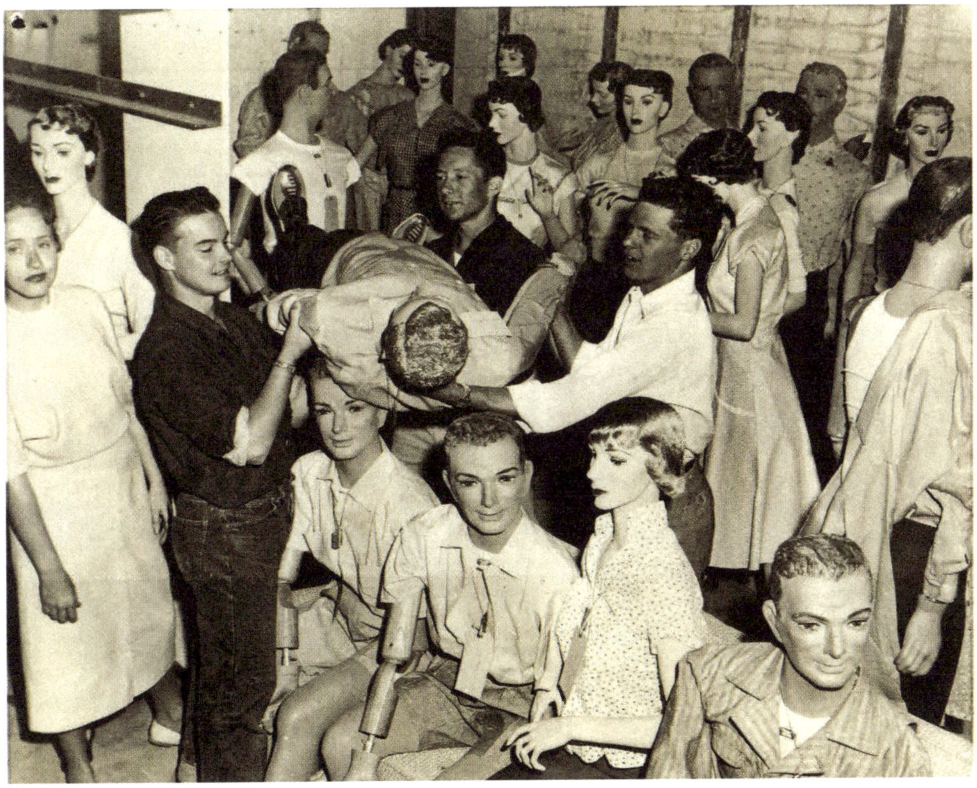

Figure 4.3: Associated Press Wirephoto. *Las Vegas High School Students Prepare Mannequins That Will Represent Them during a Nuclear Test.* 1955.

like pickin' up a box of Silly Putty in a dime store." He added, "It is the biggest non-issue of the 20th century."[2]

Douglas Coupland's portrayals of mushroom clouds are less nonchalant. [**Figure 4.2**] On a large blanket, the artist has printed a photographic image of the Nagasaki cloud, and on a second blanket the image of an explosion at the Nevada Test Site. The works are lethal security blankets that denaturalize the cataclysmic. They are soft to touch but prickly to look at. The bomb's denaturalization assumes a particularly uncanny form in a photograph of Las Vegas high school students who stand among the mannequins that will represent them during an upcoming test explosion. [**Figure 4.3**] Students and mannequins are dressed identically, making it difficult to tell the teens from their atomic doppelgängers. The picture was taken inside the facsimile of a suburban house in Survival City, also called "Doom Town" or "Doomstown," constructed in 1955 during Operation Teapot for the purpose of measuring the impact of a bomb on humans and buildings. During a guided tour of Survival City half a century later, Coupland got a splinter from a piece of wood in a house that had survived a blast and jokes that the staff was worried he might turn green from lingering radioactivity.[3]

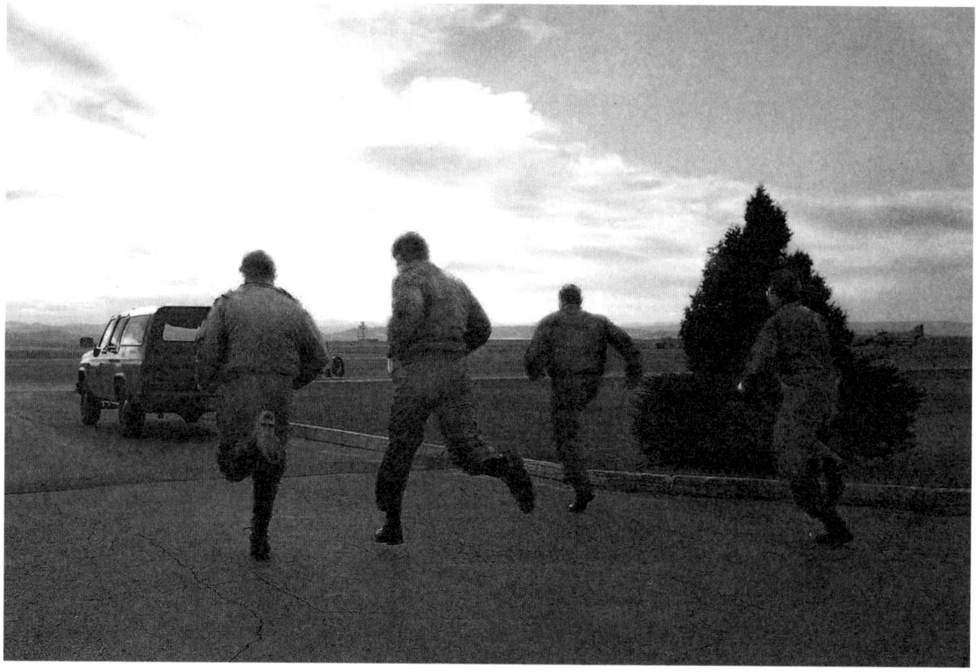

Figure 4.4: Robert Del Tredici. *The Scramble.* 1986. **Figure 4.5**: Barbara Norfleet. *Dr. J. Robert Oppenheimer, Bradbury Science Museum, Los Alamos.* 1988.

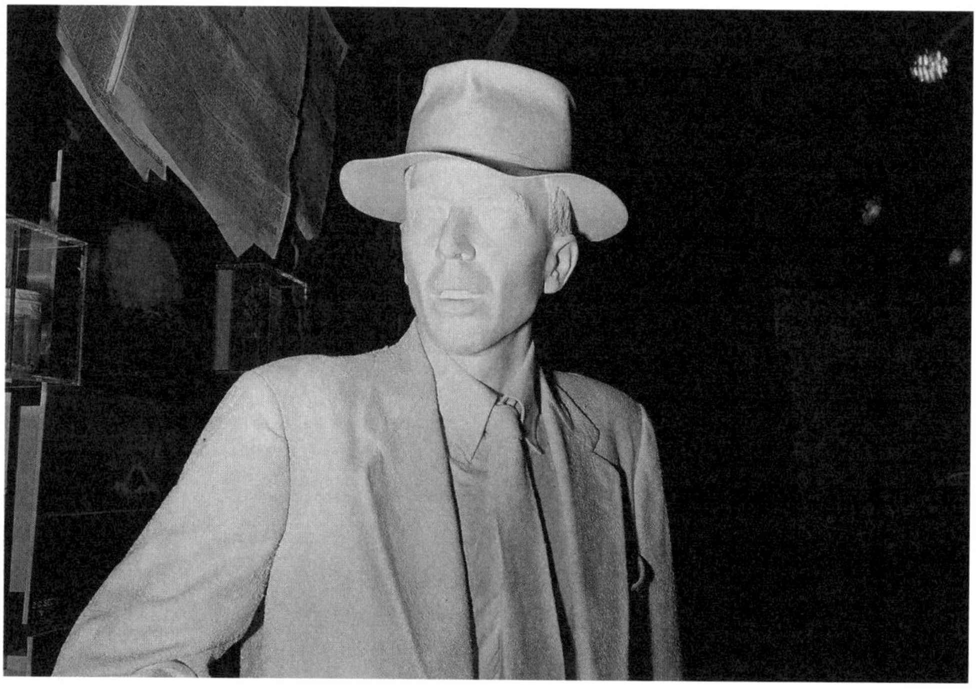

Nuclear photography, like the nuclear era itself, is haunted by questions of excess. One way to understand how atomic representation has shaped public attitudes and memory is to consider the surplus production of mushroom cloud images since the end of the Second World War.[4] The flash-and-bang imagery of mushroom clouds and fireballs has separated the spectacle from the environment in which it occurred, often generating more noise than signal.[5] Photographs taken during the attacks on Hiroshima and Nagasaki made the mushroom cloud instantly recognizable, and the atomic tests conducted by the United States at Bikini and Nevada following the end of the Second World War made it a staple of mass media and popular culture. "Our mental picture of the Bomb," writes Del Tredici in *At Work in the Fields of the Bomb,* "includes the distant mushroom cloud, a city turned to ash, rockets in the atmosphere, the tapered cone – icons worn smooth by time and use."[6] Del Tredici's photographs offer an alternative to the spectacle. In addition to images of atomic workers and their workplaces – uranium mines, ore processing plants, research laboratories, factories, nuclear power stations – his books and projects contain interviews with them about their work, ranging from mining and processing to nuclear physics and manufacturing to manning bombers equipped with nuclear weapons.[7] One photograph, *The Scramble,* shows members of a bomber crew running toward a van that will ferry them to their plane. [**Figure 4.4**] Del Tredici's photographs and writings are about the so-called Yellowcake Road – the route from the uranium mine to the mushroom cloud – which is named for the colour of the concentrate obtained from milling uranium ore after it has been mined but before enrichment. The Yellowcake Road is far less vivid in the public imagination than the Yellow Brick Road that leads from Kansas to the Emerald City. Neither is paved with real gold, though Dorothy and her companions in *The Wizard of Oz* made theirs glitter.

When the first mushroom cloud rose from the desert floor at the Trinity test site, Alamogordo, in the early hours of the morning on July 16, 1945, J. Robert Oppenheimer, often called the father of the atomic bomb for his role in its development, declared, "I am become Death, the destroyer of worlds." He was speaking in front of a motion-picture camera that was on hand to record the explosion.[8] A photograph of a statue of Oppenheimer, taken by Barbara Norfleet at the Bradbury Science Museum in Los Alamos, transmutes the scientist from a living figure into a metaphorical pillar of salt capped by his signature porkpie hat.[9] [**Figure 4.5**] Norfleet's camera freezes Oppenheimer against a black background, making him look as if he has just gazed upon the forbidden and paid a biblical price for his transgression. Her image underscores his description of himself as a "destroyer of worlds."

The iconography of nuclear explosions began with the Trinity detonation, where photographers took at least 100,000 exposures of the test.[10] At laboratories in Los Alamos, Julian E. Mack and Berlyn Brixner designed specialized cameras for recording the detonation and scientifically measuring the blast yield of the weapon. Some of the photographs give the exact moment of execution down to a fraction of a second, which is noted on the prints with time stamps (see Figure 1.7). Nuclear testing and the development of increasingly sophisticated camera and film equipment to document what otherwise could not be recorded went hand

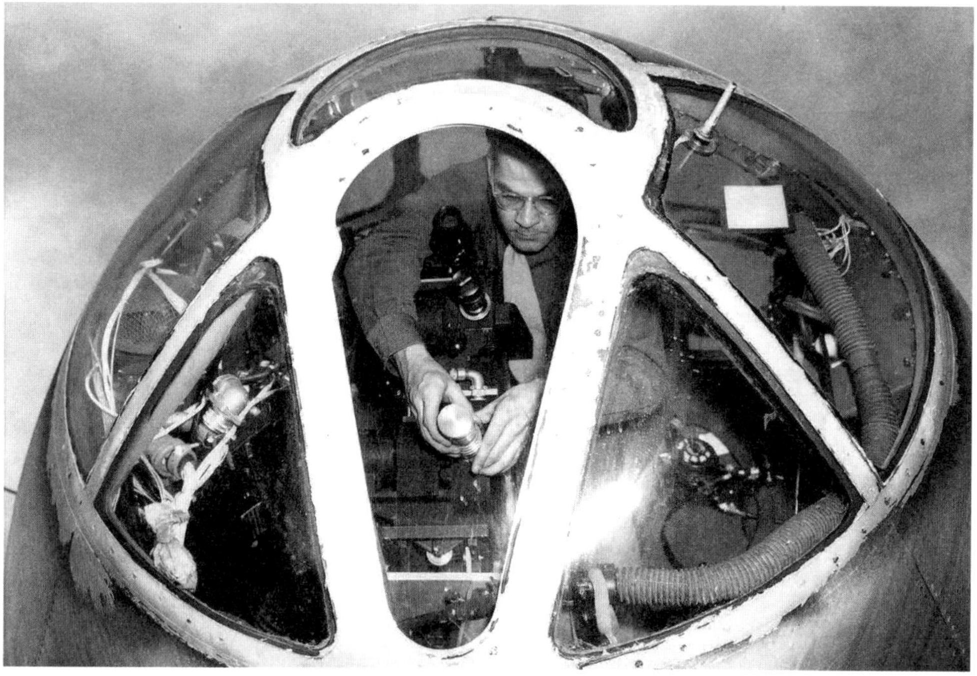

Figure 4.6: Associated Press. *Lieutenant T.H. Martin Mounting Dual Movie Cameras in the Nose of a B-29 Bomber, Bikini Atomic Bomb Tests.* 1946. **Figure 4.7**: US Strategic Bombing Survey. *Atomic Shadow Scorched into the Steps of Sumitomo Hiroshima Bank, 260 Meters from the Hypocentre of the Blast.* 1945.

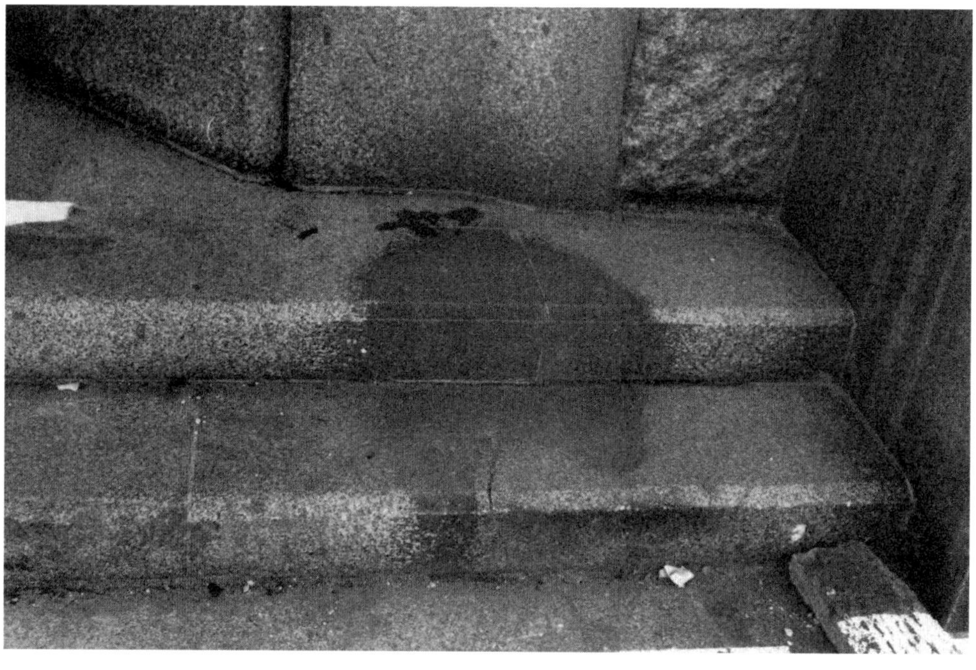

in hand.[11] A shot taken during preparations for the first Bikini test shows the mounting of dual movie cameras into the gun turrets of a B-29 aircraft undergoing conversion into a "photographic ship." [**Figure 4.6**] Here guns are turned into cameras – both aimed to shoot with deadly accuracy. "The technologies of cinema and warfare have developed a fatal interdependence," observes Paul Virilio, the French theorist of speed, technology, and warfare.[12] An extreme version of this interdependence was produced by the atomic "flash" at Hiroshima, where the violent excess of light and heat from the explosion imprinted the shadows of bodies on the steps and walls of the city. [**Figure 4.7**] Susan Schuppli, the Canadian artist and scholar, remarks that these radiographic prints – literally atomic shadows – "document life at the very moment of death."[13] A survivor described the sky over Hiroshima after the explosion as being "filled by a garish light which resembles the magnesium light used in photography."[14]

The most immediately recognizable of all mushroom clouds was produced by the Fat Man plutonium bomb dropped on Nagasaki.[15] Photographs of it exist in multiple versions.[16] [**Figure 4.8**] The shot that was released to the press is a two-tone, two-tier image taken from the air minutes after the detonation, as its caption states, and was rapidly circulated around the globe as a wirephoto. The upper part of the cloud contrasts sharply with the dark column of smoke and debris of the lower part, and is separated from it by a slender gap. The gap is highly charged, like the gap in Michelangelo's painting in the Sistine Chapel between God's forefinger and the hand of Adam, but whereas the fresco represents the creation of life, the Nagasaki photograph represents its destruction. The relationship between generative and destructive power is the subject of a nuclear-themed poem by William Carlos Williams.[17] In "Asphodel, That Greeny Flower," Williams reflects on the correlation between the mushroom cloud as a blooming flower and the mushroom cloud as a bloom of death.[18] "The bomb / also / is a flower," he writes, aware that in classical times asphodels were associated with mourning and death. Asphodels were the flowers of hell.

The gap separating the light and dark components of the Nagasaki cloud, referred to as a signature feature by those who specialize in mushroom cloud iconography, makes it unlike any other atomic image. But this has not prevented it from being identified with the Hiroshima explosion that occurred three days earlier.[19] Because Hiroshima and Nagasaki are often spoken of as a single event, and because the attack on Hiroshima preceded that on Nagasaki, the sibling cities are frequently conflated into one word: "Hiroshima" has become a sign for both. In addition to being sold for touristic purposes, a version of the Nagasaki mushroom cloud photograph was used by the US Air Force to commemorate the bombings. [**Figure 4.9**] A deluxe presentation album containing twenty-three tipped-in photographs – one shy of the number in a standard roll of film – includes the Nagasaki image. Horizontal in format, this version of the shot shows part of the wing of the observation plane from which it was taken. Written in the lower right corner is the photograph's ID number: "AEC-51-4028 58381 A.C." The overlapping of text and wing serves to emphasize, as other photographs of the Nagasaki cloud do not, the connection between the interior of the aircraft, from which

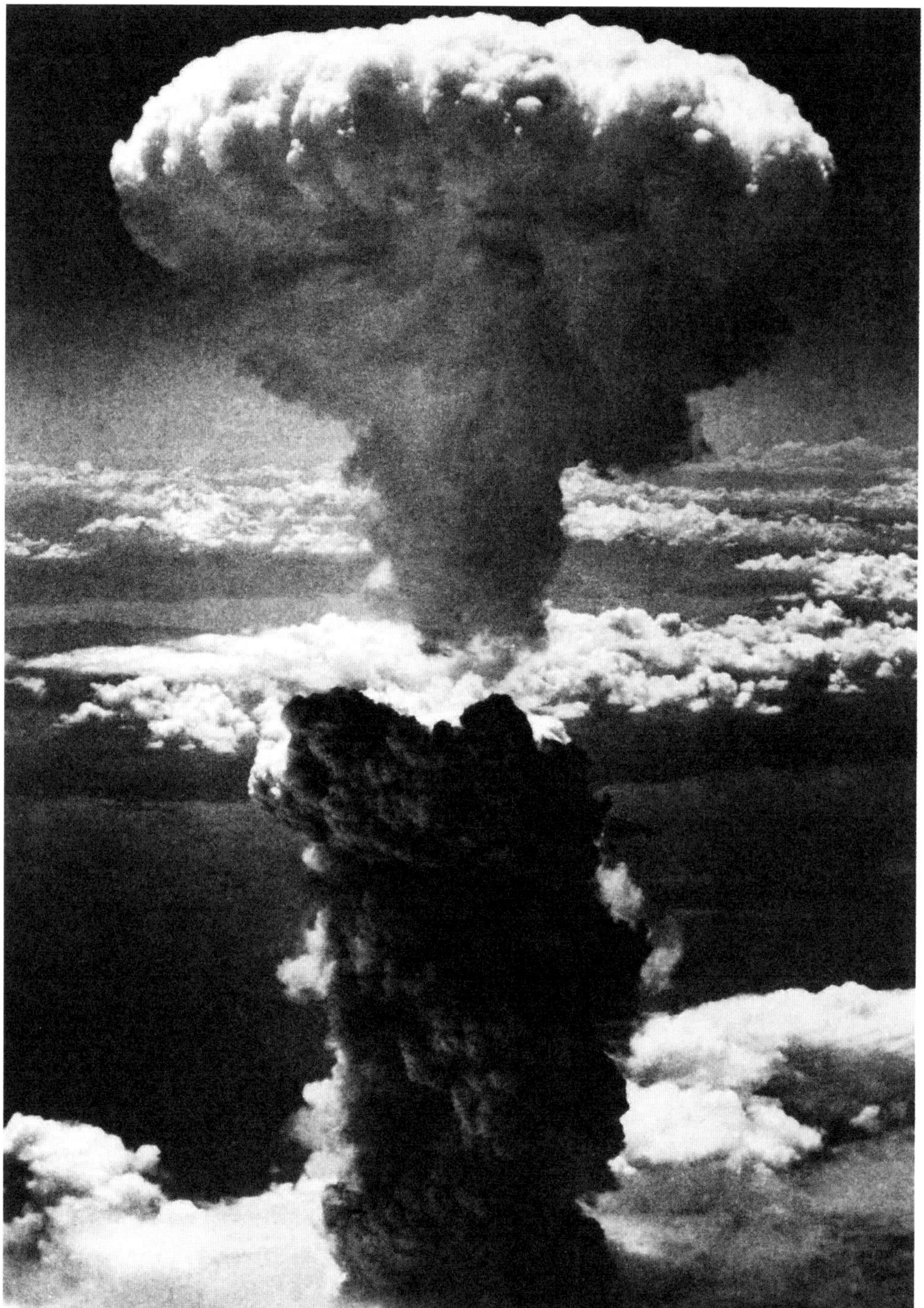

Figure 4.8: Associated Press Wirephoto. *Atomic Bomb Sends Smoke 20,000 Feet above Nagasaki.* 1945.

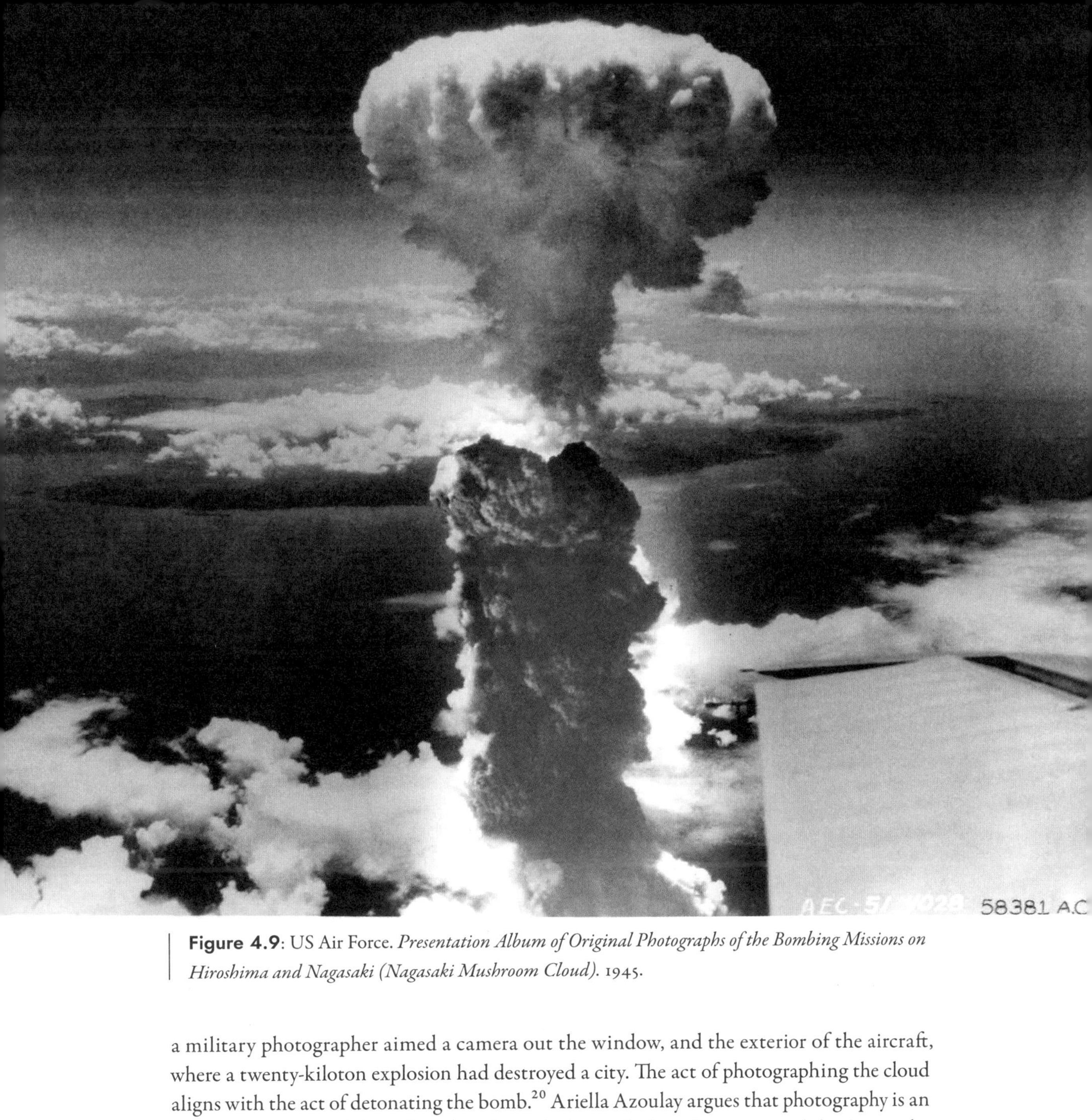

Figure 4.9: US Air Force. *Presentation Album of Original Photographs of the Bombing Missions on Hiroshima and Nagasaki (Nagasaki Mushroom Cloud)*. 1945.

a military photographer aimed a camera out the window, and the exterior of the aircraft, where a twenty-kiloton explosion had destroyed a city. The act of photographing the cloud aligns with the act of detonating the bomb.[20] Ariella Azoulay argues that photography is an *event* of a special kind that takes place through the mediation of a camera and that it must be understood as indivisible from historical events. One military aircraft delivered the nuclear weapon, another military aircraft delivered the photograph.

The Nagasaki image has also been deployed as a matrix for the production of contemporary art. Like Coupland, in *Nagasaki Blanket* (see Figure 4.2), the Korean artist Kyungah Ham repurposed the image using textiles as a medium.[21] Coupland also reflects on nuclear threat in his novels, including *Life after God* and *Generation X: Tales for an Accelerated*

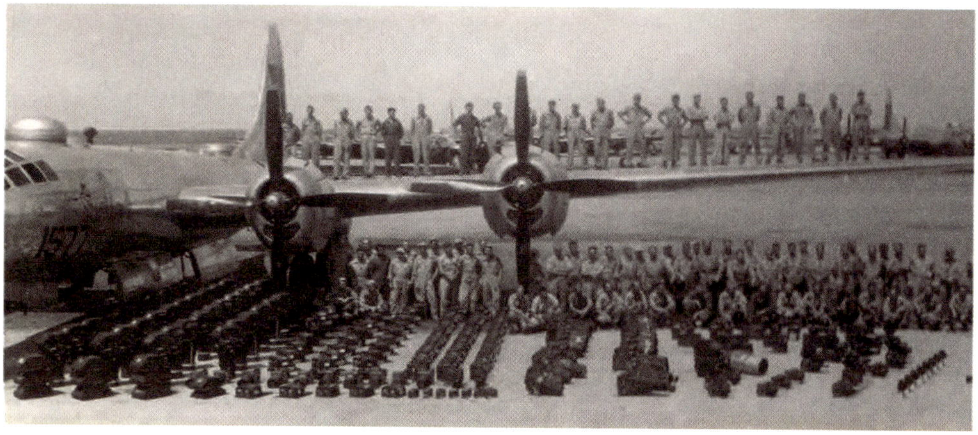

Figure 4.10: Joint Task Force One. *Mass Grouping of Cameras Used to Photograph First Bikini Tests.* 1946.

Culture.[22] In a non-fiction story from *Life after God,* titled "The Wrong Sun," Coupland recalls sitting in a McDonald's restaurant in Vancouver at a children's birthday party on November 6, 1971, the day of the Amchitka nuclear test in Alaska: "I remember sitting on my purple vinyl stool being unable to eat, gazing out the window, waiting for the flash, waiting for the cars to float up into the sky, for the Hamburglar statue to melt, for the tiled floor to break apart and expose lava."[23] The connection between McDonald's and the nuclear test, Coupland says, stayed with him and led to two conclusions: "One, that McDonald's equals evil; two, that technology does not always equal progress."[24] Coupland's nuclear blanket is not bedding for the faint of heart.

The first two atomic tests at Bikini Atoll – Test Able and Test Baker – codenamed Operation Crossroads, were conducted in July 1946.[25] They were the fourth and fifth nuclear detonations ever to take place. In part, they were scientifically managed experiments designed to assess the impact of aerial (Able) and underwater (Baker) detonations on target ships and biological specimens. At the same time, they were political theatre: carefully staged media events, with the world press in attendance, mounted to assert American nuclear superiority and to dispel reports about the dangers of radiation.[26] Photographers from the Associated Press, International News Service, Acme Newspictures, and *Life* magazine were invited to attend on the condition that they pool their images for distribution under the supervision of the US military.[27] Seven hundred and fifty cameras were employed by the news photographers and military cameramen to record the blasts, using up half the world's supply of motion-picture film in the process. A photo records a mass grouping of cameras used by the military during the tests. [**Figure 4.10**] Among them was the largest still camera then in existence, with a forty-eight-inch focal length telephoto lens and an ultra-high-speed motion camera that could shoot ten thousand frames per second. "The multiplicity of cameras was necessary to ensure full records of results," a book on the operation stated, "particularly damage results."[28]

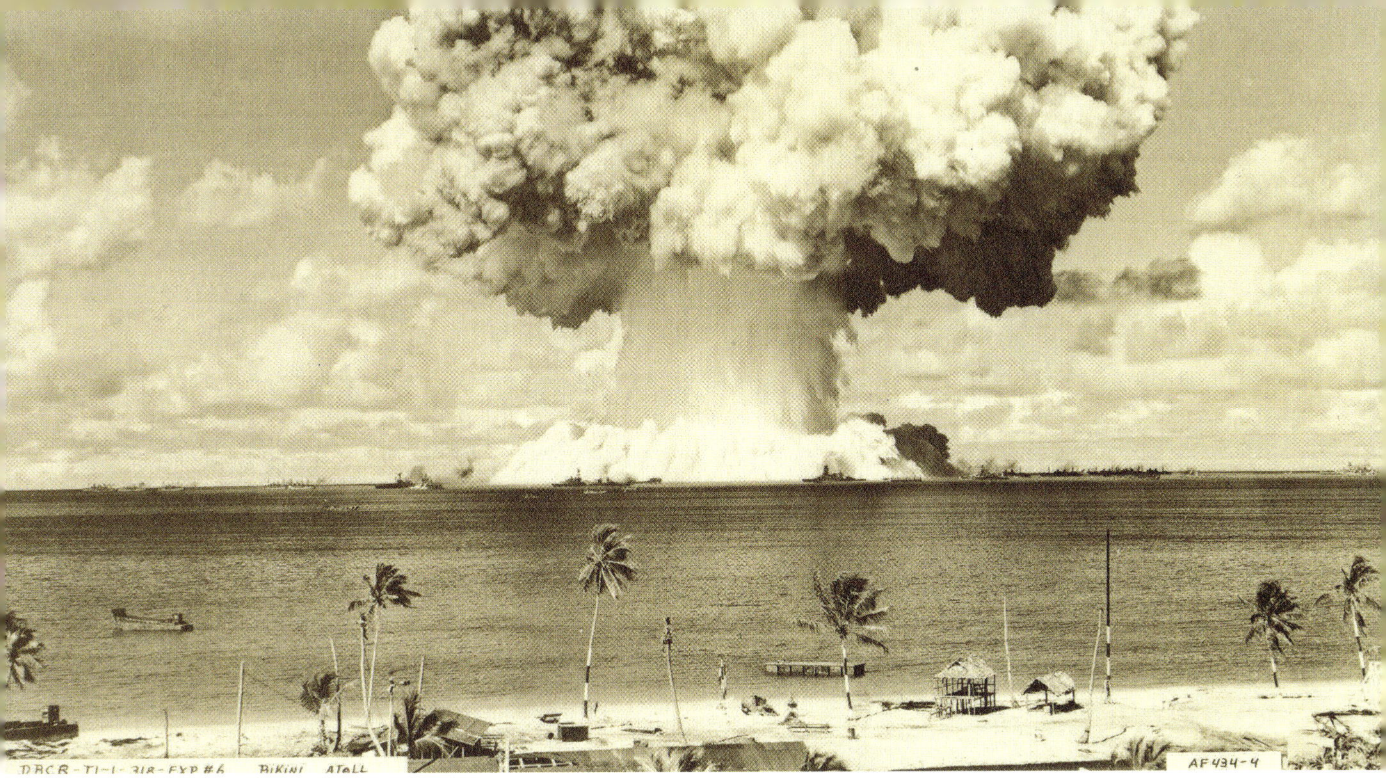

Figure 4.11: US Army Photographic Signal Corps. *Test Baker, Operation Crossroads, Bikini Atoll.* 1946.

The "damage results" from Bikini that entered most strongly into public memory were provided by pictures of Test Baker, which was detonated fifteen metres below the water. The images exist in at least as many variations as there were cameras trained on the event. Each shows a column of water rising from the blast to form a cloud of condensation that looks more like a cauliflower than a mushroom.[29] [**Figure 4.11**] The best-known example goes by several titles, including *Mushroom Cloud, Test Baker.* A description of the image in *Operation Crossroads: The Official Pictorial Record,* a book released a few months after the tests by the publisher Wm. H. Wise in conjunction with the US military, concentrates less on the shape of the cloud than on the damage wrought by the explosion. It comments on what is visible to the camera as well as what is not:

> *The wall of spray and steam at the base of the column rushes precipitantly out from the center, drenching the target ships with its thick poisonous wetness ... Small wonder that 90 per cent of the target vessel array was affected by the deadly radioactivity. Although the amount of damage done to the hulls was not very different from predicted damage, the extent of radiological hazards went beyond what had been expected.*[30]

This passage does not mince words. Ships were drenched in "thick poisonous wetness" and affected by "deadly radioactivity," while the "radiological hazards went beyond what had

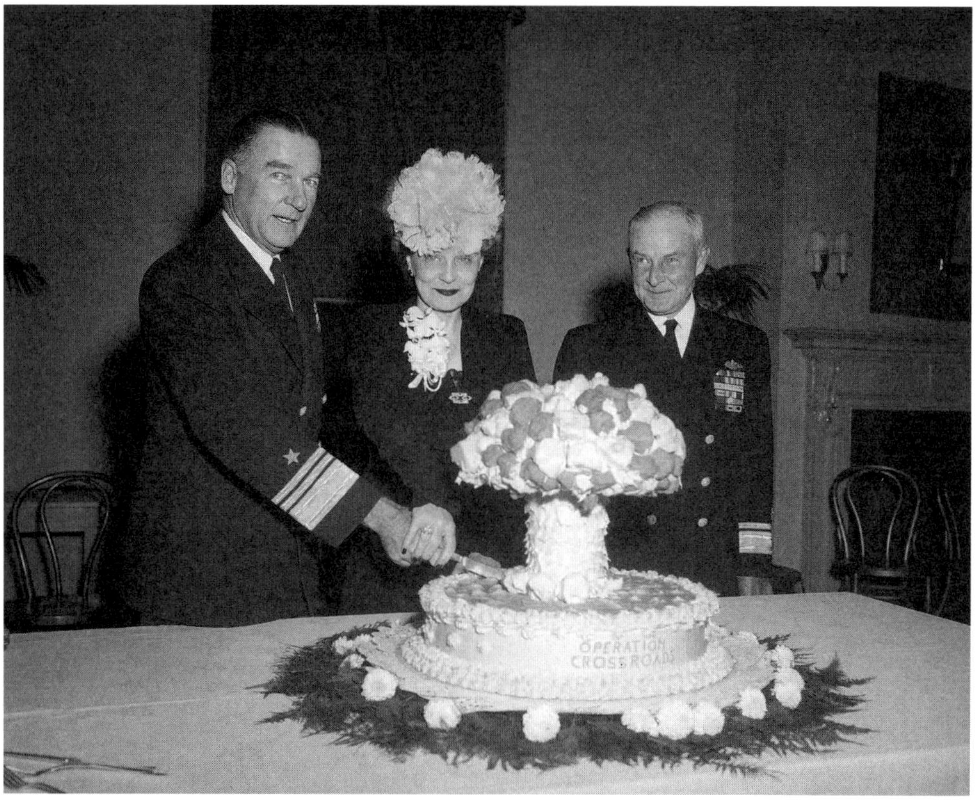

Figure 4.12: Press Agency Photo. *Vice Admiral W.H.P. ("Spike") Blandy and His Wife Cutting an Atomic Cake.* 1946.

been expected." The phrases address the dangers of radiation with a clarity and openness that was soon stifled in press communiqués issued by military authorities. After 1946, words such as "poisonous" and "deadly" were excised from the official nuclear lexicon and replaced with clichés.[31]

Among the biological specimens placed on the target ships and assessed for radiological and other kinds of harm were pigs, goats, rats, mice, and guinea pigs. Pigs were used because their skin and hair were comparable to those of humans, and goats because their weight and bodily fluids were similarly compatible. Thirty-five percent of the test animals died.[32] The circulation of photographs of irradiated goats and pigs after Operation Crossroads, like that of words and sentences describing radiological threat, was subsequently curtailed by military censors. In *Operation Sandstone: The Story of Joint Task Force Seven,* a book of photographs of tests conducted at Eniwetok Atoll in 1948, there are no images of animals.[33] The photos selected for general release by authorities are less about radiation experiments and the effects on animals and people than about the technological sublime, the flash-and-bang spectacle of fireballs and mushrooms clouds.

Mushroom Cloud, Test Baker, has supplanted most other images in constructing a public memory of the event.[34] But a shot of a cake baked to celebrate the completion of the tests falls into a special category of its own. [**Figure 4.12**] Topped with a mushroom cloud, it is being cut by Vice Admiral William "Spike" Blandy and his wife. Rear Admiral Frank J. Lowry stands beside them, with a smirk on his face.[35] Mrs. Blandy, wearing a hat to match the cake, folds her right hand over her husband's like a bird of prey. Blandy, who once denied being an "atomic playboy," was the senior officer in charge of the Bikini tests. The photograph was taken by the studio of Harris and Ewing at an officers' club in Washington, DC, and it caused an uproar when it was published in the social pages of the *Washington Post* on November 7, 1946. Reverend Arthur Powell Davies denounced it as "obscene" from the pulpit of All Souls Church. His sermon began a chain reaction of criticism that spread across North America and beyond, to Europe and the Soviet Union.[36]

In 1954, the International Nickel Company (Inco), a Canadian corporation, published a full-page advertisement in *Time* magazine, in which it transformed *Mushroom Cloud, Test Baker,* from a black-and-white photograph into a grisaille watercolour.[37] [**Figure 4.13**] The illustration, which flips the original image by 180 degrees, remains faithful to most of its characteristics, emphasizing the extremes of dark and light in the cloud as well as in the surrounding sky and water, but makes two notable alterations. It eliminates the target ships that were placed at ground zero to test the force of the bomb, and it enlarges a group of thatched huts and palm trees in the lower right foreground. The changes soften the image, transforming a representation of the atomic sublime into what I am inclined to call the "atomic picturesque." The atomic sublime can be disturbingly seductive, whereas the orderliness of the atomic picturesque is merely pleasurable. Like postcards of nuclear explosions, the Inco ad domesticates a deadly event. The bathing suit named by Louis Réard in response to the 1946 Bikini tests is a more truthful representation of the event than the advertisement; the two-piece bikini, at least, acknowledges the sexual energy associated with a nuclear discharge.

The reason for Inco's pictorial legerdemain is announced in the ad's message: "Even *this* cloud has a silver lining," the tagline reads, invoking the beneficial side of the nuclear pharmakon. Nickel alloys are needed in reactors to produce radioisotopes for medical purposes, Inco states, because the alloys are noncorrosive, heat resistant, and strong. The company is ideally situated to help in the production of radioisotopes for peaceful purposes. This message is consistent with those of other Inco ads in American magazines in 1954. One example in *Collier's* explains to readers "How Inco Nickel Is Helping to Produce Power from the Peace Atom."[38] President Dwight D. Eisenhower's recently coined slogan "Atoms for Peace" was finding an audience among military suppliers, just as the Atomic Energy Commission's (AEC) promotion of nuclear power for civilian purposes was attracting commercial interest. "While the AEC's weaponeers are at work in the Pacific," an article in *Time* observed, "the AEC is also going after peaceful nuclear power in a big way."[39] In other words, the United States was turning swords into ploughshares (while still looking for ways to build bigger bombs). The large thermonuclear test Castle Bravo took place at Bikini on March 1, 1954, just weeks

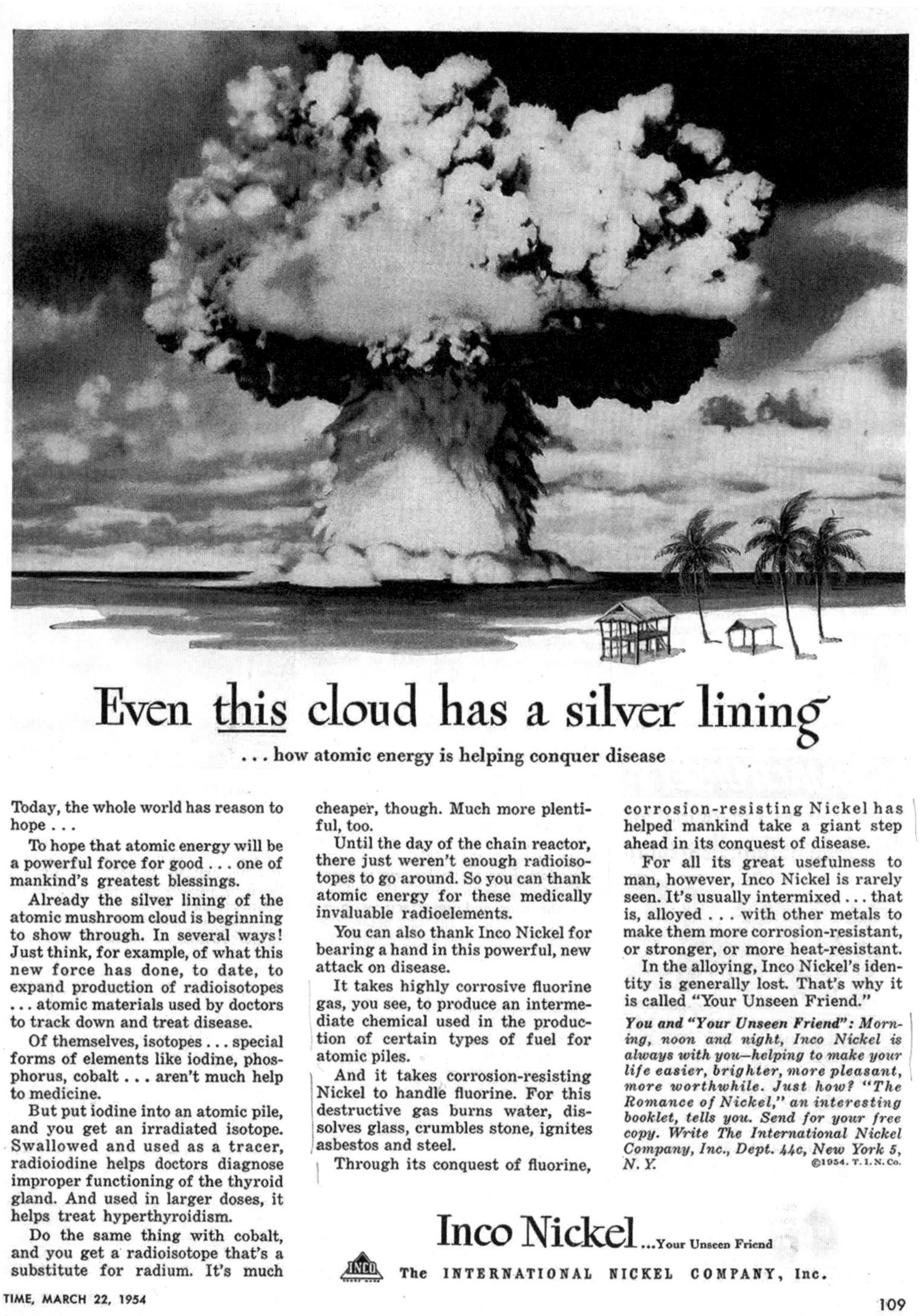

Figure 4.13: Advertisement. *Even* This *Cloud Has a Silver Lining.* 1954.

before the Inco advertisement appeared. It was the most powerful nuclear device exploded by the United States and one of the most lethal.[40]

From the tone and content of Inco's advertisement, readers might be persuaded that a significant part of the company's interest lay in the medical uses of nickel. They would be wrong. In our era of near permanent war – which shows no sign of changing more than half a century after Castle Bravo and the Inco ad – it follows that all corporate and government activities must be performed in the pursuit of peace.[41] By 1950, and the outbreak of the Korean War, Inco had established a virtual monopoly in the production and distribution of nickel in North America – it was supplying close to 80 percent of US requirements – and Congress launched an investigation into its market dominance and pricing practices. Nickel was a "true war material," the US Department of Defense declared, and it wanted the Canadian company to provide the metal to the United States at a competitive cost.[42] The construction of nuclear reactors, whether for military or peaceful purposes, demanded large quantities of nickel and aluminum, and the Americans wanted secure supplies.

The congressional investigation bruised Inco's corporate image by pointing out that it was functioning as a monopoly. To repair the damage, the company increased production to bring down prices and initiated an aggressive advertising campaign to refashion itself as a concerned corporate citizen. By 1954, it was operating at full capacity and selling all the nickel it could mine. Only a very small part of its production, however, went toward reactors that were making radioisotopes for medical purposes. Much more of its Canadian output went to feed the US war machine. A month after the Test Baker ad appeared in *Time,* the *Inco Triangle,* an in-house company publication, reported, "War or the preparation for war creates an immediate demand for all metals and destroys the usual balance between demand and supply ... We are now in such a period."[43] The nickel supplied to the United States by Inco, like the uranium supplied to the US Department of Defense by Eldorado Mining and Refining, was directed more toward making mushroom clouds than producing isotopes for medicine. But the company's gambit to present itself as a responsible corporate citizen succeeded. Inco found its "silver lining" in the Baker detonation and managed to turn a nuclear negative into a positive.

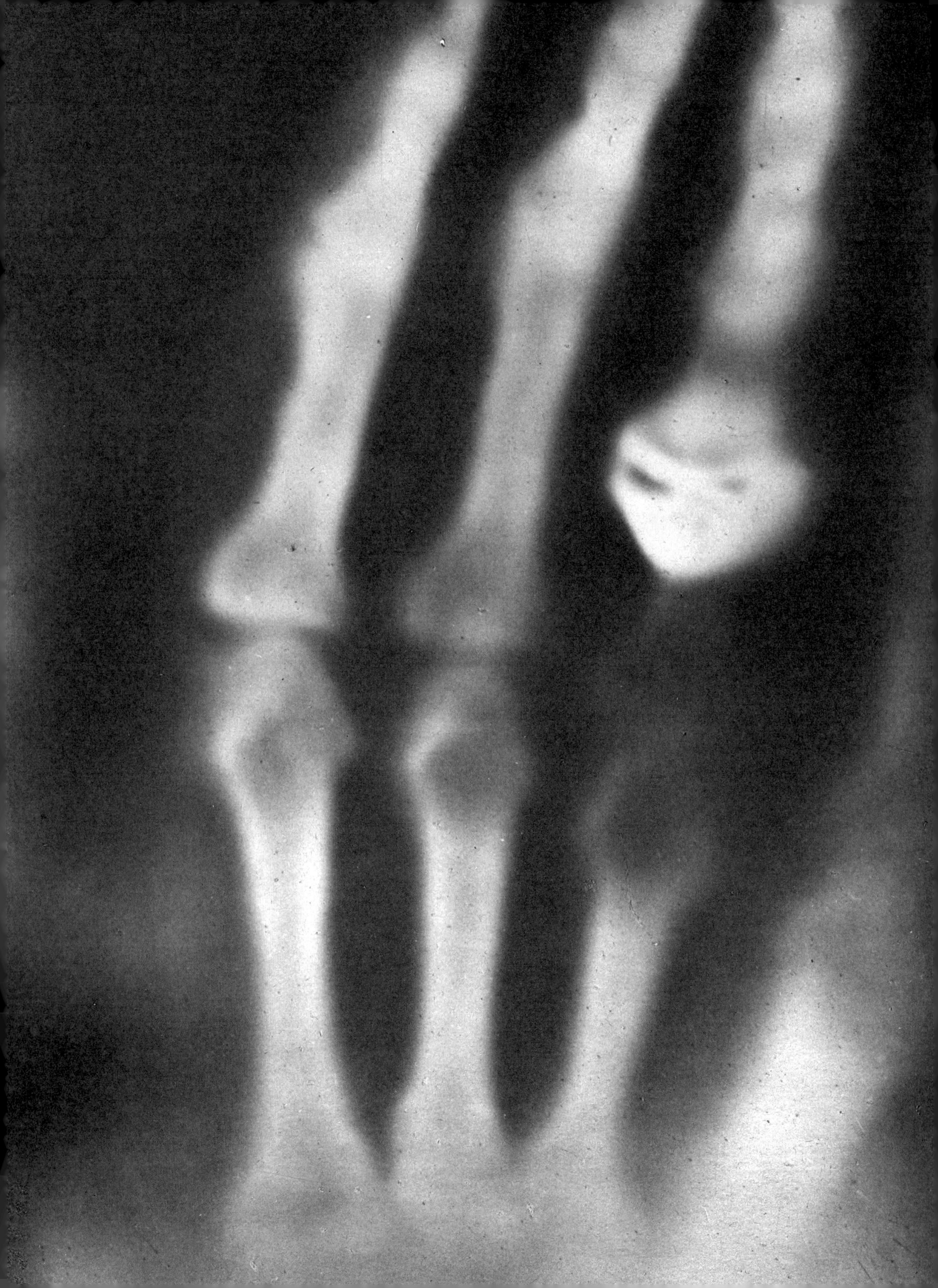

5

Atomic Soldiers

EYEWITNESS ACCOUNTS OF NUCLEAR TESTS HAVE FREQUENTLY DRAWN ON photographic imagery to describe the experience.[1] "Before the shot, we were ordered to face the tower," recounts Woodrow Whitefield Coward, a Canadian officer who participated in Operation Plumbbob at Yucca Flat in the northeast corner of the Nevada Test Site in 1957:

> *At the count of "ten" I turned around, covered my eyes with my gloved hands (it was a cold morning), and a split second after "one," even with my gloved hands covering my face, there was a flash as bright as the mid-day sun. I could see the bones in each hand detailed as clearly as if I was looking at a pink X-ray.*[2]

Coward's shock at seeing an "X-ray" of the bones of his hands parallels the shock expressed by Bertha Röntgen in 1895 upon being shown the first X-ray ever produced. [**Figure 5.1**] It was of her left hand. Her fingers, one adorned with a wedding ring, look spectral; the jewellery sits just above one of the joints of her hand. "I have seen my own death," she declared.[3] Her husband, the physicist Wilhelm Röntgen, who had invented the new photographic process, called it an "X-ray," where "X" stood for what was not yet identified or understood. X-ray photographs and atomic blasts break through skin and surface, reversing relations of inside to outside and reorganizing the economy of visuality.[4]

| **Figure 5.1**: Wilhelm Conrad Röntgen. *Bertha Röntgen's X-rayed Hand.* 1895.

As always within the compass of nuclear activities, and as demonstrated in the preceding chapters, cameras were present for the Plumbbob test series in which Coward participated. The tests took place in a broad valley surrounded by jagged mountains, where the Mojave Desert meets the Great Basin. The cameras were on hand to provide visual documentation of the explosions and of the participating troops. A photograph titled *Test Stokes, Operation Plumbbob* (August 7, 1957), shows troops, heads down and facing away from the exploding bomb. [**Figure 5.2**] The camera was rigged to set off lights pointed at the men – so the photo would register them clearly – at the exact moment of detonation.

One of the tasks assigned to photography is that of producing "official memory," which has been defined as memory propagated by an elite attempting to advance its own version of the facts.[5] In accordance with the hyper-masculinization of military activities and nuclear technology, the pictures were taken mostly in black-and-white. Colour photographs, gendered feminine, were reserved for advertising, tourist promotion, and family snapshots.[6] The Nevada Test Site was a backyard playground for scientists working on bomb design at Los Alamos in New Mexico, Lawrence Livermore in California, and Sandia in New Mexico. The test site was their "proving ground" – and photography played a central role in their work. The official memory produced in unclassified photographs taken by the US military and circulated to the media was often at odds with eyewitness accounts of the same events. Coward was deeply shaken by what he experienced and, like Bertha Röntgen, may have thought he was seeing his own death. But the photos released to the public tell a different story. For example, *Exercise Desert Rock VI, Nevada,* shows troops watching a mushroom cloud rise in the distance as if they are looking at a fireworks display. [**Figure 5.3**] Instead of representing the proving ground as a site of danger, historian and photographer Peter B. Hales notes, the image makes the Nevada desert look like "a theater for atomic performances."[7]

This chapter examines discrepancies between official and vernacular memory in the depiction of nuclear events. Canadian military personnel participated in more than two dozen nuclear weapons' operations in Nevada, the South Pacific, and Australia from 1946 until 1963, when atmospheric tests stopped.[8] Between 300,000 and 400,000 US and Canadian servicemen were involved with the testing in Nevada and the Pacific. The most notorious of the atmospheric tests was the Plumbbob series, in which Coward participated. It took place between May and October 1957, concentrating on warheads fitted to intercontinental ballistic missiles and intermediate-range missiles. Plumbbob was the most elaborate test series ever conducted in the continental United States – it comprised twenty-nine explosions in all – and its notoriety stems from the decision of US and Canadian military commanders to use large numbers of foot soldiers as nuclear guinea pigs.[9] Under the larger Exercise Desert Rock series, which ran from 1951 to 1957, troops were used experimentally in Plumbbob and other operations on simulated atomic battlefields. The other operations included Buster-Jangle,

Figure 5.2: US Army Photographic Signal Corps. *Test Stokes, Operation Plumbbob.* 1957.

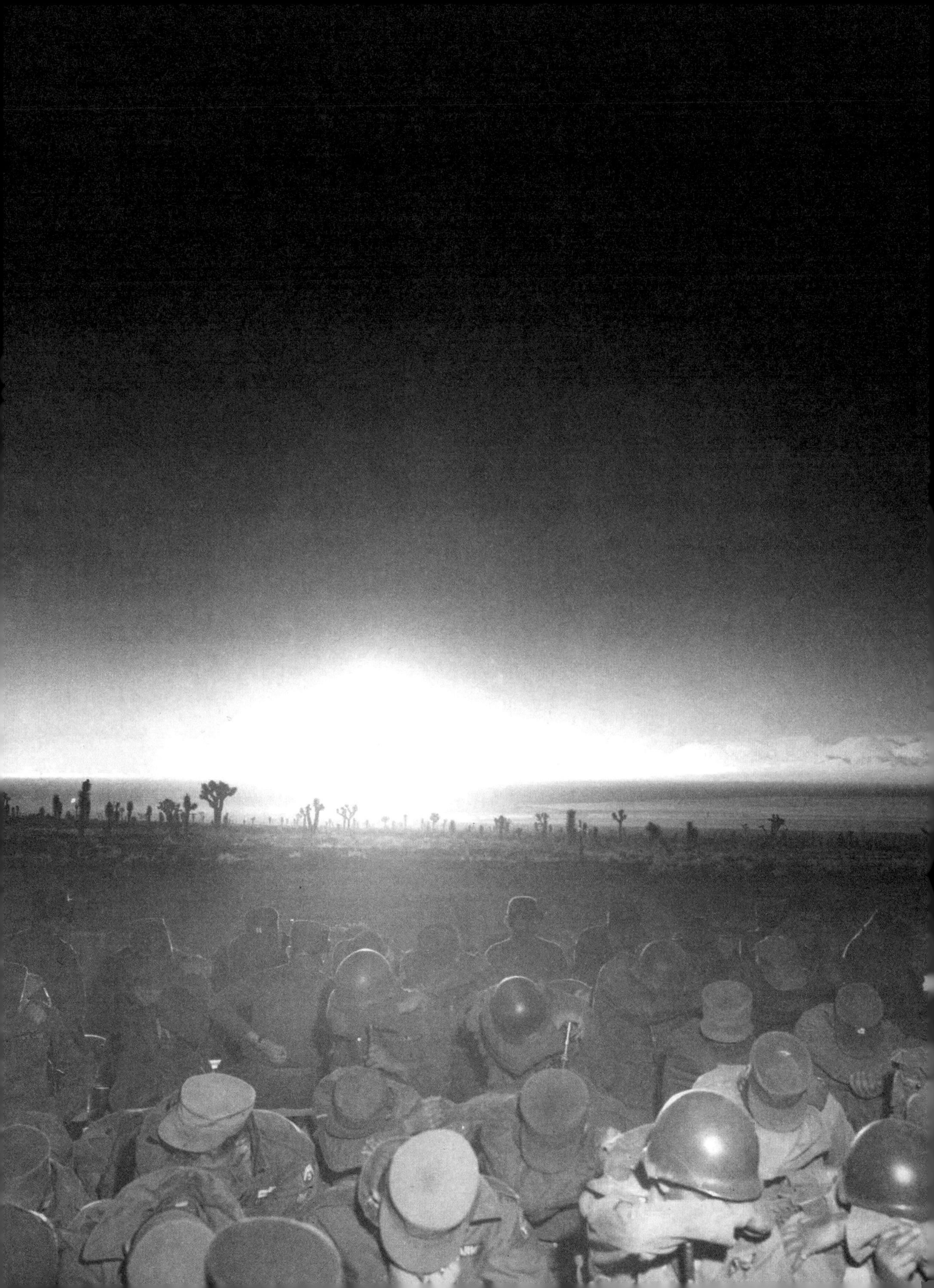

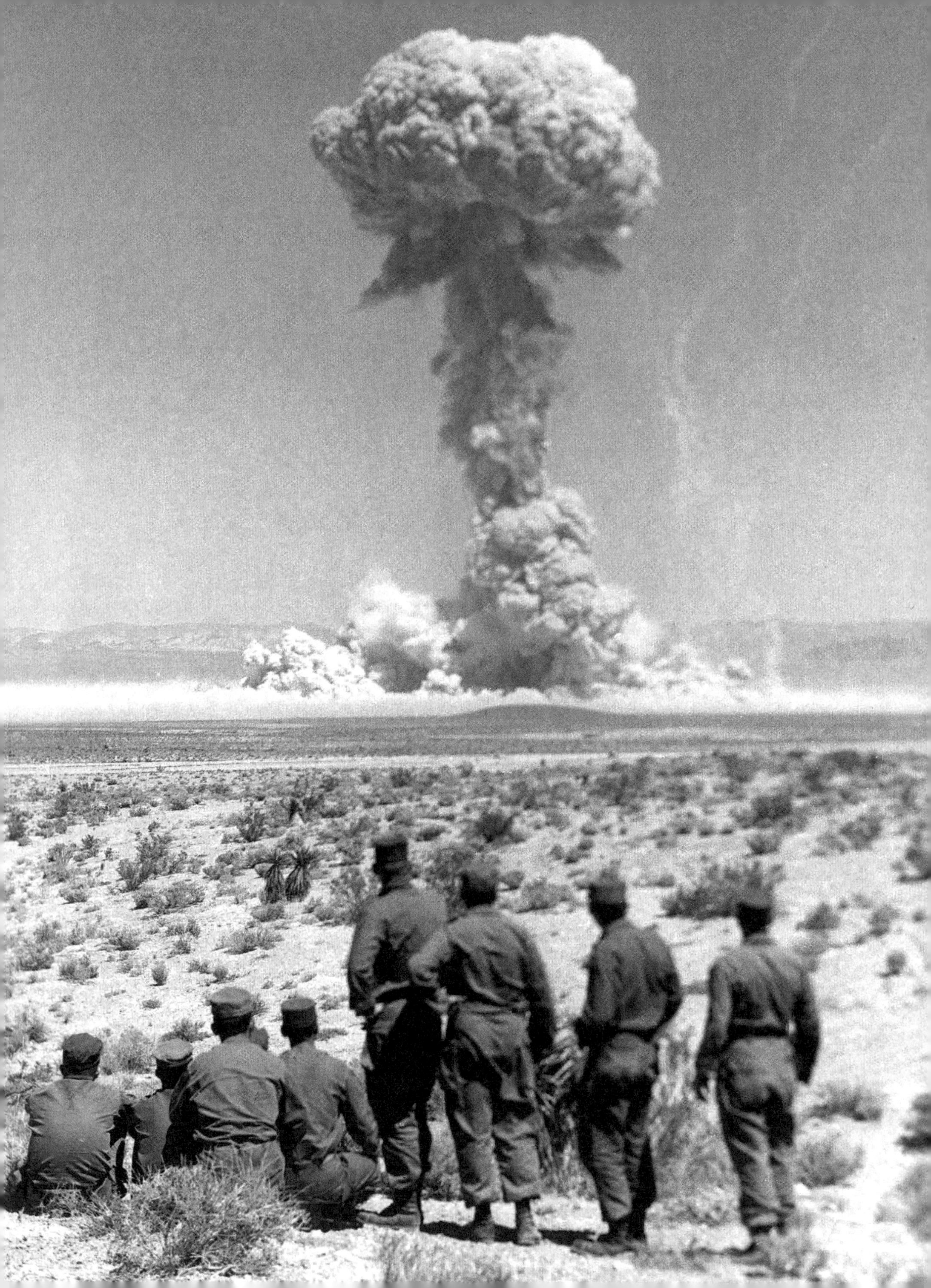

Tumbler-Snapper, Upshot-Knothole, and Teapot, but none matched the intensity of Plumbbob. Before being dispatched into a simulated war zone, soldiers were shown instructional photographs and films, briefed on the effects of nuclear weapons, and drilled in preparation for their manoeuvres.[10]

Canadian troops were engaged in six of the twenty-nine Plumbbob tests. Approximately eighteen thousand members of the combined US and Canadian armed services took part in the exercises, including seven hundred from Canada, in an attempt to discover how enlisted men would fare a nuclear war zone. With no protective gear, they were ordered to advance toward the mushroom cloud, while technicians dressed in rad-safe outfits accompanied them, monitoring for radiation. Some troops wondered why the technicians wore protective gear whereas they themselves did not, but most "trusted that their superiors knew enough to protect their safety."[11]

Their superiors did know enough to protect their safety but chose not to do so. They knew that when an atomic weapon detonates, it produces gamma radiation that passes directly through the body. They also knew that a detonation generates alpha and beta particles as well as fallout that contains a large inventory of radioactive elements, which can be absorbed internally.[12] Before Test Priscilla was held on June 24, 1957, a television program in a military series called *The Big Picture,* produced by the US Army Signal Corps, was screened for soldiers preparing to engage in the atomic test exercises. In it, a chaplain assures two soldiers that "there is no need to be worried, as the army has taken all the necessary precautions to see that we're perfectly safe here. The fireball as it ascends up into the heavens," the chaplain adds in a non sequitur that aestheticizes the test, "contains all the rich colours of the rainbow. It is a wonderful sight to behold."[13] After the test, soldiers returned to camp bleeding from ears, eyes, nose, and mouth.[14] Although the safety guidelines established by the US Atomic Energy Commission required that troops and other individuals be located eleven kilometres away from a detonation, they were sometimes placed at less than two. That was the case with Test Priscilla. The military claimed that its purpose was "to create 'hardened' troops capable of 'tactical' warfare on the atomic battlefield."[15] The American and Canadian dignitaries in attendance stayed many kilometres back from danger, watching the explosion from observation bleachers. In 1995, long before he made his series of photographs of the interior of the Diefenbunker, Mark Ruwedel produced *The Witnesses,* an image of empty and abandoned bleachers overlooking the proving ground. [**Figure 5.4**] Unlike the atomic soldiers and nearby civilians, the seated dignitaries had the wind at their backs to protect them from the fallout and were equipped with safety goggles to shield their eyes from the flash.

Canadian and US survivors of the tests subsequently formed atomic veterans associations that sued for the damage done; the legal suits included financial compensation. It was not until 2008, more than fifty years after Operation Plumbbob, that Canada settled the claims of those exposed to heavy radiation in the Exercise Desert Rock series by offering a one-time

| **Figure 5.3**: US Army. *Exercise Desert Rock VI, Nevada.* 1955.

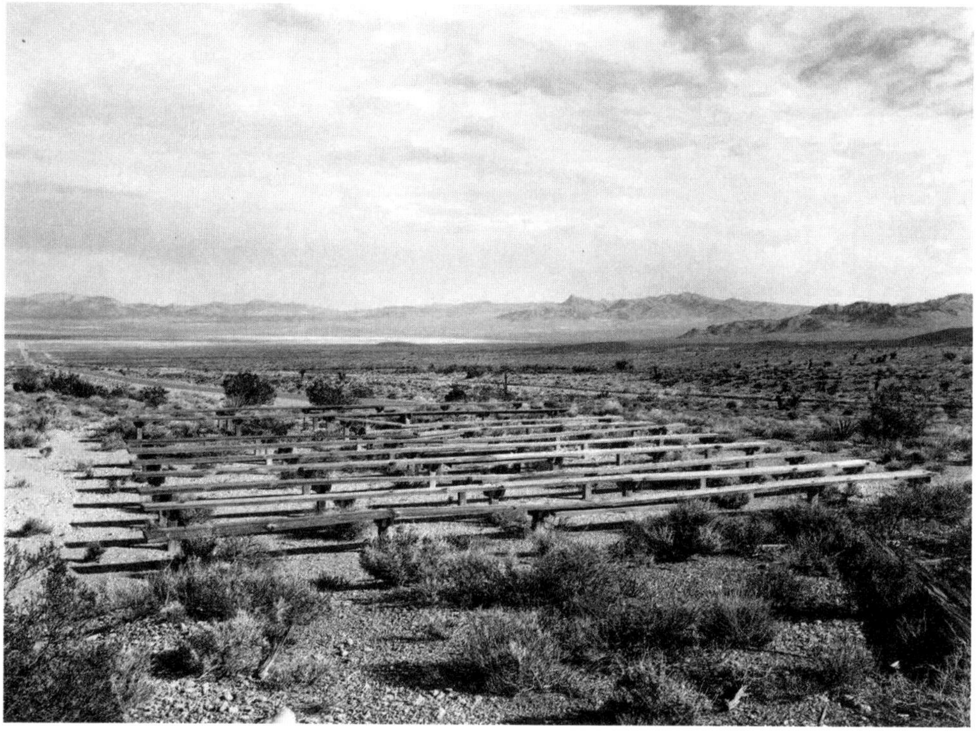

Figure 5.4: Mark Ruwedel. *The Witnesses (1), Nevada Test Site: Viewing Area for 14 Atmospheric Tests, at Frenchman Flat, 1951–1962.* 1995.

payment of $24,000 per person. Minister of National Defence Peter MacKay commented on "the tremendous contribution that atomic veterans made to the security of our country, and made with, really, little choice ... They were given an order, which they obeyed valiantly."[16] Had they known more about what they were being asked to do, the veterans might not have obeyed quite so "valiantly."

The destructive calculus of exposing troops to radiological and other dangers has been featured in board games and films as well as in still photography. The commercial board game *Risk* was launched in North America in 1959, two years after the conclusion of Operation Plumbbob. It was invented by Albert Lamorisse, the French film director responsible for the 1956 short film *Le Ballon rouge,* who called his new game *La Conquête du Monde.*[17] Played on a political map of the globe, it was the first mainstream board game ever produced about war, and it still remains popular (it is currently available as a phone app). Its goal is world domination. Although set in Napoleonic times, *Risk* is designed for the Cold War era. Over the years, it has spawned numerous variants, including *Nuclear Risk.* "Feel like RISKing everything?" asks the nuclear version, which was adapted from the original by US naval personnel in the early 1980s, during Ronald Reagan's first presidency.[18] Instead of competing against each other with armies, players of *Nuclear Risk* compete with atomic warheads represented by red tokens. The rules of engagement permit players to blow themselves up for tactical advantage, a suicidal move that parallels the destruction of the red balloon in *Le Ballon rouge* by a gang of bullies. Of course, this gambit has different implications in the real world, like the decisions taken by military brass during Exercise Desert Rock. The logic of such exercises falls into the realm of Herman Kahn, the nuclear physicist and strategist who worked for the RAND Corporation, the think tank of the US Air Force. In 1962, Kahn published *Thinking about the Unthinkable,* which used game theory expressed in layman's language to imagine whether a nuclear confrontation was winnable.[19] [**Figure 5.5**] The cover illustration depicts a mushroom cloud rising behind a red telephone, denoting one end of the nuclear hotline between Washington and Moscow. Kahn concluded that a nuclear war was winnable. To facilitate the terms of his analysis, he coined the word "megadeath" to denote a toll of 1 million dead.

"The 'nuclear age' makes for a certain type of colloquium," Derrida observed, "with its particular technology of information, diffusion and storage, its rhythm of speech, its demonstration procedures, and thus its arguments and its armaments, its modes of persuasion and intimidation."[20] An improbable participant in Derrida's "colloquium" was Stanley Kubrick. While filming *Dr. Strangelove* in 1963, Kubrick conferred several times with Kahn on the logic and illogic of mutual assured destruction. *Thinking about the Unthinkable* had turned Kahn into a celebrity – an unlikely celebrity, given his line of work – and the two men took to one another. According to scholar of scientific history Sharon Ghamari-Tabrizi, Kahn possessed the instincts of a stand-up comedian and displayed a gift for grotesque humour in his public talks on nuclear risk.[21] "One wishes to relieve the grimness of the subject matter," Kahn explained in 1962. "People in a state of horror are not good analysts or detached or objective listeners."[22] Kubrick agreed with him: "Why should the bomb be approached with reverence? Reverence can be a paralyzing state of mind."[23] Kubrick and Kahn were not the only ones to think this. As Kubrick was preparing to film *Dr. Strangelove* at Shepperton Studios outside London, the British satirical revue *Beyond the Fringe* was playing in the city. One of its sketches parodied the advice then being dispensed by British civil defence authorities on how to cope with a nuclear attack. Like Kahn and Kubrick, the makers of *Beyond the Fringe*

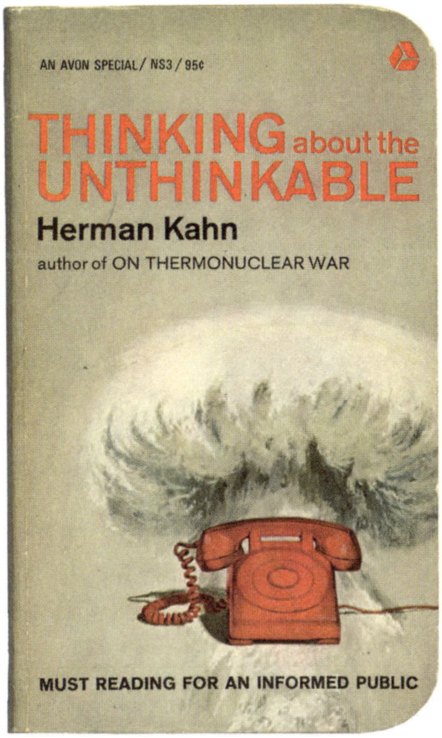

Figure 5.5: Cover of Herman Kahn, *Thinking about the Unthinkable.* 1962.

understood that irreverence could be more effective than gravitas in dealing with the topic of annihilation.[24]

Photography has informed the bomb from the beginning. It has been used to determine the explosive yields of atomic bombs, the velocity of the shock waves generated by the weapons, and the temperature progressions that occur during the blasts.[25] But photography is paradoxical in ways that nuclear science is not. At the same time that it has contributed to scientific invention, it has also showcased its threats. Some of the casualties of Exercise Desert Rock were photographers who worked for Lookout Mountain Studio, a secret facility based in Hollywood that was under contract to the US Department of Defense to record nuclear tests in motion-picture and still photography. The tensions between official and vernacular memory included the bodies of the photographers who were commissioned to document this history. A still image taken at the Nevada Test Site in the late 1950s shows four moving-camera operators on the top of a van, looking toward ground zero and equipped with breathing masks, overalls, caps, and gloves. [**Figure 5.6**] Their feet, obscured by the side of the van, would also have been protected. These men were issued with rad-safe gear for the test, but not all photographers were so lucky.

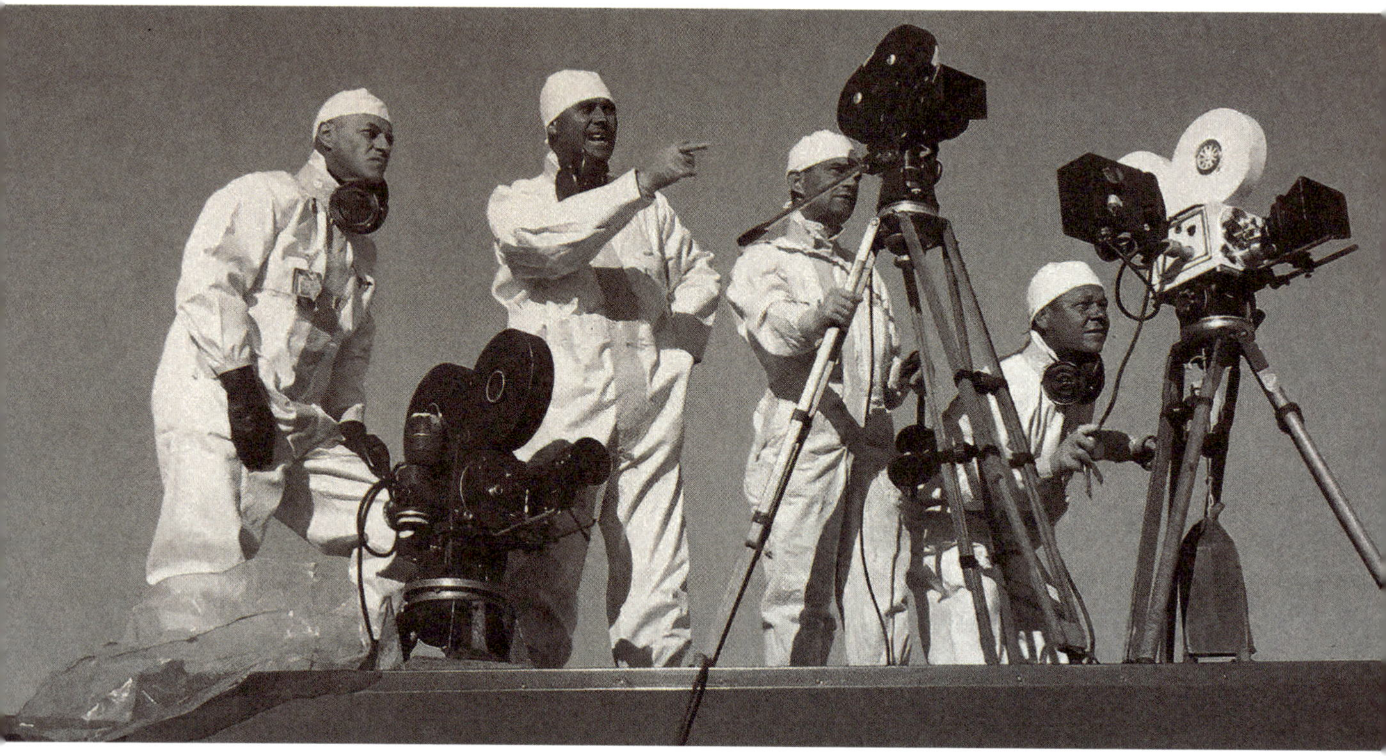

Figure 5.6: US Military. *Lookout Mountain Camera Crew in Rad-Safe Clothing, on Top of a Camera Van.* 1950s.

During a 2012 interview, photographer George Yoshitake recalled his experience with atomic testing. In July 1957, his boss at Lookout Mountain told him to travel immediately to Nevada for one of the Plumbbob detonations. He was needed out there for a "special test." His boss explained, "'Tomorrow morning you're going to go out with five other guys and you're going to be standing at ground zero.' I said, 'Ground zero??' He said, 'Yeah, but the bomb's gonna go off 10,000 feet above you.' I said, 'Well, what kind of protective gear am I going to have?' He said, 'None.'"[26] Yoshitake continues:

> *I remember I had a baseball hat, so I wore that just in case. He gave me a still camera and two motion picture cameras. These were 35 mm Eyemos. I set up the two Eyemos, and had little trip wires that I could trip with my foot starting about 5 seconds before the blast. And the still camera, I also had a trip wire so that I could trip it. I could get one exposure only. The other five guys were scientists and they volunteered to be there. I wasn't a volunteer.*[27]

Yoshitake and the five scientists all developed cancer. Only two were still alive in 2012, when the interview took place, and Yoshitake himself died the following year.[28] His shot of the five scientists has survived. [**Figure 5.7**] In the picture, his camera looks down while the men look up. Dressed in identical uniforms and standing together on the desert floor, they are peering at the first nuclear-armed air-to-air missile test, officially called Test John, detonated on July 19, 1957, over Yucca Flat. The missile's nickname was Genie.[29] [**Figure 5.8**] At a distance of 18,500 feet from the blast, the volunteers would have been unable to see anything but the explosion of the 1.5 kiloton warhead above them.

The military's rationale for placing Yoshitake and the volunteers at ground zero was to prove that the weapon was safe for use above cities and other populated areas.[30] By 1962, three thousand nuclear-tipped Genies had been produced by Douglas Aircraft Company for the US Air Force and the Royal Canadian Air Force. Genies were designed to intercept Soviet strategic bombers. In the event of an enemy attack, many of the Genies would be fired over cities in Canada and the United States. CF-101 Voodoo aircraft carried the Canadian Genies.[31] Under a dual-key arrangement, the United States supposedly kept custody of the missiles until an emergency occurred, at which point it released them to Canada – a fiction that made them more acceptable to the Canadian public.[32] In fact, Canada could use the weapons in an emergency without US approval.

Yoshitake and the scientists were no more protected from alpha, beta, and gamma radiation than the atomic soldiers at the Nevada Test Site or the residents of small towns lying to the east, such as Cedar City and St. George, both in Utah. They have been referred to as downwinders. According to the Atomic Energy Commission, populations living to the east of ground zero were entirely safe from exposure to fallout. The commission declared that the amount of radiation exposure resulting from the tests was "no worse than having a tooth X-rayed."[33] In *American*

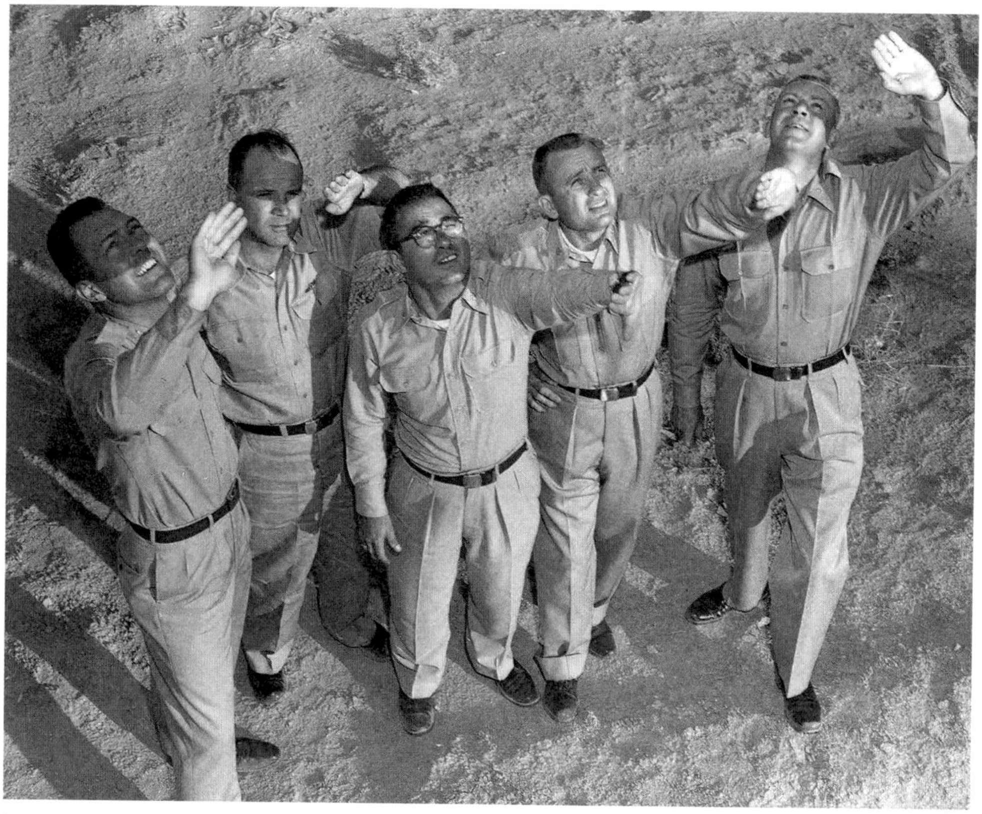

Figure 5.7: US Military (George Yoshitake). *Five Air Force Officers Are Observers at Ground Zero*. 1957.

Ground Zero, Carole Gallagher gives the lie to that assertion.[34] Using a combination of photography and text, she documents the damage done to human bodies and lives from radiation by one of the most reckless programs of nuclear experimentation undertaken in North America. Indigenous peoples were simply ignored. "Native people and their lands constituted an invisible presence in areas heavily occupied by the US military and the Department of Energy, which engaged in a consistent pattern of nuclear colonialism," Valerie L. Kuletz declares in *The Tainted Desert*.[35] The situation was no different in Canada. The Sahtu Dene at Port Radium in the Northwest Territories and the Anishinaabe at Serpent River in Ontario were similarly ignored.

If military commanders were prepared to risk the lives of hundreds of soldiers during Operation Plumbbob in 1957, Secretary of State John Foster Dulles was prepared to risk the lives of millions by engaging in nuclear brinkmanship in 1955. He was the "original misguided missile," in the opinion of Democratic senator Henry Jackson, "traveling fast, making lots of noise, and never hitting the target."[36] In a bellicose speech delivered on March 20, Dulles announced that if war with China became necessary, the United States would deploy nuclear armaments, "new and powerful weapons of precision, which can utterly destroy military targets without endangering civilian centers."[37] This was during the Formosa Crisis, when China was threatening the government of the Republic of China on the island of Formosa (now Taiwan), prompting factions within the Eisenhower administration to argue in favour of deploying nuclear warheads against an Asian nation for the second time in a decade.[38] The split between the Communist and Nationalist Parties of China had escalated into military attacks and counterattacks, and the United States was alarmed at the spiralling intensity of the conflict. Eisenhower feared that if Formosa fell, its loss would "seriously jeopardize the anti-Communist barrier ... in the Western Pacific."[39]

Dulles's disclaimer that Chinese civilians would be unharmed by nuclear weapons was a bombastic version of the Atomic Energy Commission's statement that the radiological effects of the Nevada tests were no greater than that of a dental X-ray. In the end, Eisenhower chose diplomacy over bombs. Even though China possessed no nuclear weapons at the time – it conducted its first atomic test in 1964 – the Soviet Union had an arsenal of two hundred nuclear warheads by 1955. Had the United States detonated an atomic bomb over Beijing, the attack would have sent a radioactive message to Moscow.

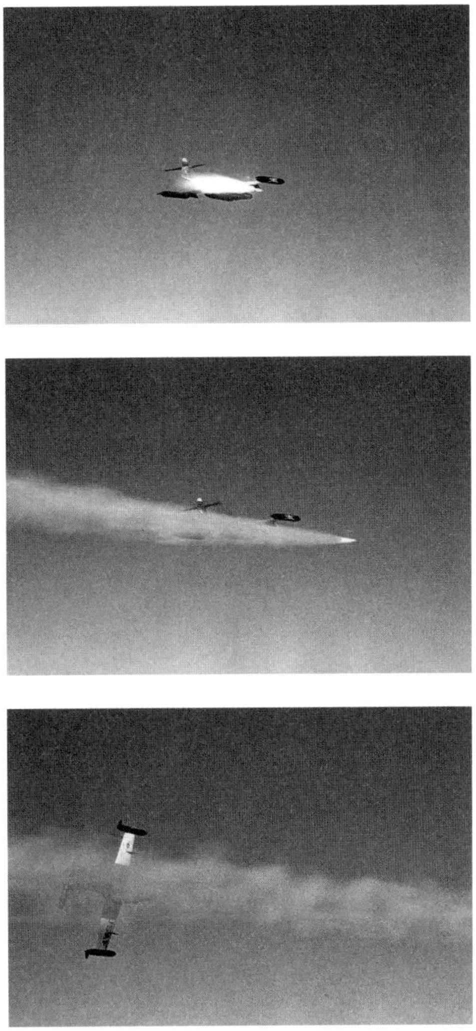

Figure 5.8: US Military. *Genie Test*. 1957.

While President Eisenhower was deciding whether or not to bomb China, the photographer Robert Frank was producing *Hoover Dam, Nevada,* a black-and-white shot of a souvenir postcard rack on a plywood shelf outside a store near the Hoover Dam in Nevada.[40] It shows a revolving rack displaying nine sets of colour postcards, only three of which are visible. The cards sell for ten cents apiece, the cost of top-quality colour cards at the time. In descending order, they offer a view of the Grand Canyon, the Hoover Dam – pictured with the American flag in the centre of the frame – and an atomic detonation at the Nevada Test Site to the west of the canyon and the dam. This perennially popular postcard of a fireball rising above the desert floor was for sale in every second drugstore and gas station in Nevada, and may have been known to Woodrow Whitefield Coward and other Canadian soldiers

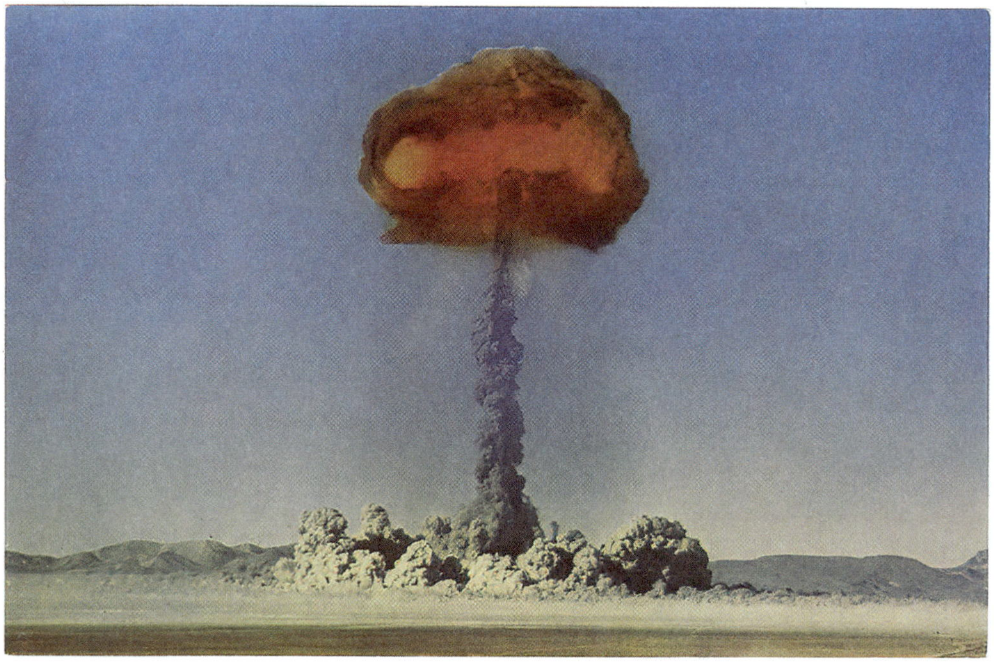

Figure 5.9: Postcard. *Atomic Explosion Frenchman's Flat, or Yucca Flat, Nevada.* c. 1955.

who participated in Operation Plumbbob. [**Figure 5.9**] On the left side of Frank's photograph, fragments of words and numbers struggle to announce the cost of a "Pictorial Tour of the Dam" (thirty-five cents), and in the background an automobile, no less a symbol of American industrial progress than the Hoover Dam, is parked on the tarmac.[41] Its right headlight and licence plate are cut off by the frame of the dam image, making it seem one-eyed (not unlike the camera recording it) and stateless (again, not unlike the camera).

Frank took the picture in 1955, two years before US and Canadian soldiers participated in Plumbbob; he was travelling across the United States on a Guggenheim grant, an excursion that resulted in his 1958 book *The Americans.*[42] He photographed other postcard stands during his journey but was so drawn to this particular rack that he shot five versions of it.[43] However, he did not include them in *The Americans,* possibly because he considered them inconsistent with the mood he wished to convey. The book is underwritten by signs and signage of everyday life, not unlike the work of Walker Evans before Frank and William Eggleston afterward, but the postcard pictures may have seemed overly redolent with both. They are what photographer and critic Nathan Lyons calls "social landscapes," which is to say, they take a self-consciously ironic and critical approach to documentary material.[44] Frank rebelled against representations of "official memory," including the photo-essays of *Life* magazine. He referred to them as "god-damned stories with a beginning and end."[45]

Frank's *Hoover Dam, Nevada,* as well as its variations, has attracted substantial commentary since it was first published in 1972.[46] This may be in part because the photographs are photographs of other photographs, sequenced in a way that invites narrative interpretation. But it is difficult to find a single, stable narrative here. Instead, the sequencing tells several narratives; the photographs are plural representations.[47] According to some readings, they are an encapsulation of the changing American West in the 1950s. They depict the impact of postwar modernity on the contemporary uses of the landscape, with the sequential organization of the cards offering a commodified touristic panorama, a source of energy for human consumption, and "the figurehead of the exuberant new Atomic Age."[48] Another reading sees them as representing nature worship, patriotism, and mushroom cloud symbolism to produce the "atomic sublime."[49] In this reading, the manufactured destructive force of the bomb and the material histories that preceded it are subsumed beneath the colours and shapes of the mushroom cloud. Yet another reading understands the three cards as a metaphor for the United States – its hopes, realities, and nightmares – that offers viewers a choice between unadulterated nature, industrial advancement, and nuclear catastrophe.[50]

An additional interpretation views the postcard sequence as an elliptical narrative of potential catastrophe that has been turned into comedy.[51] The comedy identified in the photo strikes me as pointing in a useful direction, just as Kubrick's irreverence in *Dr. Strangelove* brought a new perspective to MAD, mutual assured destruction. By having the story of nuclear annihilation played out in postcards on a plywood stage at a roadside venue, Frank's pictures become prime expressions of postwar absurdity. Samuel Beckett's Vladimir and Estragon, the embattled vaudevillians in *Waiting for Godot,* would not be out of place waiting and watching the skies at this American roadside stand. (The play premiered in English in 1955, the same year Frank took his photographs.)[52] What we see is a nuclear drama presented for the pleasure of tourists who may not appreciate the absurdity of what has been staged for them. The postcards were designed and printed in the long shadow of Hiroshima and Nagasaki, whose explosions took place only ten years earlier, while down the road at the Nevada proving grounds bombs many times more powerful than those detonated in Japan were being tested. Here in the American Southwest, history was repeating itself as farce, as Karl Marx commented about the Second French Empire of Louis-Napoleon.

Frank's photograph derives part of its complexity from the juxtaposition of postcard imagery and residual war imagery. The tension is between the message of "wish you were here" and the secondary message of "glad you are not." In the sense that a postcard connects a receiver-subject to its sender, Derrida observes in his book *The Post Card,* it is a materialized form of a Lacanian screen.[53] The image on the front of a card comes in the form of a proposition, according to Derrida, deciphering the receiver in advance of its arrival. The pre-existing group of symbols of nature, engineering, and atomic force that constitutes the "screen" projected onto viewers in Frank's photograph, by this reading, interprets viewers before they encounter it. Above all, viewers are constructed by the mushroom cloud. Mushrooms, it is worth remembering, can be either nutritious or poisonous, and the spores they throw, like the radioactive isotopes scattered by atomic explosions, are invisible to the human eye. This affinity has resonance for atomic soldiers who may have encountered the postcard.

By shooting in monochromatic film, Frank converts the compulsively overheated Kodachrome palette of the postcards into the tonal properties of a black-and-white photograph. To put it in aesthetic terms, he converts the high-chroma imagery of the American consumerist vernacular into the monochrome contrasts and shadings demanded by art photography in his era. Whatever else it was in the 1950s, art photography was not rainbow coloured, and Frank had no desire to contravene the dominant aesthetic of black-and-white. Even Walker Evans, who sometimes deployed colour in his photographic work for *Fortune* magazine after the war, decried the "screeching hues" that debased most colour shots. They were, he said, a "bebop of electric blues, furious reds, and poison greens."[54] In the years leading up to the war, colour had become coded as the photographic medium particular to Madison Avenue advertising firms and the cheaply produced popular magazines in which ads appeared. It was associated with a commercialized vision of modernity, and such a vision was anathema to serious photography.[55]

Colour television, introduced in 1954, was also associated with a commercialism mistrusted by serious photographers. The corporations involved in developing the medium, which included General Electric and Westinghouse, were also helping to build bombs. As investigative journalist Joyce Nelson points out, the ideological interests of the corporations intersected where television sets and warheads met.[56] Both devices emitted massive levels of electro-magnetic radiation, levels denied by both the corporations and the government – in other words, by official memory. Some General Electric televisions gave off 100,000 times the recommended government standard.[57] The corporations also commissioned made-for-TV films on the benefits of nuclear power, which were provided free of charge to domestic television stations. In the decade or so after President Eisenhower's "Atoms for Peace" speech in 1953, which promoted "the sunny side of the atom," nuclear postcards and promotional films such as *The Atom Comes to Town* (1957) and *The Atom and Eve* (1965) were common fare. The latter film promoted nuclear-generated electricity for the home by following a housewife through the now-made-easy tasks of her electrified day – presented in living colour.[58]

Hoover Dam, Nevada, draws colour and monochrome photography together, just as it draws popular and art photography together. It also unites postcard imagery with the lives of atomic soldiers to produce a form of vernacular memory. Official images from Exercise Desert Rock, by contrast, embody the sanctioned memory of military activities. [**Figure 5.10**] A photograph of a collapsed dirigible, a ZSG-3 airship under the command of the navy, which was destroyed during Test Stokes in 1957, epitomizes the representation of official memory.[59] The dirigible was unmanned to avoid loss of life in case of failure, authorities stated. But elsewhere on the battlefield, the same authorities were exposing soldiers to ionizing radiation by ordering them to advance on foot toward ground zero.

Figure 5.10: International News Photo. *Shockwave of the Blast from Test Stokes, Operation Plumbbob, Collapses Dirigible.* 1957.

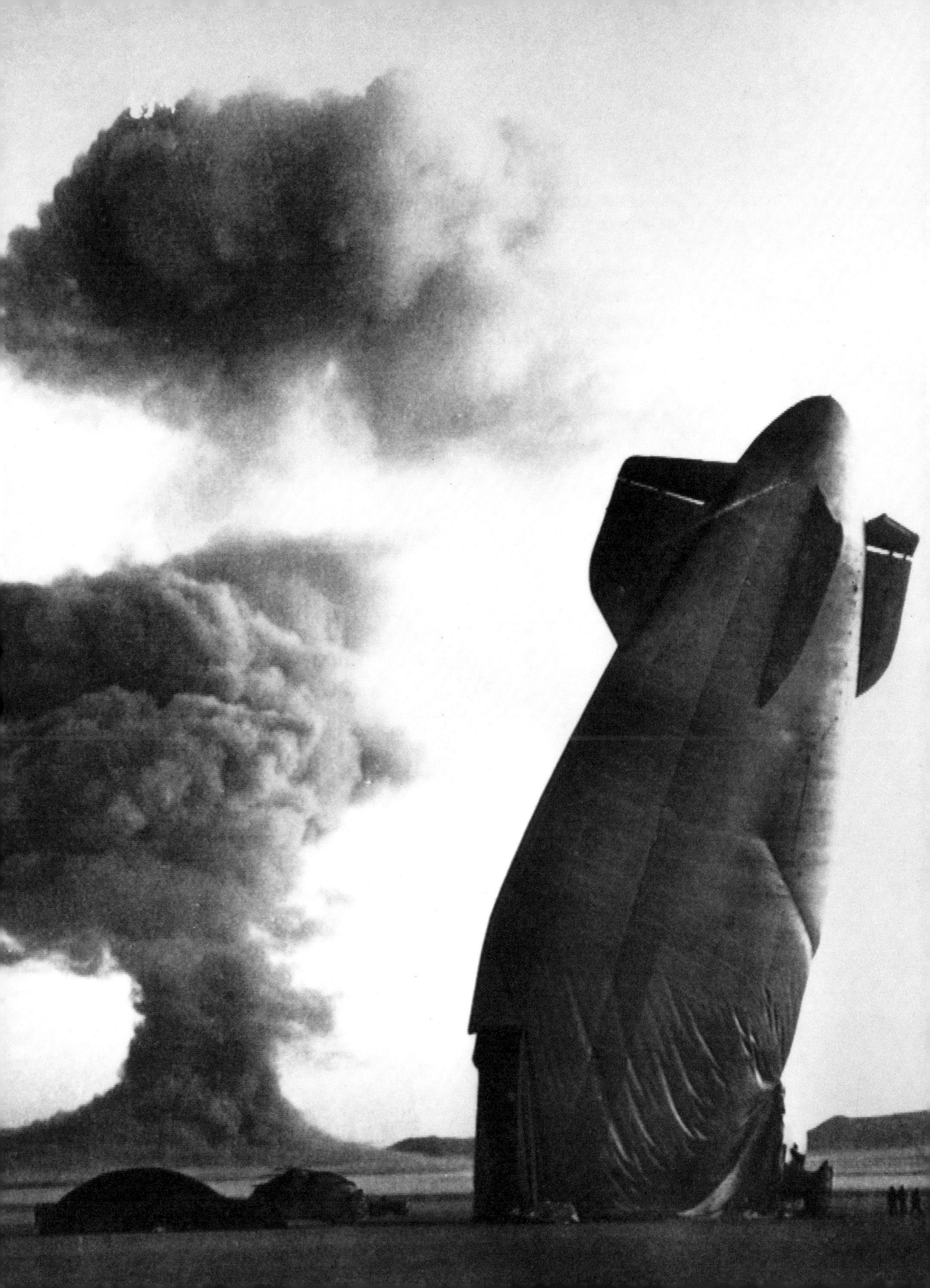

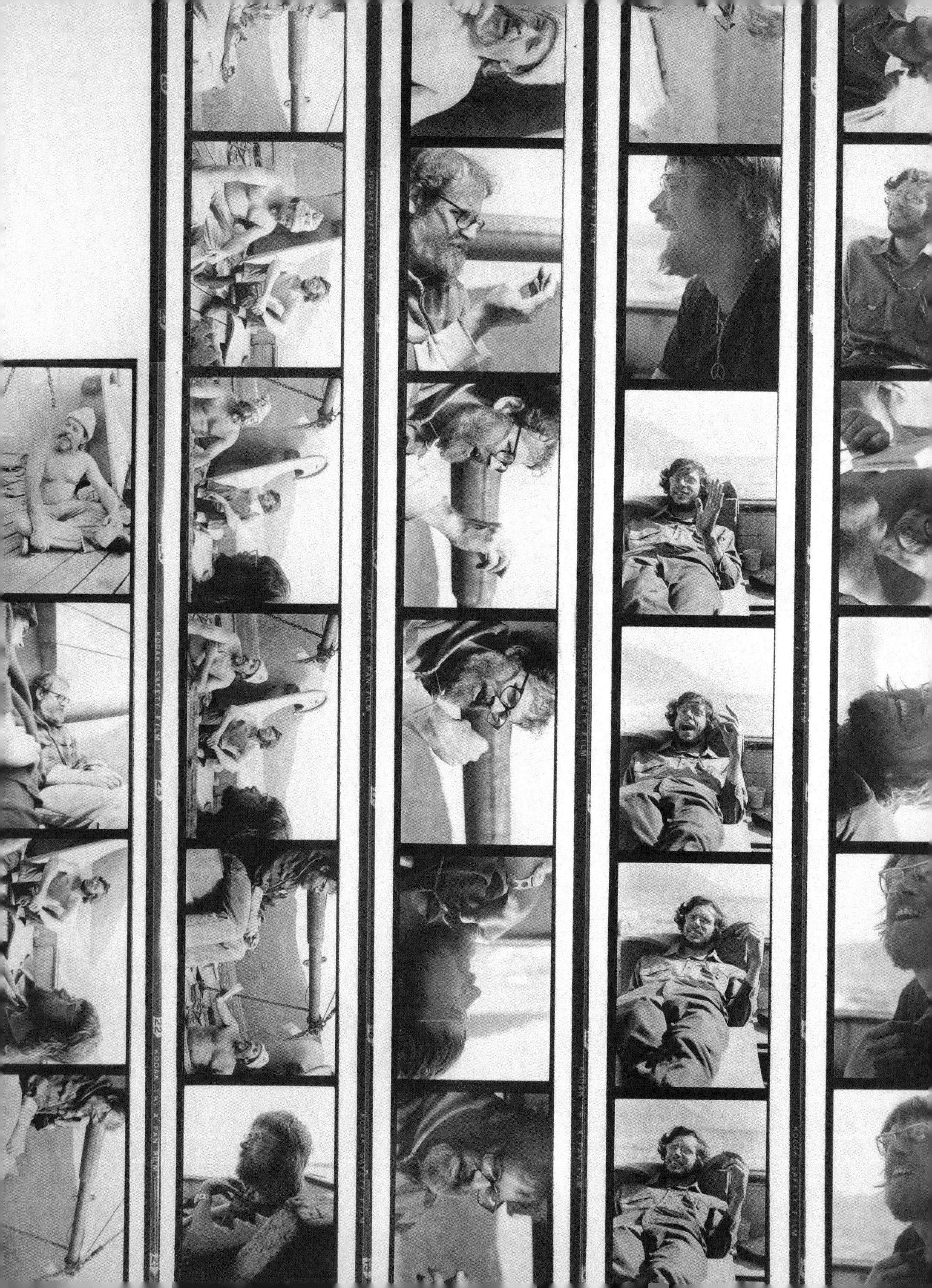
KODAK SAFETY FILM
KODAK TRI X PAN FILM

6 Mass Protest

GREENPEACE WAS FOUNDED IN VANCOUVER IN 1971 TO PROTEST US NUCLEAR testing at Amchitka in the Rat Islands of the Aleutian chain, Alaska. Amchitka Island is tectonically unstable. A battered fishing boat, renamed *Greenpeace* for the occasion, was chartered to disrupt the testing, and among those on board was the photographer Robert Keziere. During the boat's voyage north from Vancouver to blockade the Amchitka test, Keziere shot dozens of rolls of film, providing a visual record of the journey from above and below deck.[1] A contact sheet produced from one of the rolls shows some of the directions in which he pointed his camera. [**Figure 6.1**] One was toward other members of the crew, another toward the boat itself. Keziere's photo *US Navy Surveillance* shows an observation aircraft tracking *Greenpeace,* and in turn being tracked by the photographer. [**Figure 6.2**] As the vessel progressed north, the US Navy kept a close eye on it. In the photograph, the surveillance aircraft is viewed through a cat's cradle of nautical rigging, and the second half of the word "Greenpeace" is visible on the sail, between the symbols for ecology and peace. The peace symbol was designed in the late 1950s as the logo for the Campaign for Nuclear Disarmament in Britain, before being adopted by anti-nuclear activists around the world. During the Cold War, it was often accompanied by the slogan "Ban the Bomb."

| **Figure 6.1**: Robert Keziere. *Greenpeace Voyage to Amchitka.* 1971.

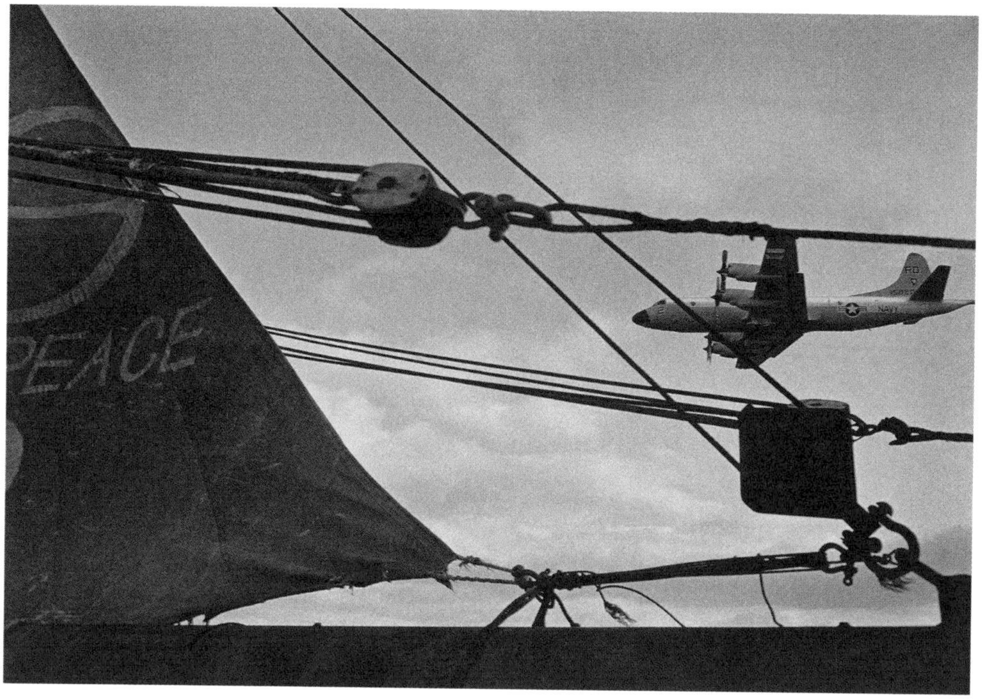

Figure 6.2: Robert Keziere. *US Navy Surveillance.* 1971. **Figure 6.3**: Unknown. *Protest against Amchitka.* 1971.

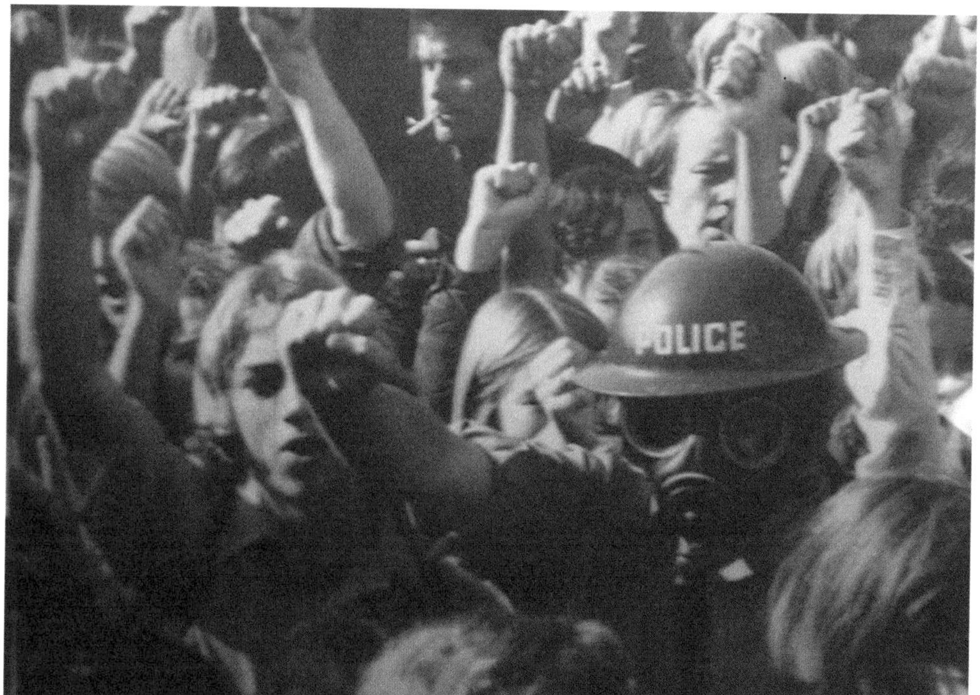

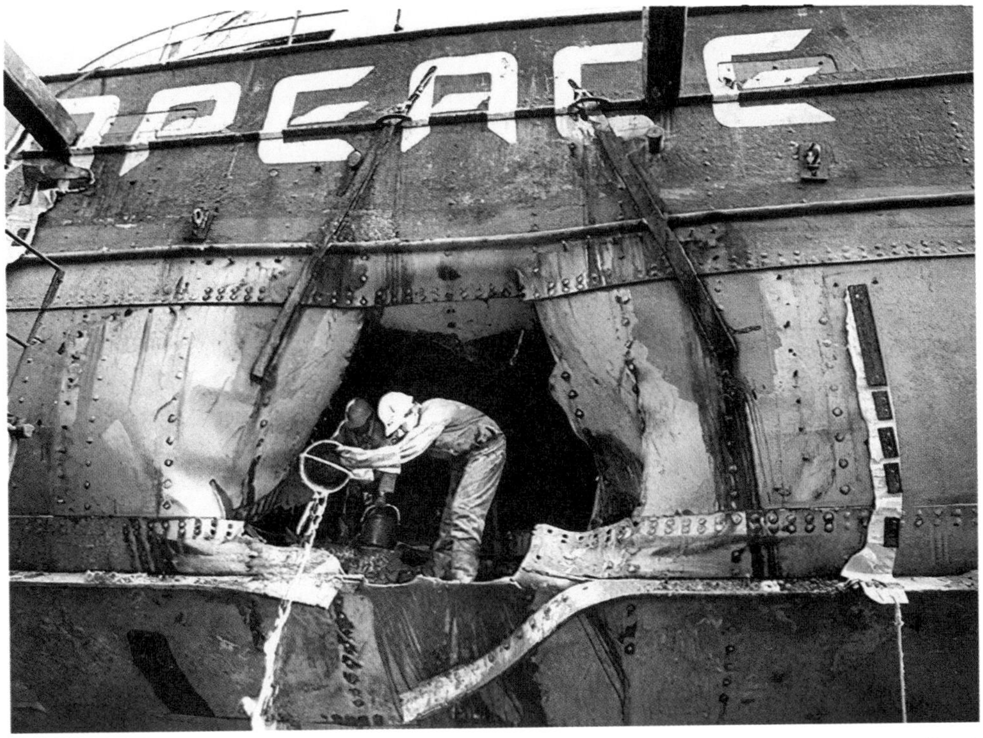

Figure 6.4: Unknown. *French Secret Service Agents Sink Greenpeace's Rainbow Warrior, Auckland Harbour.* 1985.

In preparation for the Amchitka voyage, Greenpeace developed a publicity machine. According to Bob Hunter, a Greenpeace cofounder who also filed a regular newspaper column with the *Vancouver Sun* during the trip, the crew wanted to fight a "McLuhanesque 'war of the icons'" by firing off images at the media like missiles.[2] The strategy brought international attention to Greenpeace and helped to generate mass protests in Canada and elsewhere. A photograph shows Vancouver protesters with their fists raised against the Amchitka test. [**Figure 6.3**] One wears a gas mask and a metal helmet with the word "police" written across it. Although the *Greenpeace* intervention failed to halt the Amchitka test, the Nixon administration cancelled further testing for "political and other reasons."[3] One reason was the negative publicity being generated by Greenpeace. Amchitka Island eventually became a bird sanctuary. I cannot think of many mass protests whose outcomes have been more successful.

A year later, in 1972, members of Greenpeace travelled to Moruroa to demonstrate against France's nuclear-testing program in the South Pacific. On orders from the French government, a minesweeper rammed their boat when it entered the test zone. The violent response of the French state to the intervention of Greenpeace was repeated with greater force in 1973. French sailors boarded the protesters' ship, even though it was in international waters, and beat the crew with truncheons before throwing their cameras overboard.[4] As this chapter demonstrates,

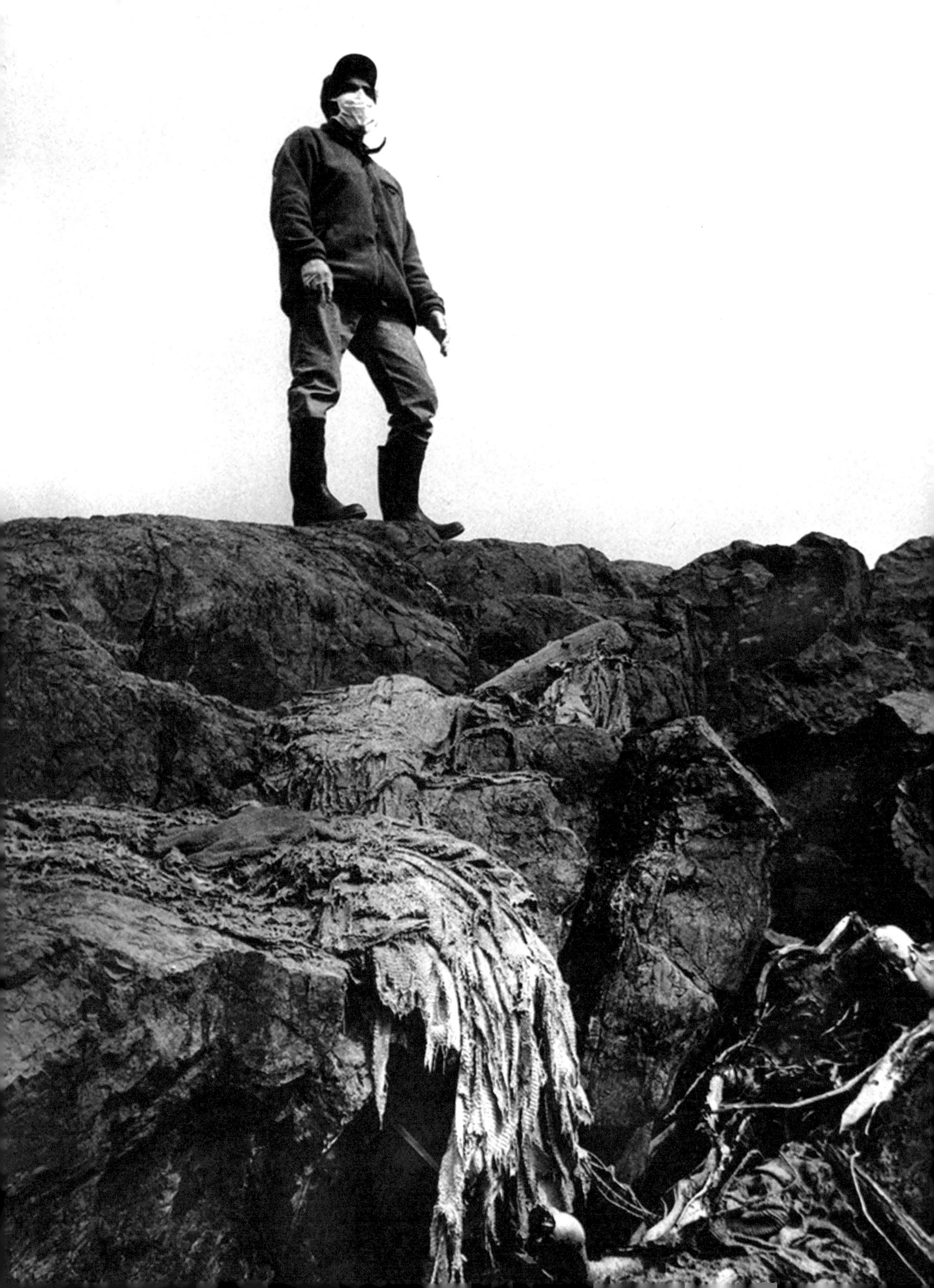

cameras record acts of violence and provide images that can mobilize public opinion; they are rightly seen as dangerous by those in power. The images challenge those made available by military authorities for public consumption on nuclear testing, as discussed in the previous two chapters. When French secret service agents sank Greenpeace's *Rainbow Warrior* in Auckland Harbour in 1985, in a covert attack codenamed Opération satanique that ripped a hole in the side of the vessel, the blast killed Fernando Pereira, a photographer.[5] [**Figure 6.4**] The attack was a public relations disaster for France. Photographs of the wrecked boat, like the one illustrated here, spurred further action. New Zealanders responded by sending a flotilla of private yachts into the test area, where they succeeded in halting the nuclear testing.[6]

In *On Photography* (1977) and *Regarding the Pain of Others* (2003), Susan Sontag asks whether photographs have the power to communicate the suffering of others. Can they convey the pain of those who experienced the atomic bombings in Japan, or those in the Northwest Territories who suffered radioactive poisoning from transporting uranium ore for the Little Boy and Fat Man bombs, or those whose lives were damaged by nuclear testing in the South Pacific? Can a photograph of Joe Blondin Jr., seen wearing a protective mask and looking over a rocky knoll at discarded burlap sacks used to carry the uranium ore, convey the suffering from thyroid cancer, leukemia, and other cancers that was experienced by the Sahtu Dene who transported them? [**Figure 6.5**] Cultural theorist and poet David Levi Strauss is on the side of those who think that photographs can communicate suffering and alter attitudes, even though the communication may fall short. Photographs must "work in the fissures, the wounds, of the social," he writes, a position with which I concur.[7] Sontag is on the other side of the debate. *On Photography,* in particular, is skeptical about the capacity of photographs to incite action that might lead to political change. Sensationalist photography has become commonplace, she argues, and visual representations of suffering have become trite. If photographs once had the power to shock, now they aestheticize suffering. Although they continue to generate emotional force, Sontag maintains, they are incapable of generating an ethical response. In *Regarding the Pain of Others* and a related 2004 article entitled "Regarding the Torture of Others," however, Sontag modifies her former conclusions and concedes that photographs of violence can sometimes be an invitation for viewers to think through the power structures responsible for mass suffering.[8]

The work of Strauss and Sontag is valuable in explaining why those in power, such as the French in Moruroa, fear the visual representation of violence and atrocity. Photographs taken in the aftermath of the atomic bombings in Japan were censored by the United States until 1952, when the occupation ended, because American authorities feared that images of burned and irradiated bodies might nauseate not only Japanese viewers but also American viewers, challenging the US assertion that the bombs ended the war and saved lives.[9] The authorities feared photographs like *Atomic Bomb Victim No. 1, Kiyoshi Kikkawa,* which shows corrugated

| **Figure 6.5**: Robert Del Tredici. *Abandoned Uranium Sacks.* 1998.

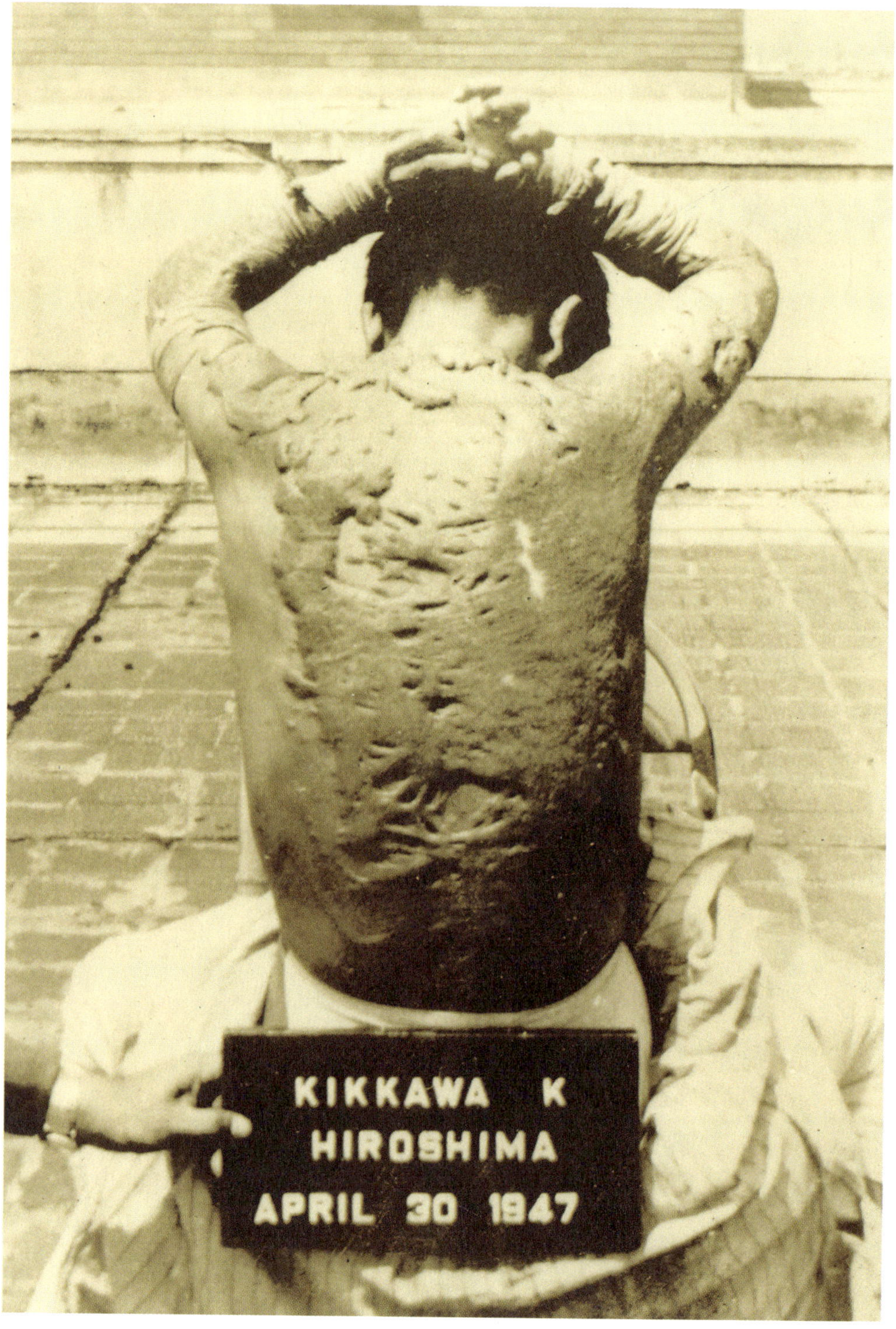

Figure 6.6: Unknown. *Atomic Bomb Victim No. 1, Kiyoshi Kikkawa*. 1947.

keloid scars from thermal burns on the back of a hibakusha. [**Figure 6.6**] A handmade sign held by an anonymous arm is written in English, indicating that the picture was intended for a non-Japanese audience. Photographs of tortured prisoners at Abu Ghraib that were leaked and circulated in 2006 did sicken and enrage people, and Secretary of Defense Donald Rumsfeld expressed his frustration at the ubiquity of the electronic communication devices, such as "digital cameras [and] a global internet without inhibitions," that were responsible for disseminating the pictures.[10] Digital images are not easily controlled, he was saying, making the fighting of wars and the maintenance of empire more troublesome enterprises than they used to be. Photography *is* difficult to constrain, a characteristic emphasized by the photo historian and theorist John Roberts in his deliberations on "the uncontrollable volatility of the photograph."[11] President George W. Bush voiced his frustration more directly than Rumsfeld. It was regrettable, he said, "that people seeing the [Abu Ghraib] pictures didn't understand the true nature and heart of America."[12]

The dangers of nuclear testing began to enter public consciousness in 1954, after the Castle Bravo thermonuclear test on Bikini went out of control and poisoned Marshall Islanders as well as the crew of the Japanese fishing boat *Daigo Fukuryū Maru* with fallout. Many young people at this time supported anti-nuclear causes. Following the Castle Bravo disaster, a campaign against thermonuclear weapons was quickly mounted in Japan, launching a petition signed by 32 million people – one-third of the population.[13] In Canada, anti-nuclear peace movements, inspired by those in the United States and Britain, started attracting members a short while later. The opinions of Canadians on the health hazards of atomic testing and the arms race were heavily influenced by the US Student Peace Union and the National Committee for a Sane Nuclear Policy (SANE) as well as by the Campaign for Nuclear Disarmament (CND) in Britain.[14] Anti-nuclear movements were also influenced by the work of Canadian scientist Ursula M. Franklin, who helped to discover high levels of bone-seeking strontium-90 in children's teeth. Strontium-90 was produced from the fallout of atmospheric nuclear tests.[15]

Bertrand Russell was a prominent member of both CND and SANE, and Martin Luther King Jr. spoke for the policies of SANE in Gandhian terms. "I don't think the choice is any longer between violence and non-violence in a day when guided ballistic missiles are carving highways of death through the stratosphere," he said in a 1961 interview. "I think now it is a choice between non-violence and non-existence."[16] King knew how to command the attention of the media, whether his cause was civil rights or nuclear disarmament. So did the CND. In 1958, it began to mount a series of non-violent mass demonstrations demanding that the British government unilaterally disarm. The demonstrations took the form of annual marches on Easter weekend between Aldermaston, a nuclear research centre in Berkshire, and Trafalgar Square, London. In the 1962 photograph *Aldermaston March with Peace Flag in Foreground,* a black peace flag imprinted with a white anti-nuclear logo is whipped by the wind, making the logo look like a skull. [**Figure 6.7**] The march of that year attracted a record 150,000 demonstrators.

In histories of nuclear protest, it is sometimes forgotten that the first Aldermaston march did not actually occur in 1958, but six years earlier. The 1952 march, called Operation Gandhi,

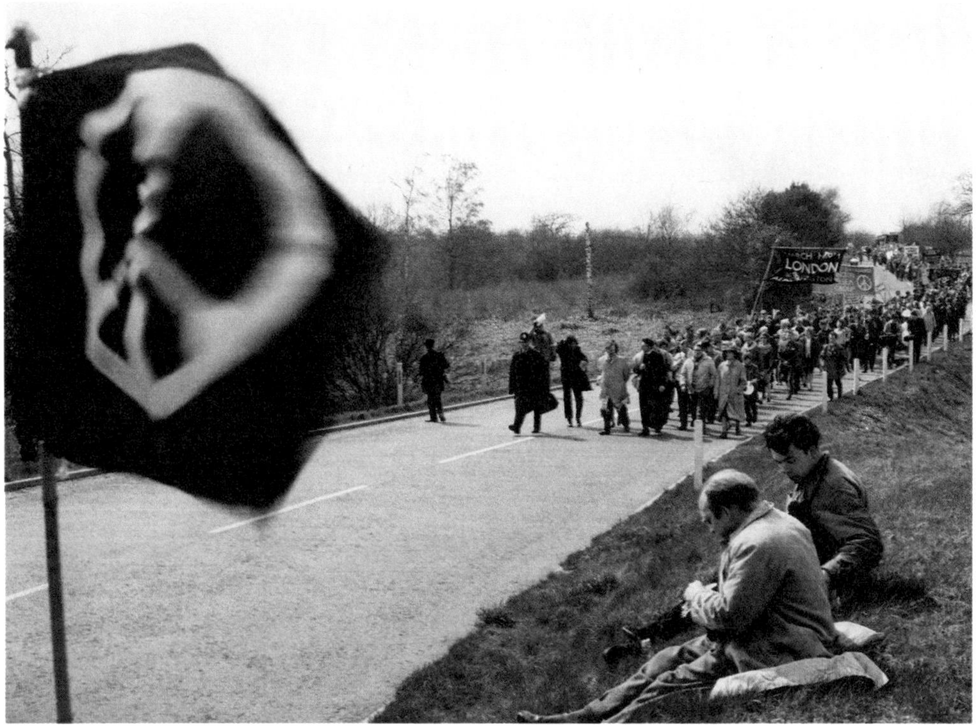

Figure 6.7: London Express News. *Aldermaston March with Peace Flag in Foreground.* 1962.
Figure 6.8: Frank Grant. *Ban the Bomb Protest, Toronto.* 1961.

attracted only a few dozen people. Drawing on the philosophy and practice of *satyagraha,* non-violent civil resistance, the actions of a few demonstrators provided a moral compass for what followed.[17] They pointed the way to the later Aldermaston marches and then, as those marches successfully raised awareness among the British people of the dangers of atmospheric testing and the arms race, to anti-nuclear marches and protests elsewhere in the world, including Canada. Greenpeace likewise recognized that its moral stature and ability to bring about change derived from the principle of satyagraha. By insisting on non-violence, it captured international attention. At the same time, this position cost it some members, including Paul Watson, whose radicalism and sometimes violent tactics crossed its line of acceptability. Watson was eventually voted off the board.[18] Photographs of large anti-nuclear marches and demonstrations were frequently given space on the front pages of newspapers. "As a political tactic," the sociologist Becki L. Ross observes, street demonstrations like the 1961 Ban the Bomb protest that took place on a rainy November day in Toronto, "express anger, empathy and identification with those who suffer locally and globally. 'Demos' invoke unpredictable movements ... in collective disobedience."[19] [**Figure 6.8**]

In the years leading up to the Cuban Missile Crisis in 1962, the number of Canadians who opposed American atmospheric nuclear testing rose sharply, with fully one-half the

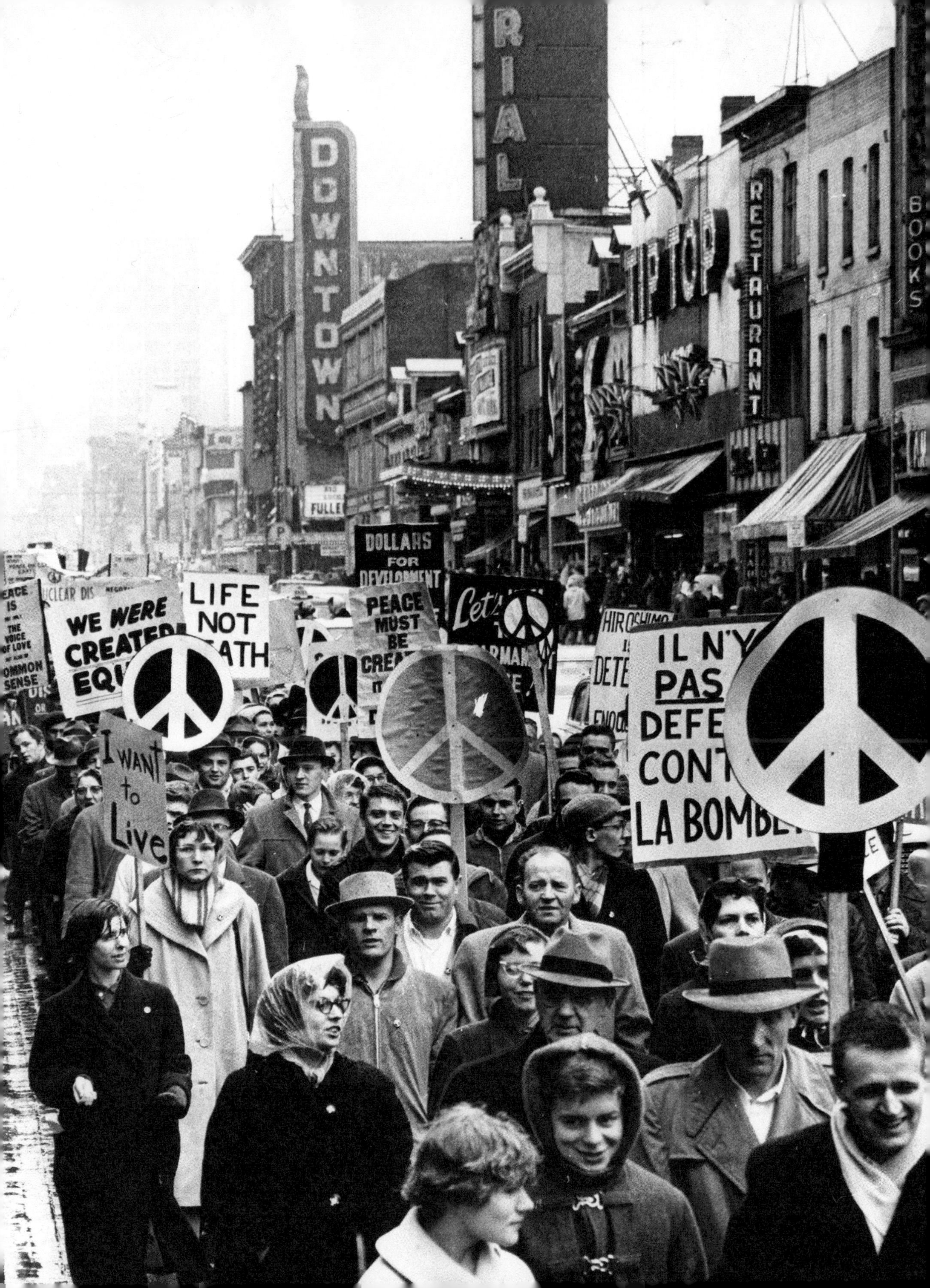

RIAL
DOWNTOWN
TIP TOP
RESTAURANT
BOOKS
DOLLARS FOR DEVELOPMENT
LIFE NOT
WE WERE CREATED
PEACE MUST BE
I WANT to Live
IL N'Y PAS DEFE CONT LA BOMBE
COMMON SENSE

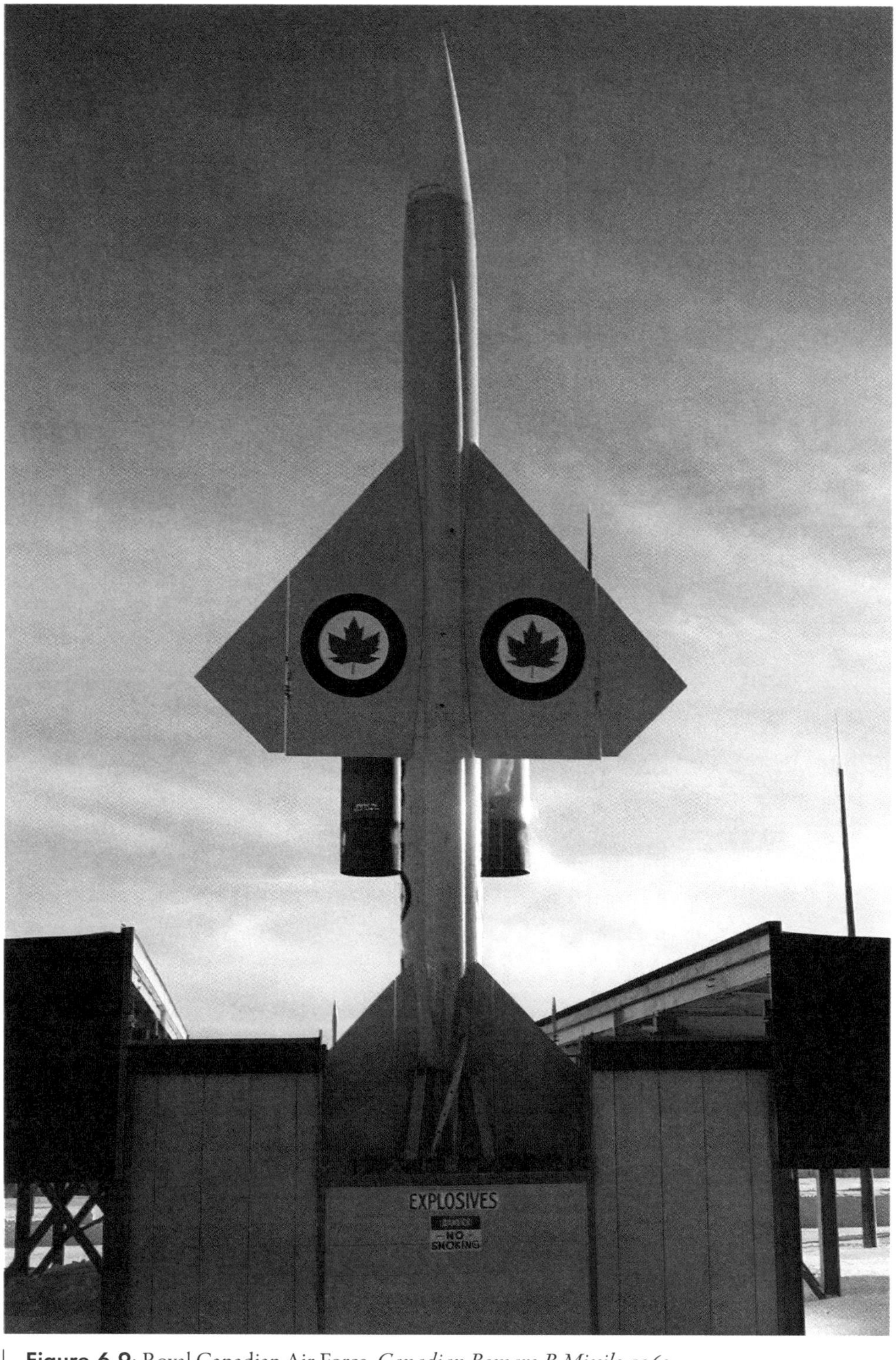

Figure 6.9: Royal Canadian Air Force. *Canadian Bomarc-B Missile.* 1963.

population in favour of banning it.[20] In the same year that Martin Luther King Jr. spoke out against death-dealing missiles in the stratosphere, Prime Minister Diefenbaker was confronted by a large anti-nuclear protest on Parliament Hill as well as by a petition coordinated by the Canadian Committee for the Control of Radiation Hazards. The petition was signed by between 140,000 and 180,000 Canadians who opposed the installation of nuclear-tipped Bomarc-B missiles in Canada.[21] The Bomarc-B system, a long-range surface-to-air missile network designed for use over North America, had been purchased by Canada from the United States on the understanding that the missiles would be armed with nuclear warheads. This would fulfill a commitment made under the North American Air Defense Command (NORAD), the integrated defence system formed by the two countries in 1958.[22] *Canadian Bomarc-B Missile* shows the weapon, with Royal Canadian Air Force roundels on its wings, ready for launching. [**Figure 6.9**] The petition captured the attention of the media, as did the demands of other protest movements such as the Voice of Women for Peace (VOW), a Canadian organization formed under the impetus of journalist Lotta Dempsey. VOW called for women to become leaders in the peace movement. The protests caused an already irresolute Diefenbaker to waver even more about accepting Bomarc-B missiles.[23] In the end, the Bomarc-B missile controversy and Diefenbaker's indecisiveness – he could not make up his mind one way or the other about accepting any type of nuclear weapons in Canada – helped to bring down the Conservative government. As historian John Boyko remarks, Diefenbaker wanted to "maintain his nuclear virginity while sleeping with the neighbor," and it ended badly for him.[24]

The controversy about Bomarc-B missiles was coming to a head while I was an undergraduate student in political science and economics at the University of Toronto. It coincided with the October 1962 Cuban Missile Crisis. Scarcely more than a month into my first term at the university, President Kennedy went on television to announce that Soviet nuclear missiles had been smuggled into Cuba and that the US military had gone on high alert. In the vocabulary of war readiness, the military had moved to DEFCON 3, meaning that it was in a position to mobilize within fifteen minutes. The Diefenbaker government did not follow suit, which deepened the rift between the two administrations.[25] Evidence of the Cuban missile bases was presented in aerial photographs taken by high-altitude surveillance planes operated by the US Air Force – another instance of photography's documentation of nuclear events. [**Figure 6.10**] In *Medium Range Nuclear Ballistic Missile Site at Sagua La Grande,* two missile launchers and a missile-ready tent are identified. Throughout the next few days, the world came as close to nuclear destruction as it has ever been. Along with others, I retain a flash memory of where I was standing on October 22, when Kennedy revealed the threat on television. I was in a house on the edge of the university's downtown campus, later knocked down to make way for Robarts Library. In my imagination, the library is a monument to the occasion, its Brutalist architecture a grim reminder of nuclear threat.

Following the Cuban Missile Crisis, the mood in Canada became more hawkish, and an increasing number of Canadians became receptive to the idea of Bomarc-B missiles being

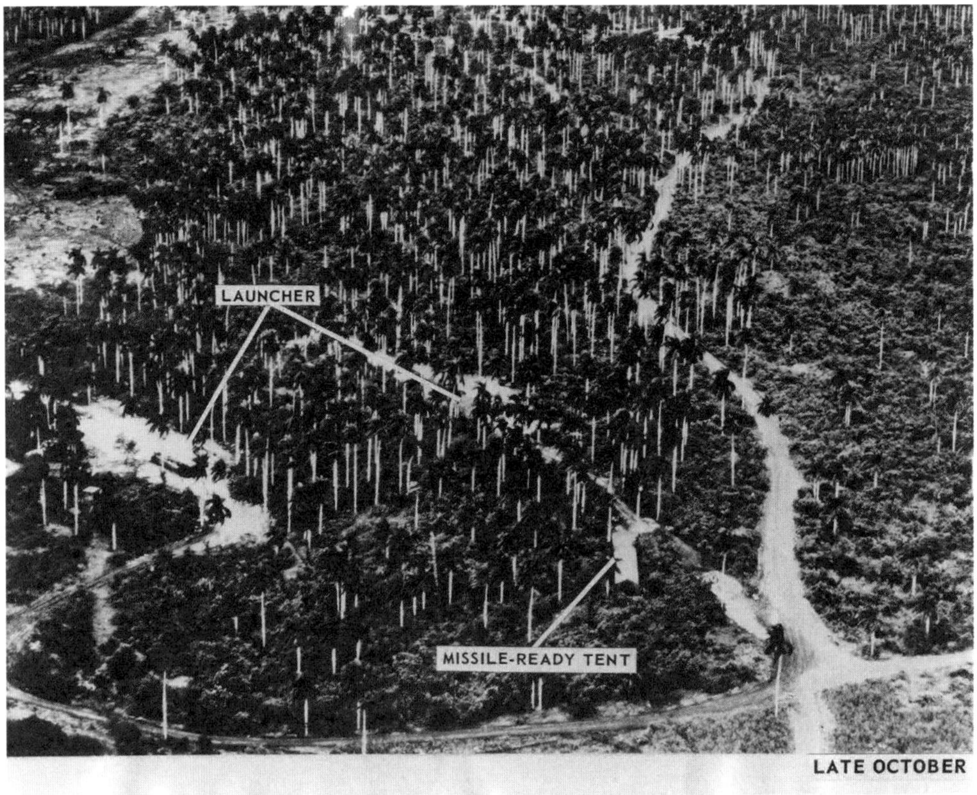

Figure 6.10: Associated Press. *Medium Range Nuclear Ballistic Missile Site at Sagua La Grande, Cuba.* 1962.

installed within the country's borders. A poll commissioned in 1963 by the Liberal Party under Lester B. Pearson determined that 54 percent supported the acquisition of nuclear weapons, with only 34 percent opposed.[26] If the Liberals wished to win the upcoming election – if Pearson wished to achieve something that had so far eluded him, to become prime minister – a change in the party's nuclear stance was necessary. But few expected Pearson, winner of the Nobel Peace Prize and an outspoken opponent of nuclear arms, to perform a U-turn in Parliament and come out in support of nuclear weapons for Canada. Pearson's wife, Maryon, was quickly forced to resign her honorary membership in VOW. On December 31, 1963, nuclear warheads for the Bomarc-B system were delivered to North Bay, Ontario, and La Macaza, Quebec. Organized protests took place in the months leading up to the deployment, but they did not attract the same attention from the media as protests held before the Cuban Missile Crisis.[27] The weapons operated under a two-key system, which had less to do with the maintenance of Canadian sovereignty than with the prevention of an accidental launch. (The pretense of independent control was politically convenient for the Liberals.) Also in 1963, the United States provided Canada with Voodoo aircraft armed with nuclear-tipped

Genies – the same type of missile detonated above George Yoshitake and the five volunteer scientists during Test John in 1957 – for the purpose of intercepting Soviet bombers. At bases in Europe, Canadian forces also acquired Honest John missiles. The launch platform for the Honest John was the back of a truck. Pierre Trudeau, then a professor of law and a critic of the Liberal reversal on nuclear weapons, dubbed Pearson "the unfrocked priest of peace."[28]

After the election, Prime Minister Jawaharlal Nehru of India was more generous than Trudeau toward Pearson and issued an invitation for him to visit.[29] In 1956, Canada had provided India with a nuclear research reactor modelled on its Chalk River NRX, on the condition that it be used for "peaceful purposes" only.[30] Although the CIRUS, as it was called, was capable of producing weapons-grade plutonium, Canada naively chose not to attach inspection requirements to the agreement.[31] By the time Canada had committed itself to the sale of additional reactors as well as of uranium ore in the 1960s, India had already developed a secret plan to use spent reactor fuel to build an atomic bomb.[32] In the 1968 National Film Board production *Juggernaut,* director Eugene Boyko follows the journey of a seventy-tonne calandria, the core of a CANDU nuclear reactor, from a port on the Arabian Sea to its final destination at a nuclear plant in Rajasthan, the state bordering Pakistan.[33] The enormous piece of machinery is shown trundling inexorably through the small villages of Rajasthan. The word "juggernaut," from the Sanskrit, refers to an unstoppable destructive force and suggests within the context of the film that what has been exported from Canada to the subcontinent could not be contained. The reactor's slow march from the sea to the interior metaphorically reversed Mahatma Gandhi's Salt March of civil disobedience from the interior to the sea. NFB films were distributed widely, and *Juggernaut* was screened internationally upon its release in 1968.

When India detonated a nuclear weapon at Pokhran in 1974, in an underground test codenamed Smiling Buddha, there was a public outcry in Canada. Smiling Buddha was the first successful atomic test in India, and news of it spread quickly. Photographs of it also spread quickly, including an image of four soldiers standing in an artificially lit trench, looking skyward as if they were gazing at a mushroom cloud. [**Figure 6.11**] The Canadian government cut off nuclear exports to India, and relations between the two countries cooled. The coolness lasted for more than three decades, until June 2010, when the Conservative government of Stephen Harper signed a new Nuclear Cooperation Agreement, which allows for the resumption of Canadian shipments of uranium and nuclear technology to India. The agreement contains assurances that the technologies and materials exchanged will be used for civilian non-explosive purposes only, and upon signing Prime Minister Manmohan Singh remarked that both governments had put foolproof systems of control in place: "We have complete civilian control of our own nuclear facilities, and as such there is absolutely no scope whatsoever of the nuclear materials and nuclear equipment supplied to India being used for unintended purposes."[34] His words were evasive, and both countries knew it. In a legalistic sense, the prime minister was right, but in an ethical sense, he was wrong. India is engaged in a regional nuclear arms race with Pakistan, and uranium exported by Canada for so-called civilian purposes frees up India's own supplies of uranium for use in the production

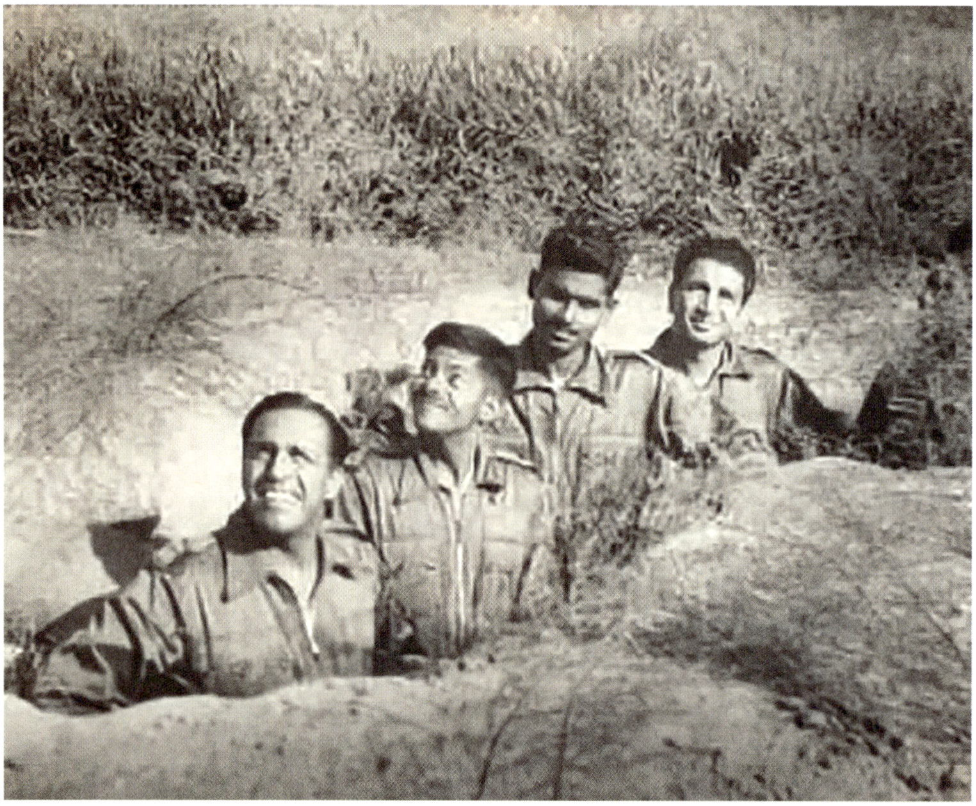

Figure 6.11: *Soldiers Posing at the Pokhran Nuclear Test Site.* c. 1974.

of weapons-grade fissile materials.[35] With the complicity of the Canadian government, India now has the means to produce additional weapons of mass destruction while maintaining its production of civilian nuclear power. Little public protest followed the signing of the Nuclear Cooperation Agreement between Canada and India, even though the risk of nuclear conflict in South Asia escalated as a result. Guided ballistic missiles were carving a new highway of death across the subcontinent.

In 1979, a partial meltdown occurred in the reactor core of Unit 2 at the Three Mile Island nuclear-generating plant near Harrisburg, Pennsylvania. The accident was a significant turning point in protests against nuclear power as well as against the arms race.[36] The number of reactors under construction fell worldwide after the incident, while the number of groups opposed to nuclear power and weapons rose.[37] Environmental and anti-nuclear groups began to coordinate their activities. When the Canadian media reported in 1982 that the Liberal government under Trudeau was negotiating with the United States to allow the testing of cruise missiles in Canada, the public response was loud. "Refuse the Cruise" was chanted by demonstrators across the country.[38] Designed to carry nuclear warheads or conventional weapons, the cruise was a six-metre-long, 1,450-kilogram missile that operated under the code name Tomahawk. A National

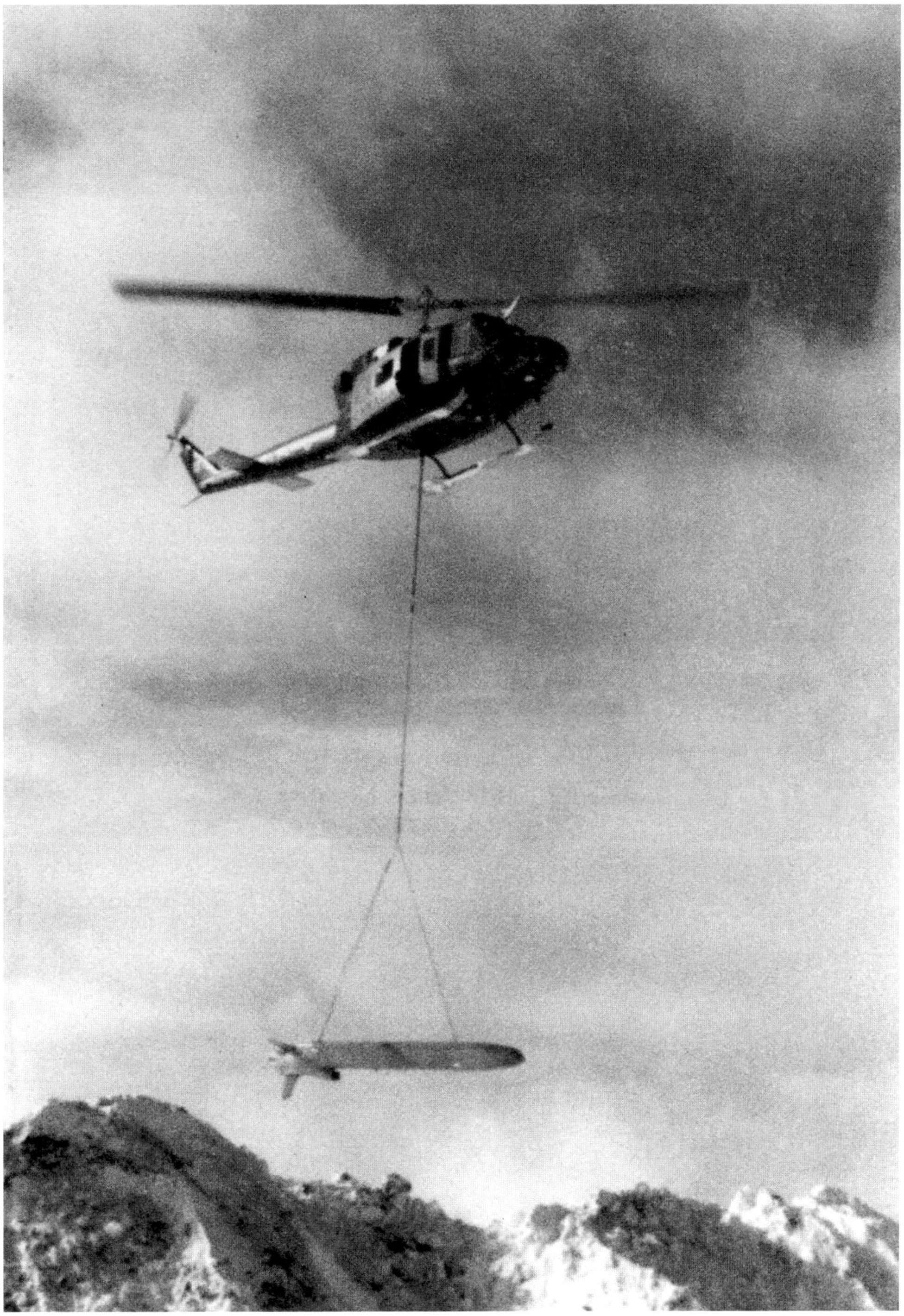

Figure 6.12: National Defence Canada. *Cruise Missile Airlifted after Being Tested at Primrose Lake Weapons Range, Alberta.* 1985.

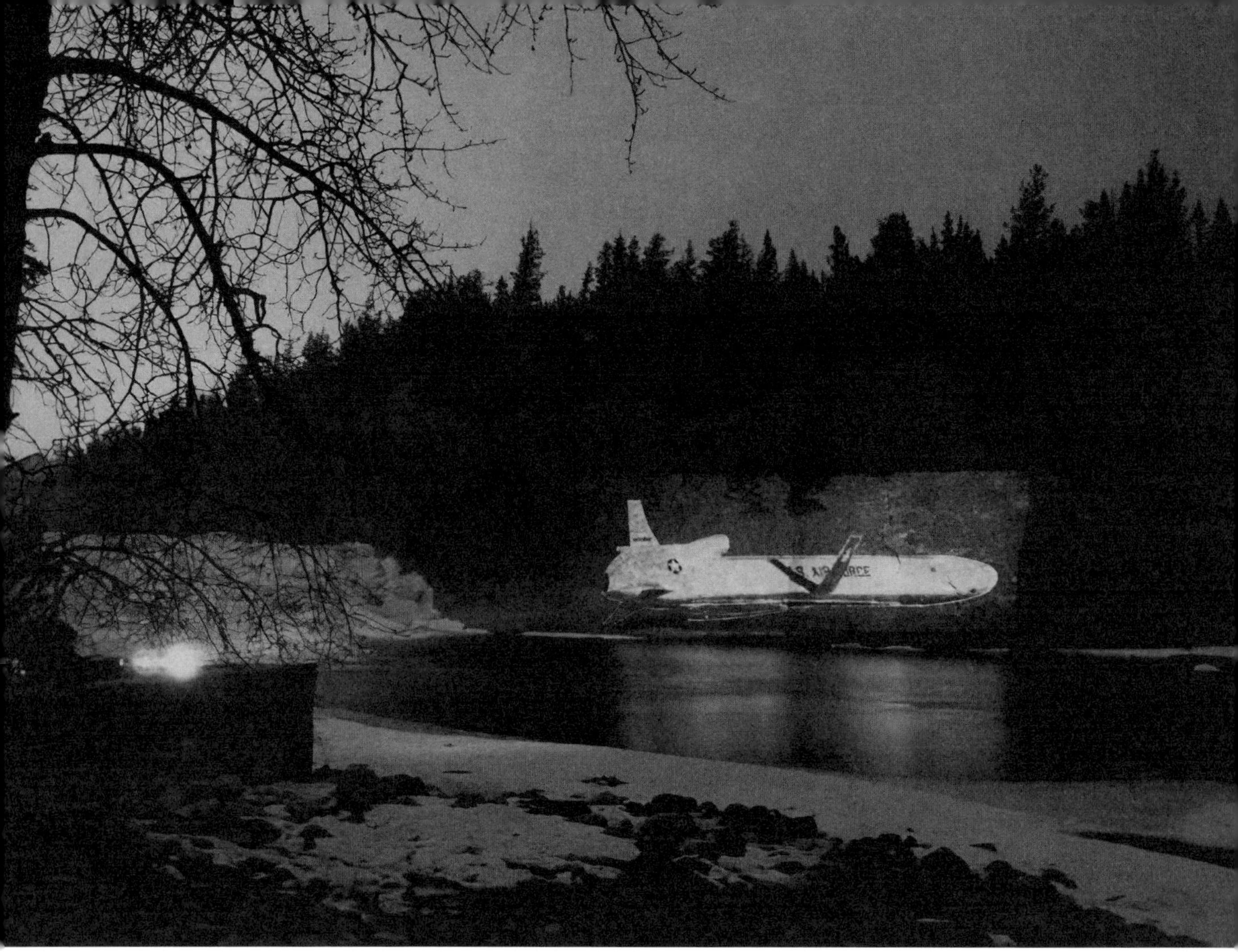

| **Figure 6.13**: Krzysztof Wodiczko. *Bow Falls Projection.* 1983.

Defence photograph shows a spent cruise being airlifted by helicopter at the Primrose Lake Weapons Range, Alberta, after testing. [**Figure 6.12**] The missile could travel at low altitudes to avoid detection and hit targets up to 2,600 kilometres away with an accuracy measureable in metres, a level of precision made possible by a computerized navigation system. Launched from a B-52 bomber, the cruise could be programmed to attack any target on earth with deadly accuracy. Because Canada's terrain resembled that of the Soviet Union – in particular, the Arctic and the Russian steppes – the US military wanted to test the Tomahawk north of the 49th parallel under the terms of NORAD.[39] Alberta was considered the ideal location.

Prime Minister Pierre Trudeau, who had belittled Pearson as a plaster saint for his *volte-face* on the Bomarc-B missile, turned plaster saint himself. He acquiesced to the demands of President Reagan and the US government by permitting the testing of cruise missiles on Canadian territory. Subservience to the United States prompted direct action by five Vancouver-area militants with left-wing and anarchist convictions. They had lost patience with sit-down protests and peace marches. The Squamish Five, as they were dubbed by the media after the town near their training grounds, stole more than a tonne of dynamite from a British Columbia Highways depot for the purpose of blowing up the Litton Systems Canada factory in Toronto. With support from the federal government, Litton was producing navigational guidance components for the cruise missile. The Squamish Five understood the degree to which big data was becoming more and more implicated in nuclear threat – as the artist James Bridle suggests, computers are also war machines – and they wanted to damage the Litton plant.[40] They placed a stolen van loaded with dynamite beside the factory and detonated it on October 14, 1982. But the action went wrong. The explosion occurred ahead of schedule and injured ten people, some permanently. The activists were tracked down, charged, and jailed. On the day following the detonation, the president of Litton declared – without irony – that "bombing is madness."[41] He was right. Few responses to the nuclear arms race have been more succinct.

On the evening of March 22, 1983, the Polish Canadian artist Krzysztof Wodiczko projected an image of a cruise missile against the side of Tunnel Mountain at Bow Falls in Banff National Park, Alberta, southwest of the Primrose Lake testing range. [**Figure 6.13**] He used three projectors and three slides to compose a single image of the missile, which he had appropriated from a photograph released by the Los Alamos National Laboratory. The image represents the missile in flight. Wodiczko had previously projected slides onto the sides of buildings but not onto the landscape. He used the same procedure for his landscape projection as for his urban interventions: "The attack must be unexpected, frontal, and must come with the night when the building, undisturbed by its daily functions, is asleep ... By introducing the technique of an outdoor slide montage and the immediately recognizable language of popular imagery, the Public Projection can become a communal, aesthetic counter-ritual."[42]

"The attack must be unexpected," Wodiczko emphasized, just as a cruise missile attack is unexpected. Wodiczko wanted to reveal something hidden in the landscape, and what was hidden in the Rocky Mountains at Banff was a nuclear weapon. In *Bow Falls Projection,* the photograph on the side of Tunnel Mountain presented the degree to which the militarization of everyday life in Canada had infiltrated every corner of the country. The presence of the weapon at Banff turned the town into a nuclear site, much as Chalk River and Los Alamos are nuclear sites. "In a way the missile was already in that canyon," Wodiczko later reflected about his projection protest, "a lost, romantic missile, a weapon moving along."[43] I take his reference to the missile as "romantic" to be ironic, a comment on Canadian identity and the idea of a pristine, empty landscape. The landscape is not pristine and empty. Cruise missiles fly through it.

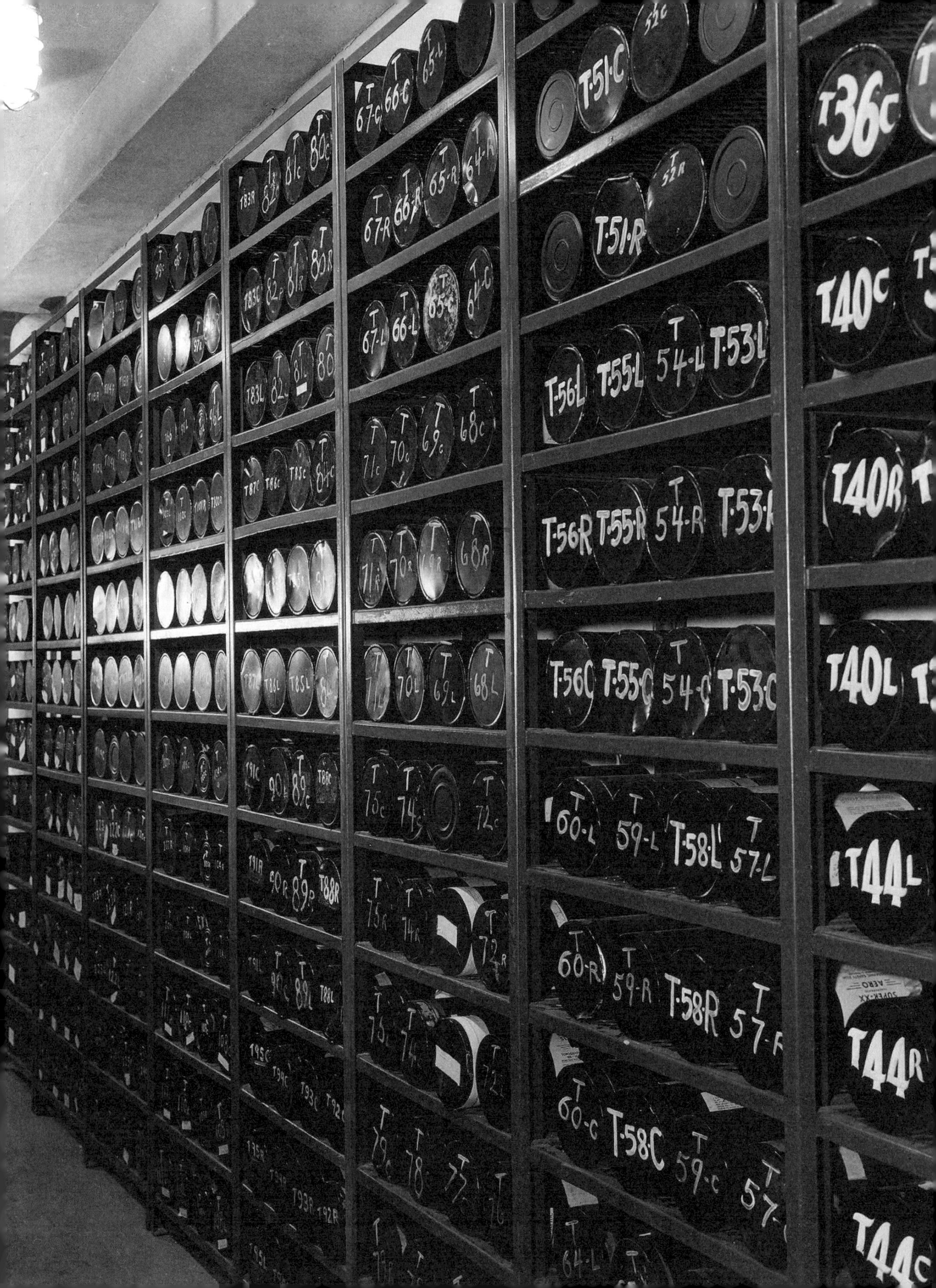
T36C
T·51·C
T·51·R
T40C
T·56·L
T·55·L
T·53·L
T40R
T·56·R
T·55·R
T·53·R
T40L
T·56C
T·55·C
T·53·C
T·58·L
T44L
T·58·R
T44R
T·58C
SUPER-XX
AERO

7
Camera Targets

AS AN AERIAL PHOTOGRAPHER WITH THE ROYAL CANADIAN AIR FORCE IN THE 1950s, the Canadian novelist and poet George Bowering shot film through a peephole in the wing of a reconnaissance plane, adding to an extensive archive produced by the RCAF. Negative rolls of film were stored by the RCAF in metal canisters in a fireproof vault in Ottawa. [**Figure 7.1**] "It was always a movie of one thing, / a target," Bowering has remarked.[1] In modern warfare, especially in aerial reconnaissance, as discussed in Chapters 4 and 5, photography provides pictures of targets. It is a sighting technology that complements the deployment of troops and weapons of mass destruction.[2] But it is also an expressive medium for art. In this chapter, I want to relate the imagery of five art photographers who have worked in Canada to a 1965 book of poems by Bowering, *The Man in Yellow Boots.*[3] The threat of nuclear war, which is a recurring theme in both the poems and the photographs, makes a target of everyone.

There is a close affinity between Bowering's nuclear poems in *The Man in Yellow Boots* and photographs taken by Roy Kiyooka, Mark Ruwedel, Yousuf Karsh, Richard Harrington, and Kristan Horton. There is also an affinity between the poems and Stanley Kubrick's film *Dr. Strangelove,* as I will discuss toward the end of the chapter. Kiyooka was commissioned

Figure 7.1: Royal Canadian Air Force. *No. 1 Photo Establishment: Negative Rolls Stored in Fireproof Vault, Rockcliffe, Ottawa.* c. 1948.

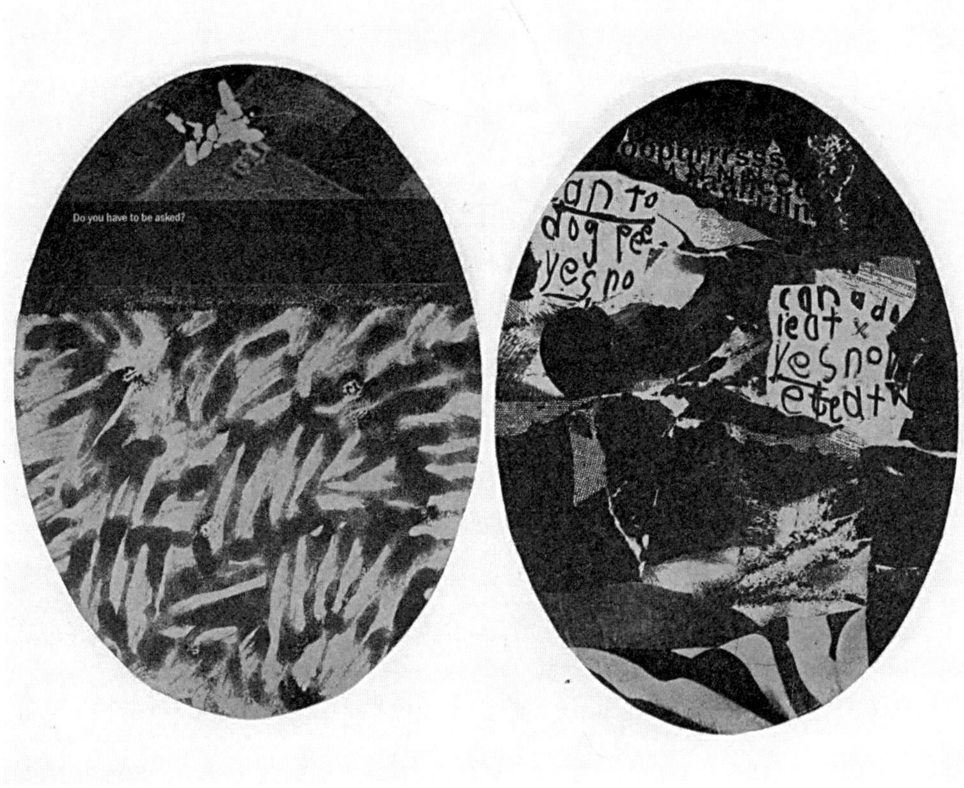

| **Figure 7.2**: Roy Kiyooka. *Untitled.* 1965.

to produce black-and-white photo-collages for Bowering's book, which are glued into the centre of the volume. [**Figure 7.2**] The twelve collages are untitled and appear mostly as double-page spreads. When Bowering invited Kiyooka, who, like Bowering, was a poet, to contribute to the project, he would have known that Kiyooka planned to cut and paste images from magazines and newspapers to make photo-collages.[4] He also would have known that Kiyooka often worked in series.[5] After spending two weeks with the abstract expressionist painter Barnett Newman at the Emma Lake Artists' Workshop in Saskatchewan in 1959, organized that year by Kiyooka and Art McKay, Kiyooka began producing large abstract paintings serially. The hoarfrost series came first, then the hard-edged elliptical paintings.[6] Kiyooka continued to work serially for the rest of his career, long after he had rejected the modernist certainties championed by Newman.

Bowering's biographer Eva-Marie Kröller identifies four poems in *The Man in Yellow Boots* as being more politically charged than the others: "Her Act Was a Bomb," "The Good Prospects," "The Day before the Chinese A-Bomb," and "Vox Crapulous (Alternate Title: J. Edgar Hoover)."[7] I would add a fifth. Although Kröller characterizes Bowering's "Old Time Photo of the Present" as an elegy, possibly a lament for the deaths of his father and his friend

Red Lane, to whom the book is dedicated, it also reads as intensely political.[8] The elegiac is in dialogue with the political, and the poem oscillates between the two:

> *I shall clamber out filthy*
> *from the wreckage*
> *of collapsing universities!*
>
> *There is not much time*
> *not much time ...*
>
> *... fingers pasting my picture*
> *in the photo album of death!*[9]

At the time he was preparing the book, Bowering was waging war against the University of Calgary for its attempts to censor his work and teaching.[10] "Now listen, George Bowering," the poet admonishes himself in "Old Time Photo of the Present," "dont write poems in an office / yell poems of destruction."[11] In 1964, he had a lot to yell about, starting with Western Canada. "Alberta is a terrible place for anyone who wants to say fuck without going to the firing squad," he wrote in a letter to Sergio Mondragón and Margaret Randall, the Marxist editors of *El Corno Emplumado / The Plumed Horn,* who were publishing *The Man in Yellow Boots* as a special issue of the journal.[12] He also wanted to yell about other parts of North America. J. Edgar Hoover was still giving orders to armed FBI agents in Washington, DC; President Kennedy had been murdered in Dallas; Malcolm X was soon to be gunned down in the Audubon Ballroom, New York. The big gun that everyone feared, the one that could go off at any time, was the atomic bomb. The principle of mutually assured destruction, in which Soviet and American firing squads faced one another while trying not to sneeze, was no guarantee against nuclear Armageddon.

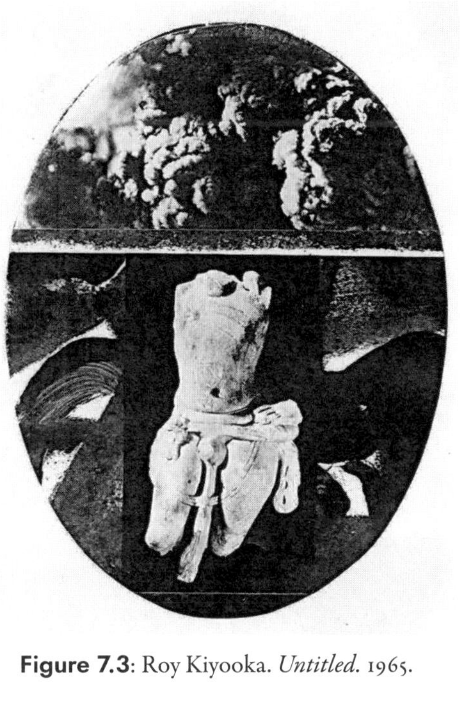

Figure 7.3: Roy Kiyooka. *Untitled.* 1965.

Some of the photo-collages made by Kiyooka for the book also yell of destruction. They would not be out of place in the "photo album of death" that Bowering imagines to be waiting for his own image. A mushroom cloud occupies the top portion of the first collage in the book, spreading above an antique Japanese sculpture of a male torso, its wooden limbs truncated, like the crown of a spreading tree. [**Figure 7.3**] In another photo-collage, reproduced on the left side of Figure 7.2, a parachutist tumbles into a void, his chute still unopened, his fall underwritten by a printed question cut from a magazine: "Do you have to be asked?" In two more collages, crowds look expectantly up at the sky, perhaps transfixed by a falling figure – or by an atomic explosion.

Kiyooka later called this series the *Zodiac* collages, for their elliptical shapes and horizontal bands. As he told curator Chris Varley, they were his first attempts at "illustration."[13] The word is used loosely, referring more to figurative representation than to illustration as it is commonly understood. Kiyooka asked Bowering not to scatter the collages throughout *The Man in Yellow Boots* but to group them in the middle. He wanted them to complement (or interrupt) the poems but not to illustrate them. The photo-collages combine cuttings from newspaper and magazines with abstract brush marks in black enamel, using an oval shape to frame and hold disparate parts together. When Varley asked about his choice of an oval format, Kiyooka replied that the clouds he was painting while living in Regina, before moving to Vancouver in 1961, wanted to turn into ellipses after the move. This may have been the case, but the answer was disingenuous. Kiyooka was an attentive student of art history, and he knew that art comes from art more often than it comes from prairie clouds. Pablo Picasso and Georges Braque used oval frames in their cubist works to upset traditional hierarchies of pictorial space. The Dadaists and surrealists did the same. An ellipse has the formal effect of redistributing the weight and significance of an image's parts so they become equal, disrupting visual connections of a more traditional kind. These disruptions, and the resulting visual ambiguities, were what Kiyooka was after.[14]

The mushroom cloud in Kiyooka's collage is connected to the human figure. The wooden torso, positioned vertically on severed legs and with missing arms, is Japanese. Kiyooka visited Japan for the first time shortly before collaborating with Bowering, though he had learned a broken form of Japanese while growing up in the prairies and continued to speak it throughout his life. As a nisei, or second-generation Japanese Canadian, he began in the 1960s to engage more fully with his parents' country of origin. He called his first book of poetry *Kyoto Airs* (1964) and saw himself standing, as he later remarked, "athwart east and west" – in art as in life.[15] Kiyooka opted for multiplicities rather than singularities, a predilection reflected in his multipartite practice as a visual artist, poet, and musician. His mushroom cloud photo-collage is also multiple rather than singular. It refers to the toxic impact of atomic detonations everywhere, though the scissored photograph of the Japanese sculpture suggests that Kiyooka was focusing specifically on Hiroshima and Nagasaki while making the piece. He was also focusing on Hiroshima and Nagasaki when he represented Canada at Expo '70 in Osaka, as discussed in the Preface.

Kröller observes that the twin of collage is montage. She is struck by the elements of montage in the poem "Her Act Was a Bomb," pointing out that three synchronous but unrelated events are juxtaposed, two of them banal and one catastrophic.[16] The poem reads,

All over America I know
people are switching off the sound
when Sophie Tucker appears
on the Ed Sullivan reruns.

It is of course
an honest gesture, severe
perhaps. But in Las Vegas
I saw an old heron woman

pull down the lever
on a cafe slot machine
and fifteen miles away
on the desert, America

dropt a Bomb on Nevada.[17]

The montage in the poem stitches the unconnected events together by means of conjunctions and enjambments between the stanzas, and the sequencing produces a chain reaction that culminates in the dropped bomb. The act of muting Sophie Tucker, known to her fans as the "Last of the Red Hot Mamas," on *Ed Sullivan Show* reruns leads to the arm movement of the old heron woman in Las Vegas, and the lever on the slot machine leads to the "red button" that detonates a bomb at the Nevada Test Site. The conjunction is between the forces of consumer capitalism and those of the military-industrial complex. In Japanese folklore, the heron woman has transformative powers. The wife of a poor fisherman, she turns herself into a white heron to weave bolts of exquisite cloth from her feathers. When the fisherman discovers her at work, the spell is broken and she is free to leave.

Also set in Las Vegas, an artist's book by Mark Ruwedel offers an even closer visual parallel to "Her Act Was a Bomb" than Kiyooka's mushroom cloud photo-collage. Ruwedel created *Lili's Wedding* in 2007, featuring striptease artist Lili St. Cyr.[18] [**Figure 7.4**] When St. Cyr performed her act at Ciro's in Hollywood, she was billed as the "Anatomic Bomb," and from 1944, when the Gayety Theatre reopened in Montreal, until the early 1950s, when her career took off in the United States, she was considered "Montreal's most famous woman, the city's femme fatale, a person whose name invoked sophistication, mystery, sin."[19] Given the hostility of the Catholic clergy to sophistication, sin, and art – the same clergy that attacked Les Automatistes painters for their 1948 manifesto, *Le Refus global,* and forced their founder Paul-Émile Borduas into exile – the stripper's success in Montreal became legendary.[20] Born Willis Marie Van Schaack in Verdun, Quebec (some biographies list her birthplace as Minneapolis), Lili St. Cyr took her stage name from the *École spéciale militaire de Saint-Cyr,* France's foremost military academy, whose motto is "They study to vanquish." She rose from

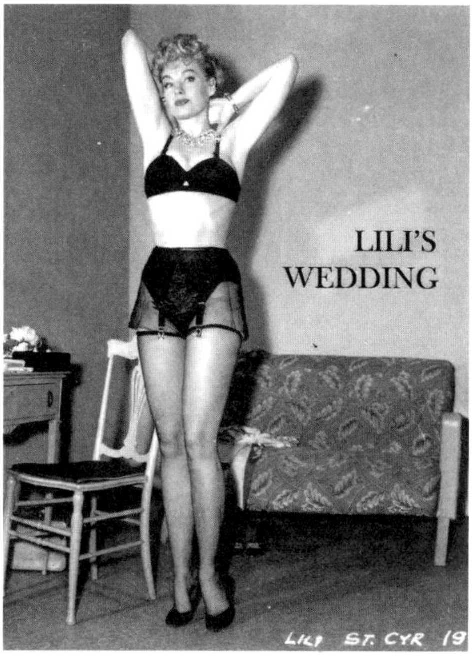

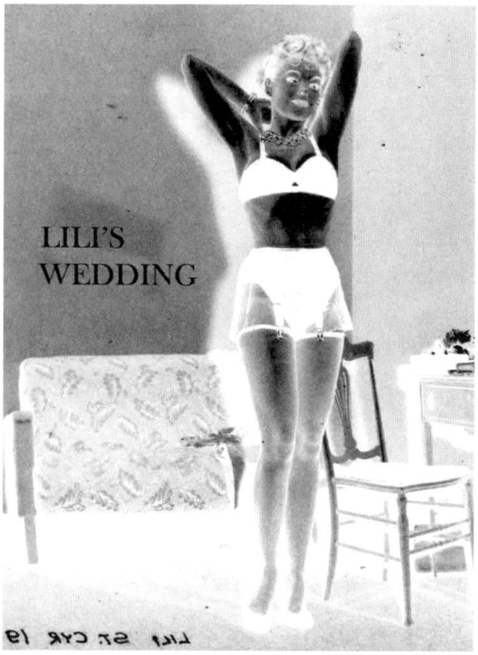

Figure 7.4: Mark Ruwedel. Front cover of *Lili's Wedding*. 2007. **Figure 7.5**: Mark Ruwedel. Back cover of *Lili's Wedding*. 2007.

a burlesque chorus line to become, along with Gypsy Rose Lee and Ann Corio, one of the most recognized striptease stars in the 1940s and 1950s.[21]

In 1955, St. Cyr postponed her Las Vegas wedding to actor Ted Jordan three times.[22] According to a note in Ruwedel's book, she wanted the marriage ceremony to coincide with the explosion of an atomic bomb at the Nevada Test Site. The detonation was cancelled three times in a row because of bad weather, but the bride had her hand on the lever and waited for the right moment to pull it. When Test Moth finally exploded on February 22, in Area 3 of Yucca Flat, with a yield of two kilotons and the creation of unusually large amounts of radioactivity, wedding guests were presented with slices of a cake baked in the shape of a mushroom cloud.[23] The Québécoise had dropped her own bomb on Las Vegas.

The front cover of Ruwedel's book features a photograph of the stripper from the 1950s posing with her arms bent behind her head and wearing a black bra and panties embroidered with a fleur-de-lys. It was taken by Irving Klaw, whose best-known model was Bettie Page. Ruwedel's back cover shows the same picture printed in negative. [**Figure 7.5**] In this inverted image, the shadow projected by St. Cyr on the dressing-room wall behind her is no longer dark but light, and her frame glows as if she has been irradiated, caught by the camera at the exact moment of a nuclear explosion. The book includes two appropriated colour images of mushroom clouds, a postcard image of the Nevada Test Site and a telephoto image released by the US Atomic Energy Commission, as well as two double-page spreads of black-and-white photographs taken by Ruwedel. By functioning indirectly, the two Ruwedel photographs are analogous to Kiyooka's collages in *The Man with Yellow Boots*. The first image shows the city of Las Vegas, shot from the desert floor at a distance, looking insubstantial in the harsh light. [**Figure 7.6**] The scene is devoid of life, with just a few white plastic bags tangled in the scrub brush, like the final scene of a post-apocalyptic film. The second image represents the Nevada Test Site, with some abandoned buildings and pieces of equipment. Again, lifeless. Only Lili is alive in the book – until she turns into a negative of herself and is not.

A few months after the stripper from Quebec got married in Las Vegas to the accompaniment of an atomic blast, the Montreal aircraft manufacturer Canadair published an advertisement in *Maclean's* magazine. [**Figure 7.7**] Depicting a new long-range Soviet bomber, the ad asked, "Do we actually *know where* to face Communism?" The suggestion was that Canada did not know where its enemy was, while sitting squarely in the crosshairs of the bomber. The plane symbolized "Communism's bid in the great race for world technological supremacy." The photo in the ad was by Yousuf Karsh – "Photographed especially for Canadair by Karsh," the tagline announces – and represents five aeronautical engineers and scientists poring over a technical drawing of the bomber.[24] It was one of eight taken by Karsh for a Canadair ad campaign organized by Walsh Advertising Company. Although Walsh claimed that the text in the ad flowed from the shot, in part to trade on Karsh's reputation, James Opp has discovered that the process took place the other way around: Karsh made his picture to complement the copy.[25]

| **Figure 7.6**: Mark Ruwedel. *Lili's Wedding (Las Vegas Seen from the Desert Floor)*. 2007.

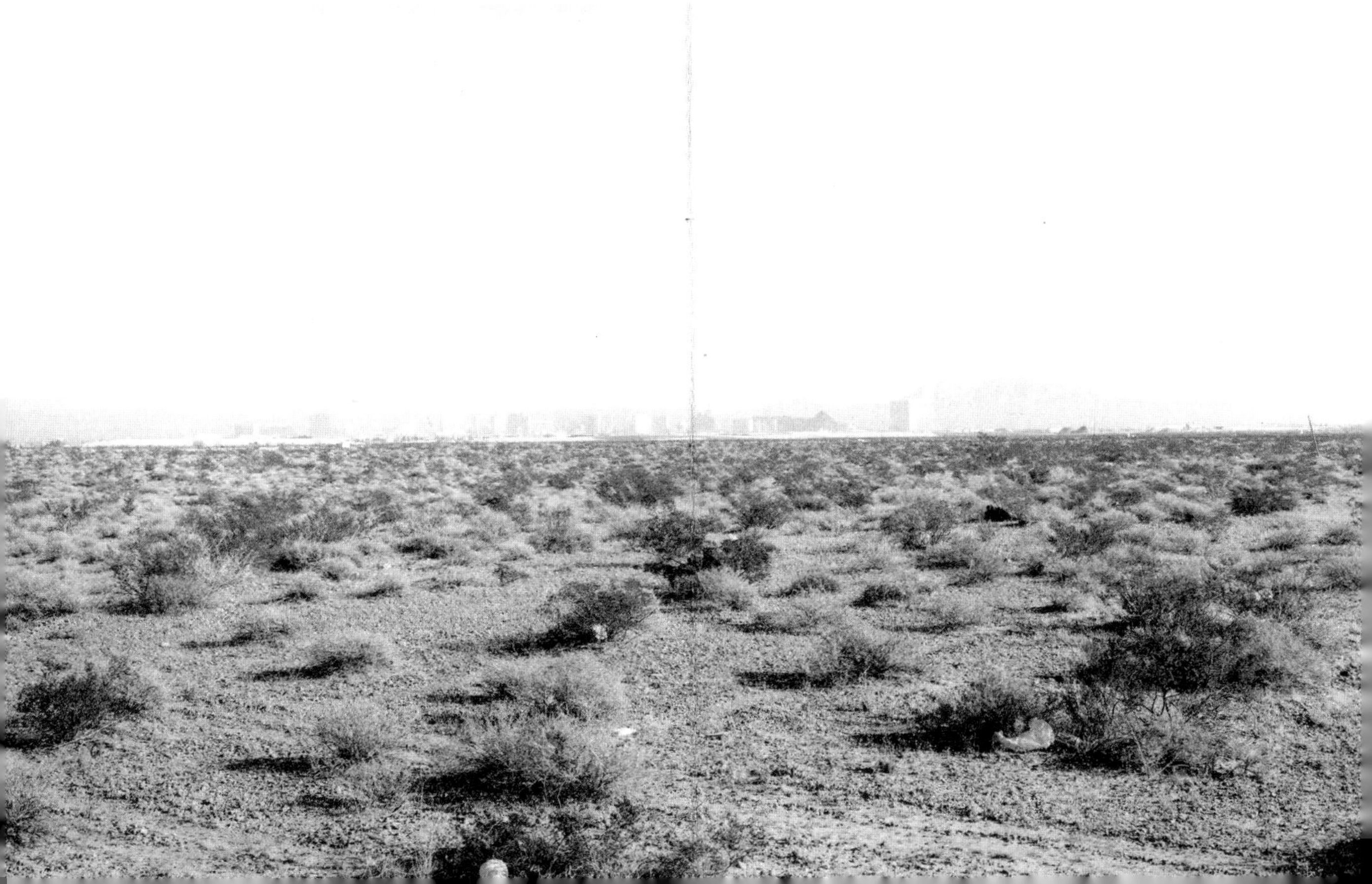

Do we actually __know__ __where__ to face Communism?

If you could use reprints of this message for friends, staff, or associates, just write Dept. AA.

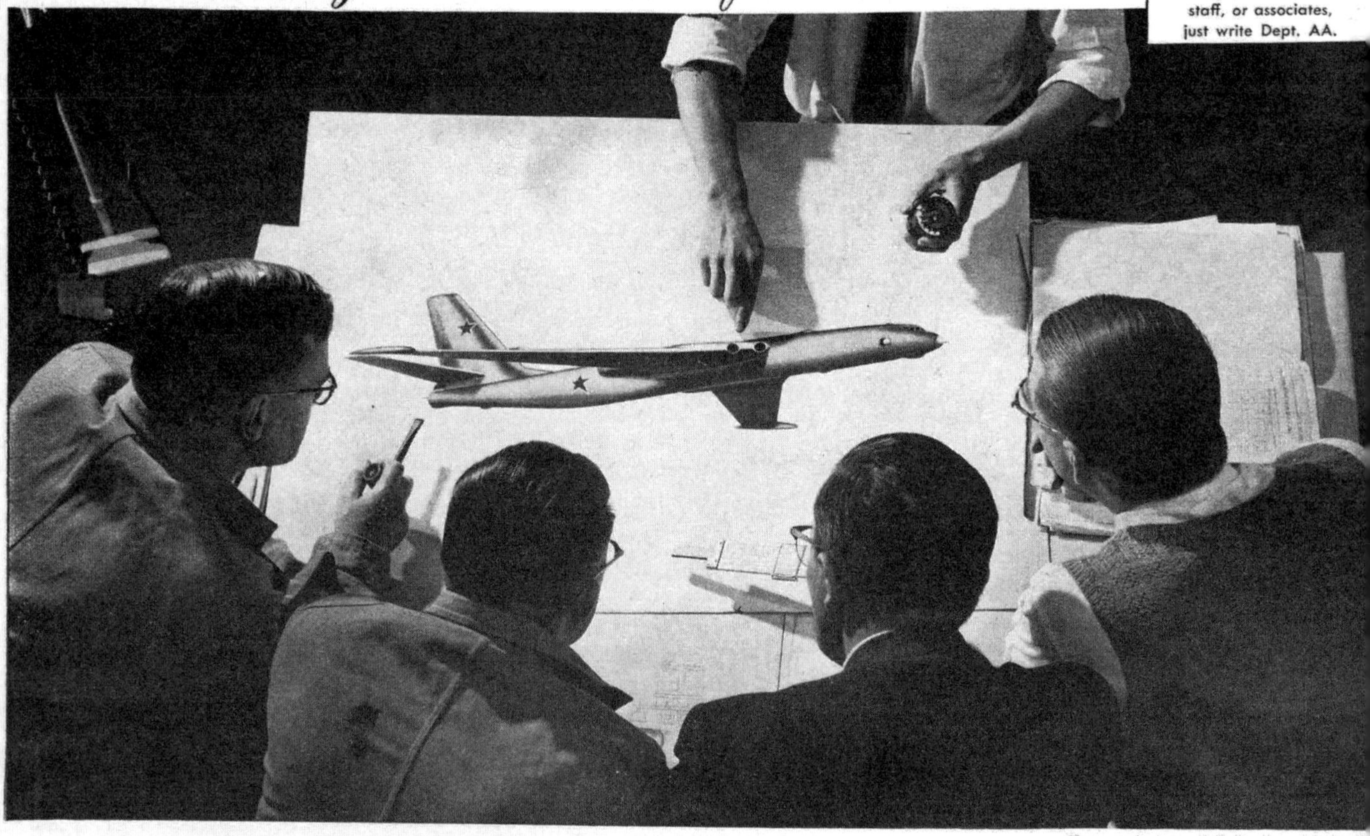

Photographed especially for Canadair by K

Communism's Technical Ability

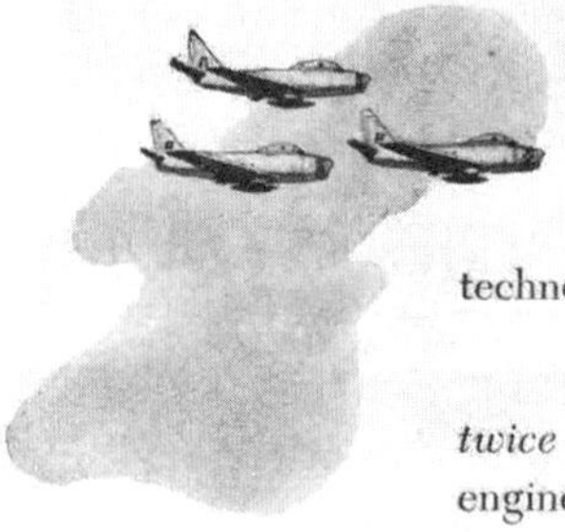

The jet engines on the huge Russian bomber these Canadian scientists are studying may well be — so say competent authorities — the most powerful in the world . . . a vivid symbol of Communism's bid in the great race for world technological supremacy.

For Communism is in dead earnest. Communist countries will graduate this year *twice* as many engineers as Canada and the United States together — may have *more* engineers and technicians than all the Western world by 1960.

Our old exploits are no guarantee of future achievements . . . we must encourage more young people to study science . . . to become engineers, technicians, graduate scientists, engineering professors. We need these "front-line soldiers" to throw the Red Challenger back — to make sure that scientific leadership remains in the hands of free men.

C-L CANADAIR LIMITED

CANADAIR

— AIRCRAFT MANUFACTURERS —

LIMITED, MONTREAL

CAS

YOUNG MEN! YOU CAN SERVE YOUR COUNTRY NOW IN THE AIR FORCE

When China first detonated an atomic bomb on October 16, 1964, it christened the explosion, just as the United States christened each of its nuclear tests. At the suggestion of the Lop Nur basketball team, at the nuclear weapons base in Xinjiang, the test was named Tou Lan, or Shoot the Basket. News of its success was immediately relayed to Chairman Mao Zedong and Premier Zhou Enlai by telephone, during which the reporting officer exclaimed, "We have seen the mushroom cloud!"[26] Bowering did not see the Chinese mushroom cloud – it was not broadcast live in China or abroad – but he did see the St. Louis Cardinals beat the New York Yankees to win the World Series on the same day. He wrote about the coincidental events, and two others as well, to Mondragón and Randall in Mexico City:

> *Oct 16/64 ... What a lovely day yesterday! The Cardinals batter the Yankees for the 1964 World Series; the Laborites knock off the Tories in Britain, Khruschev is deposed as bossman in Russia, and the Chinese drop their first A bomb. What are the Americans going to do now a country they dont recognize has detonated the Bomb? Theyll have to keep it a secret from America, or walk around wondering, where did that noise come from? It cdnt have come from that part of Asia because there aint no country there, you can see for yrself.*[27]

The conjunction of political and sporting events that caught Bowering's attention might have excited him even more if he had known that the Chinese test was named by a basketball team. At the beginning of the poem "The Day before the Chinese A-Bomb," Bowering repeats the events that he listed in the letter. Each represents a cleansing, a throwing out of the old. "The Great Purge / begins today, the Cardinals are / World Champions," the poem's final lines read.[28] Eight years would elapse before the United States responded to the noise produced by the Chinese test. President Richard Nixon did not meet with Chairman Mao until February 1972. By comparison, Canada had established trade relations with China as early as 1961, starting with a wheat sale agreement.

The Canadian photographer Richard Harrington, best known for his work in the Arctic among the Inuit, visited China several times in the years after the atomic test but before the visit of President Nixon.[29] His Chinese photos concentrate mostly on street life and signage, but he also photographed Chairman Mao at a National Day banquet and took three "nuclear" shots during his visits.[30] In one of the nuclear images, two children look at an anti-American billboard. [**Figure 7.8**] It depicts a cartoonishly drawn American soldier, with a missile slung under his arm and his nose in a magazine, while three disembodied hands toss made-in-America bombs at Korea, Cuba, and Vietnam. "Attack Korea," "Attack Cuba," "Attack Vietnam," the bomb casings repeat in unison. Another sentence on the billboard translates

| **Figure 7.7**: Yousuf Karsh. *Communism's Technical Ability*. 1955.

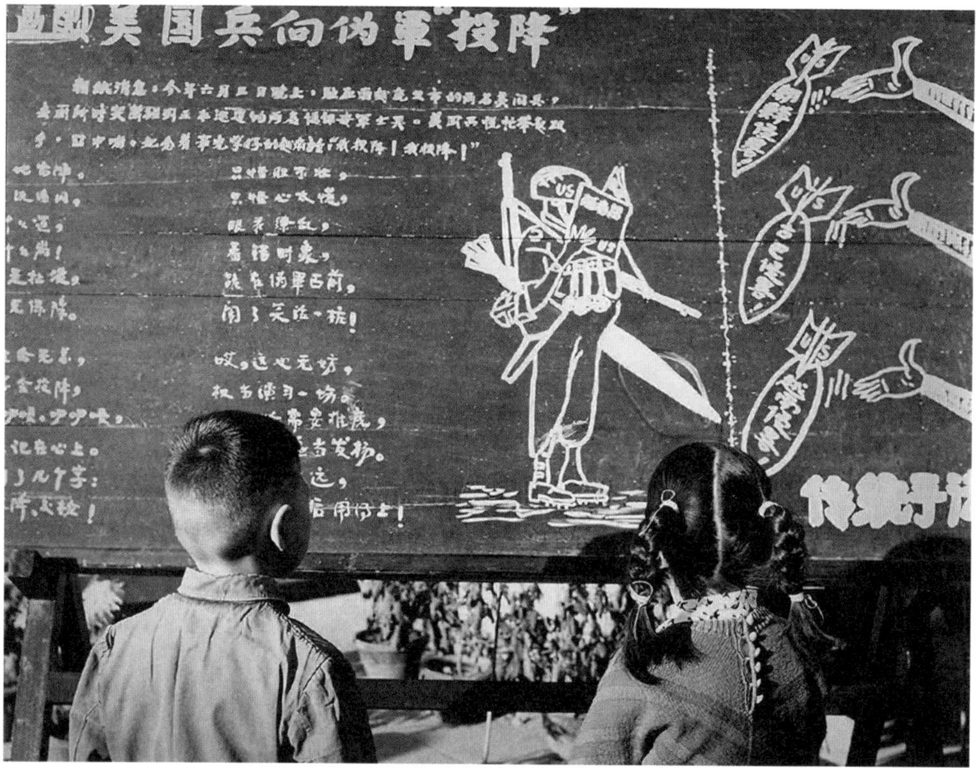

Figure 7.8: Richard Harrington. *Beijing, China (Children Reading Anti-American Sign).* 1966.
Figure 7.9: Richard Harrington. *Beijing, China (Movie Billboard with Mushroom Cloud and Soldier Wearing Anti-Radiation Gear).* 1966.

as "American Soldiers Surrender to the Chinese Army." The other two nuclear photographs are of movie billboards in pedestrian areas. [**Figure 7.9**] They show Chinese soldiers, equipped with guns and wearing anti-radiation gear as protection, moving toward the viewer, while mushroom clouds painted in a traditional brush-and-ink style *(guó huà)* rise nearby. (An ink-on-paper scroll showing a mushroom cloud, painted in 1965 by the celebrated Chinese artist Wu Hufan, is in the collection of the Shanghai Institute of Chinese Painting.)[31] In the busiest of the two photographs, taken in front of a movie theatre, the poster dominates the top half of the image while people on bicycle and foot fill the bottom half. "Let Us Cheer the Country for Its Three Successful Atomic Tests," Chairman Mao is quoted as saying, indicating that the film was propaganda on the bomb, made for domestic consumption.

When Bowering was starting to write *The Man in Yellow Boots* during the summer of 1964, *Dr. Strangelove,* first released by Columbia Pictures on January 29 of that year, was still playing in some theatres. Kubrick began shooting the film after the Cuban Missile Crisis had raised the stakes in the Cold War game of thermonuclear extinction. The nihilistic satire concludes with a vision of nuclear apocalypse unfolding to the refrain of "We'll Meet Again," sung by

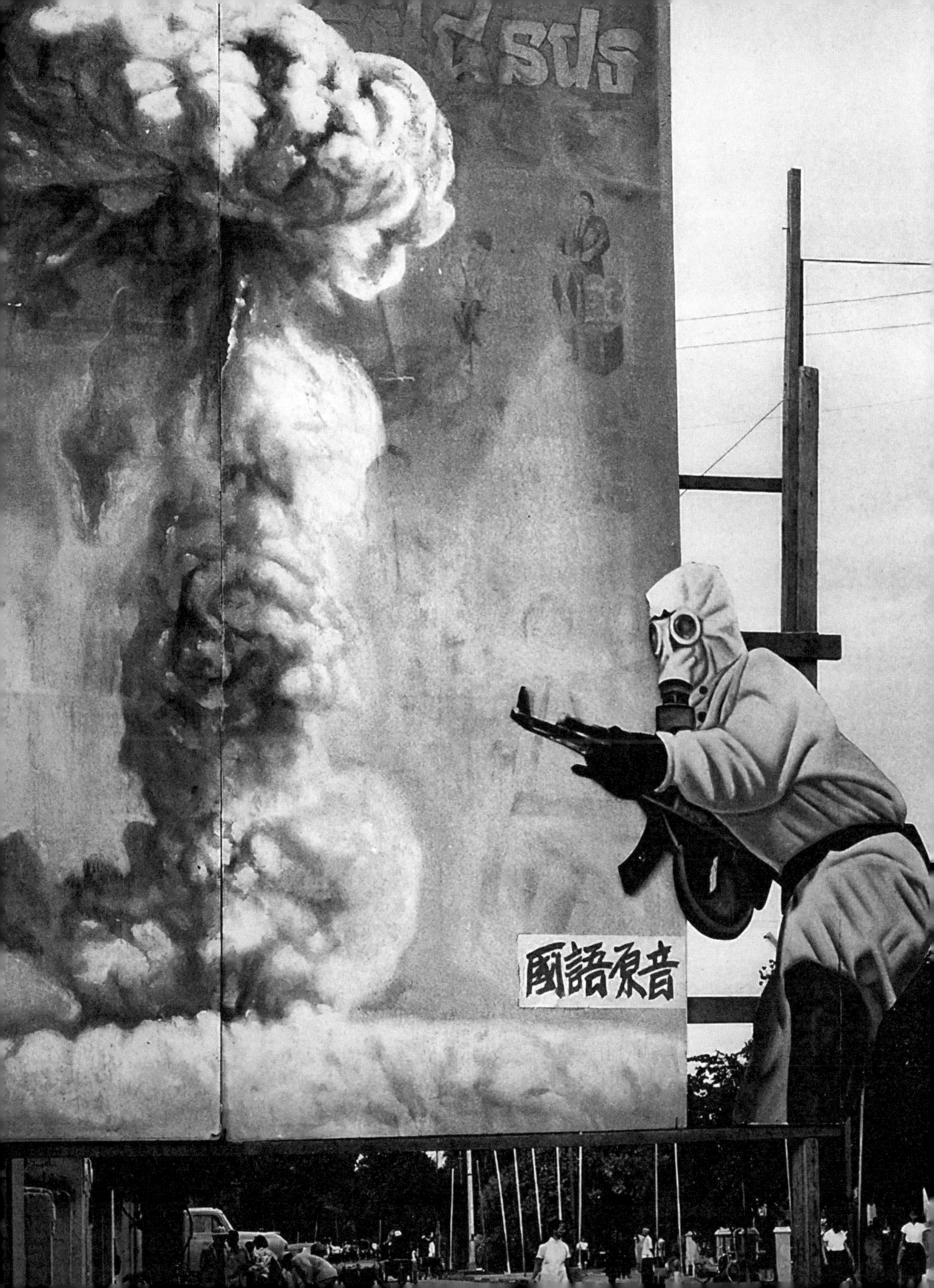
國語原音

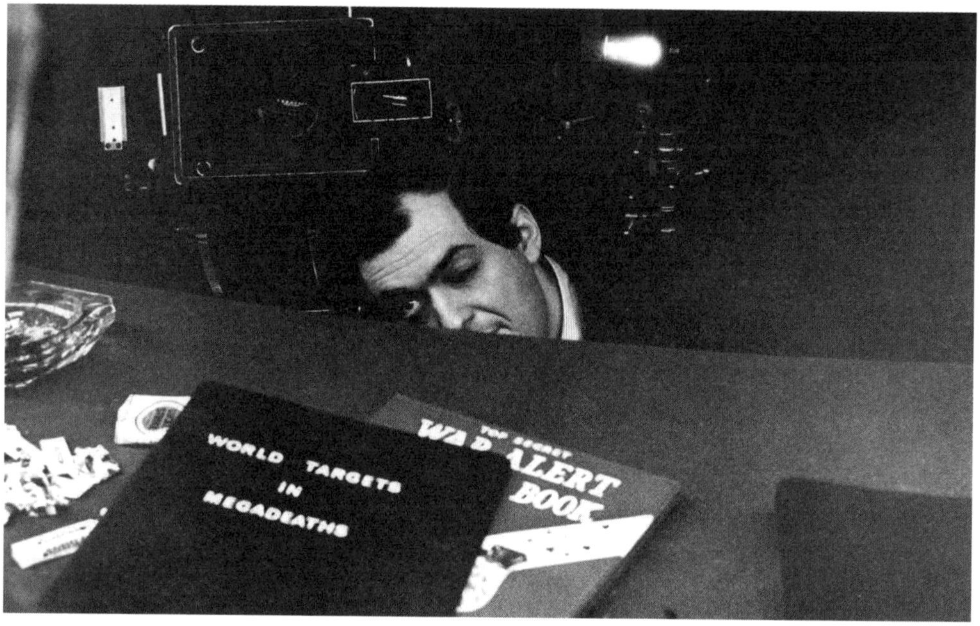
WORLD TARGETS
IN
MEGADEATHS
ALERT
BOOK

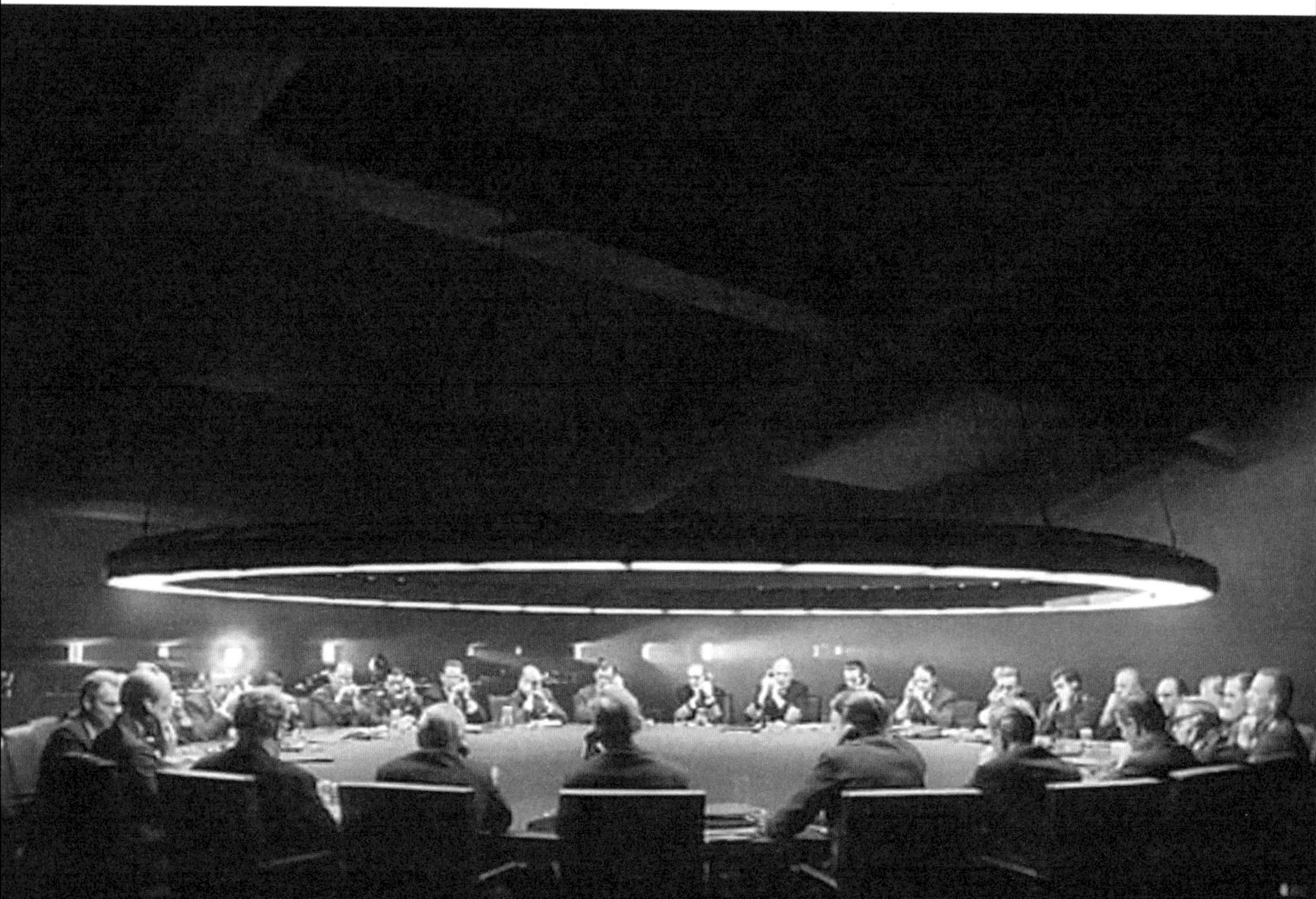

Vera Lynn. "The greatest message of the film is in the laughs," Kubrick remarked, confirming his irreverent approach to the bomb. "The most realistic things are the funniest" – which is another way of saying that humorous things are sometimes the most horrific.[32] Kubrick once thought of living in Vancouver.[33] He feared the destruction of American cities and assembled a small library of books on nuclear war and its risks. He believed that Vancouver was less vulnerable to nuclear attack than New York, where he lived until 1961, before deciding to move to London just as he was beginning work on *Dr. Strangelove.*

The film imagines a scenario where no place on earth is exempt from conflagration and nuclear winter.[34] The lower half of a photograph of Kubrick taken by Dmitri Kasterine, hired to provide publicity stills for the film, is dominated by a black binder entitled "World Targets in Megadeaths."[35] [**Figure 7.10**] The picture is taken from a low camera angle, producing an off-balance image marked by broken lines. Kubrick peers over the edge of the table of the War Room set, with his head cocked to one side.

Figure 7.10: Dmitri Kasterine. *Stanley Kubrick Lining Up a Shot on the Set of Dr. Strangelove or: How I Learned to Stop Worrying and Love the Bomb.* 1964. **Figure 7.11**: Kristan Horton. *Dr. Strangelove Dr. Strangelove.* 2003–06.

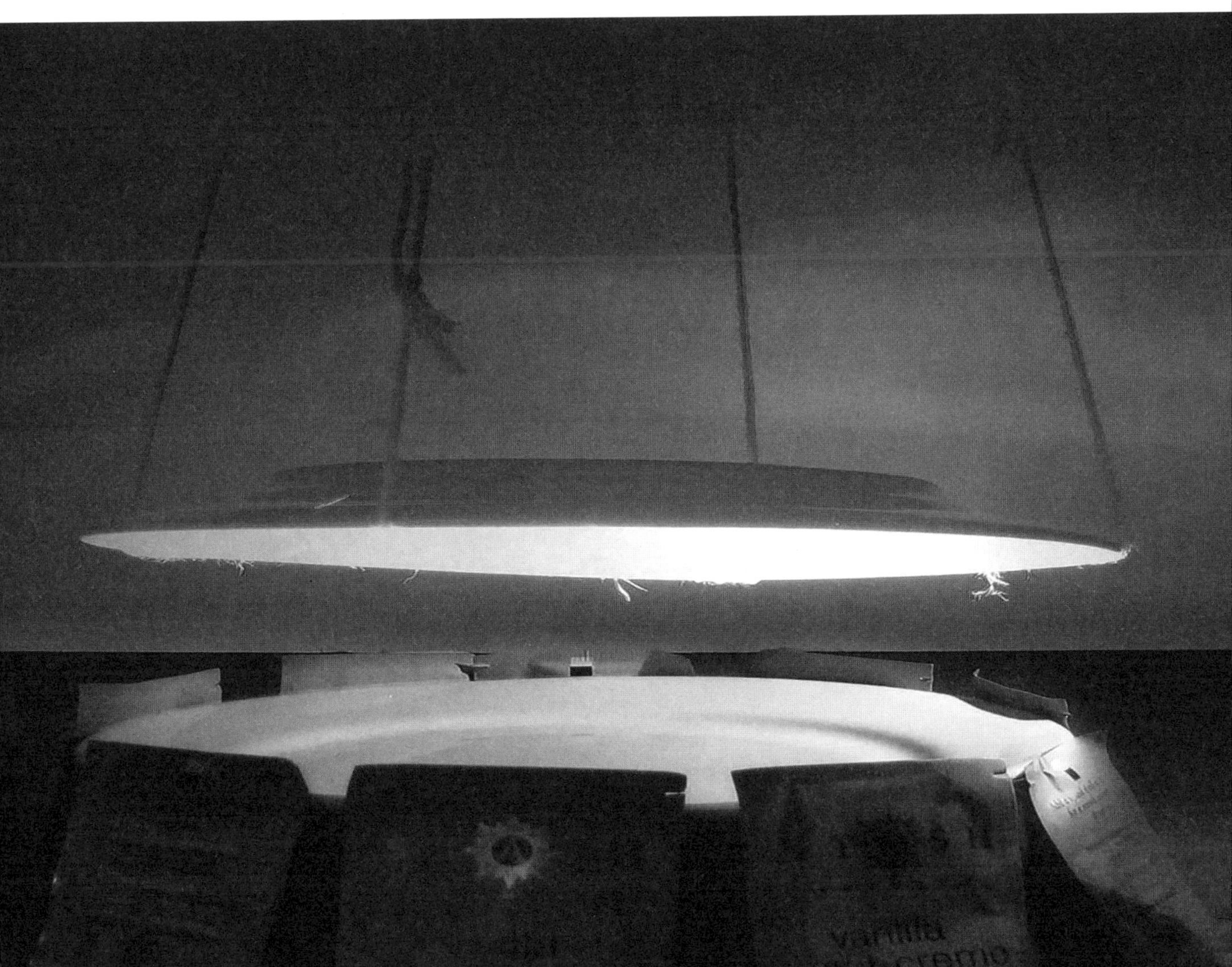

Bowering does not refer to *Dr. Strangelove* in *The Man in Yellow Boots,* but the film's presence is felt in both the atomic poems and the letters. A note beneath "The Good Prospects" informs readers that the poem was written on the occasion of a meeting between the United States and the Soviet Union in Moscow to discuss a test ban treaty. The poem and the note are reminiscent of a conversation in the *Dr. Strangelove* War Room between US president Merkin Muffley and Soviet premier Dimitri Kissov: "Now then Dimitri. You know how we've always talked about the possibility of something going wrong with the bomb. The bomb, Dimitri. The hydrogen bomb."[36] The artist Kristan Horton also knows about the possibility of something going wrong. The threat of the hydrogen bomb is present in all two hundred images, including one of the Kubrick War Room, in his 2007 book project *Dr. Strangelove Dr. Strangelove.*[37] [**Figure 7.11**] Horton's doubling of the film's title is the first clue to his concept for the book. From commonplace objects in his studio – plastic bags, ceramic bowls, glue sticks – he produced improvised constructions of scenes from the film and then took pictures of what he had built. He placed the photographs adjacent to stills from the movie, bringing the pairings together in a series of single printed images. Such humorous mimicry of the horrific is what Kubrick found so potent in representing the bomb.

The "Good Prospects" announced in the title of Bowering's poem have turned to cinders by the end of it:

> *Because*
> *there may be*
> *half erected*
> *superstructures*
>
> *left unfinished*
> *done*
> *with ash*
> *falling on them.*[38]

Kubrick gives his characters in *Dr. Strangelove* scatological names to match their personalities. The effeminacy of President Merkin Muffley earns him two colloquialisms related to female genitalia; the attraction of the title character to atomic devices becomes a sobriquet for sexual perversion; and the obsession of General Buck Turgidson with bodily functions is reflected in a moniker relating to defecation.[39] When Bowering writes to Mondragón and Randall "a last letter in case Goldpiss gets in as prez," he is working the same ground as Kubrick.[40] Goldpiss, better known as Barry Goldwater, won the Republican nomination in 1964 despite calling for military action to overthrow communism that would have risked escalation into nuclear war. Hence Bowering's ironic comment about sending "a last letter" to the editors of his book. "I would remind you that extremism in defense of liberty is no vice," Goldwater declared in his acceptance speech at the Republican convention.[41]

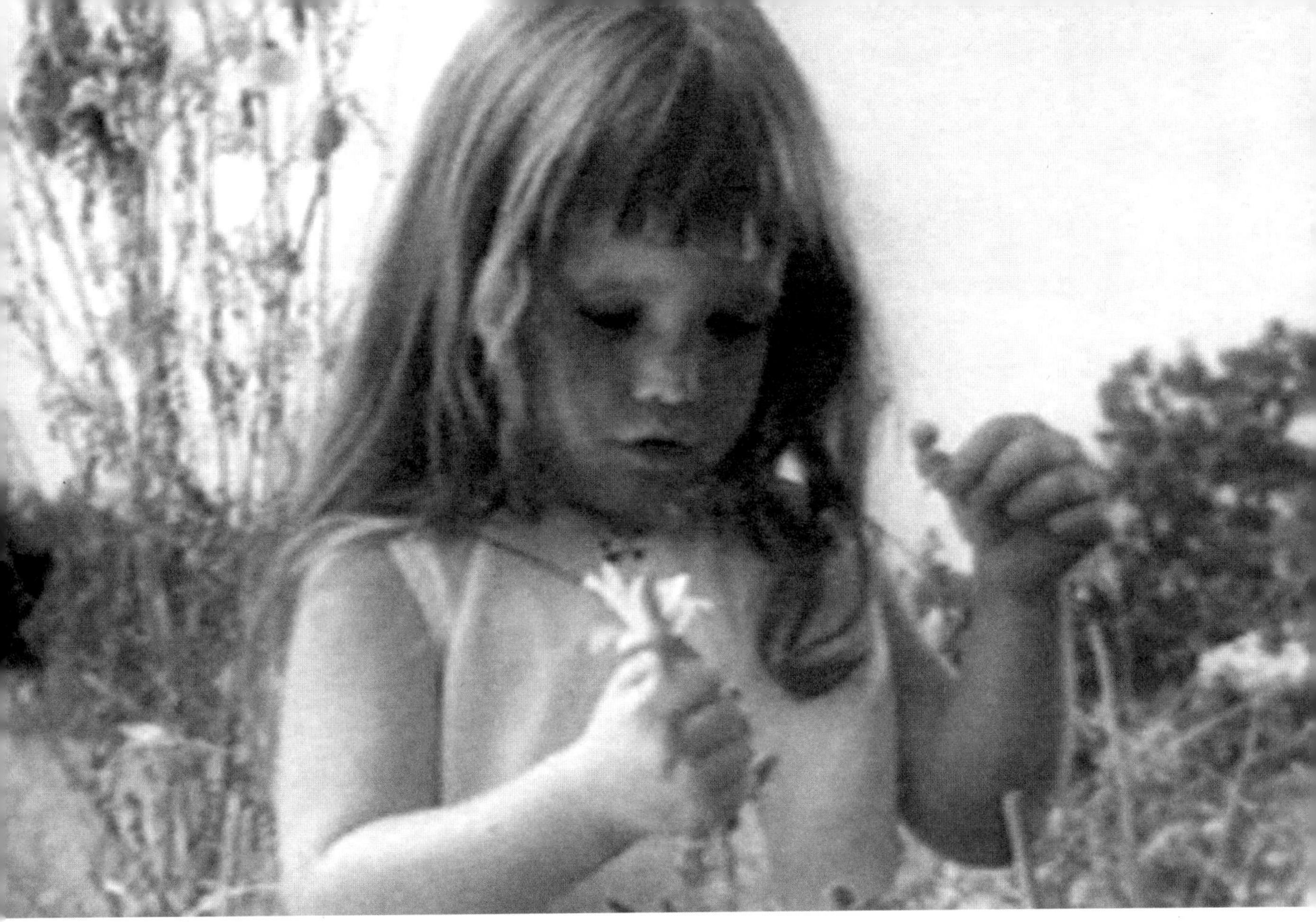

| **Figure 7.12**: Tony Schwartz with Doyle Dane Bernbach. *Peace Little Girl ("Daisy Ad")*. 1964.

President Lyndon B. Johnson and the Democrats wasted no time in exploiting widespread public anxiety about the risk of an atomic showdown that might be triggered by Goldwater if he became president. They commissioned a forty-three-second television ad in which a little girl counts daisy petals. [**Figure 7.12**] She counts hesitantly from one to ten before a male voice, reversing the order of the numbers, counts decisively down to one, as Lili St. Cyr had done ten years earlier at her wedding. As the numbers decline toward the inevitable atomic explosion, toward the ultimate camera target, the lens zeroes in on the girl's frozen face, producing a series of still photographs that belong in Bowering's "photo album of death." The "Daisy Ad," officially called "Peace Little Girl," aired only once before getting pulled as too inflammatory; but the damage had already been done.[42] Goldwater, like the antique Japanese sculpture in Kiyooka's photo-collage, had himself become a nuclear target. All the Republicans' horses and all the Republicans' men could not put him back together again. He lost the election.

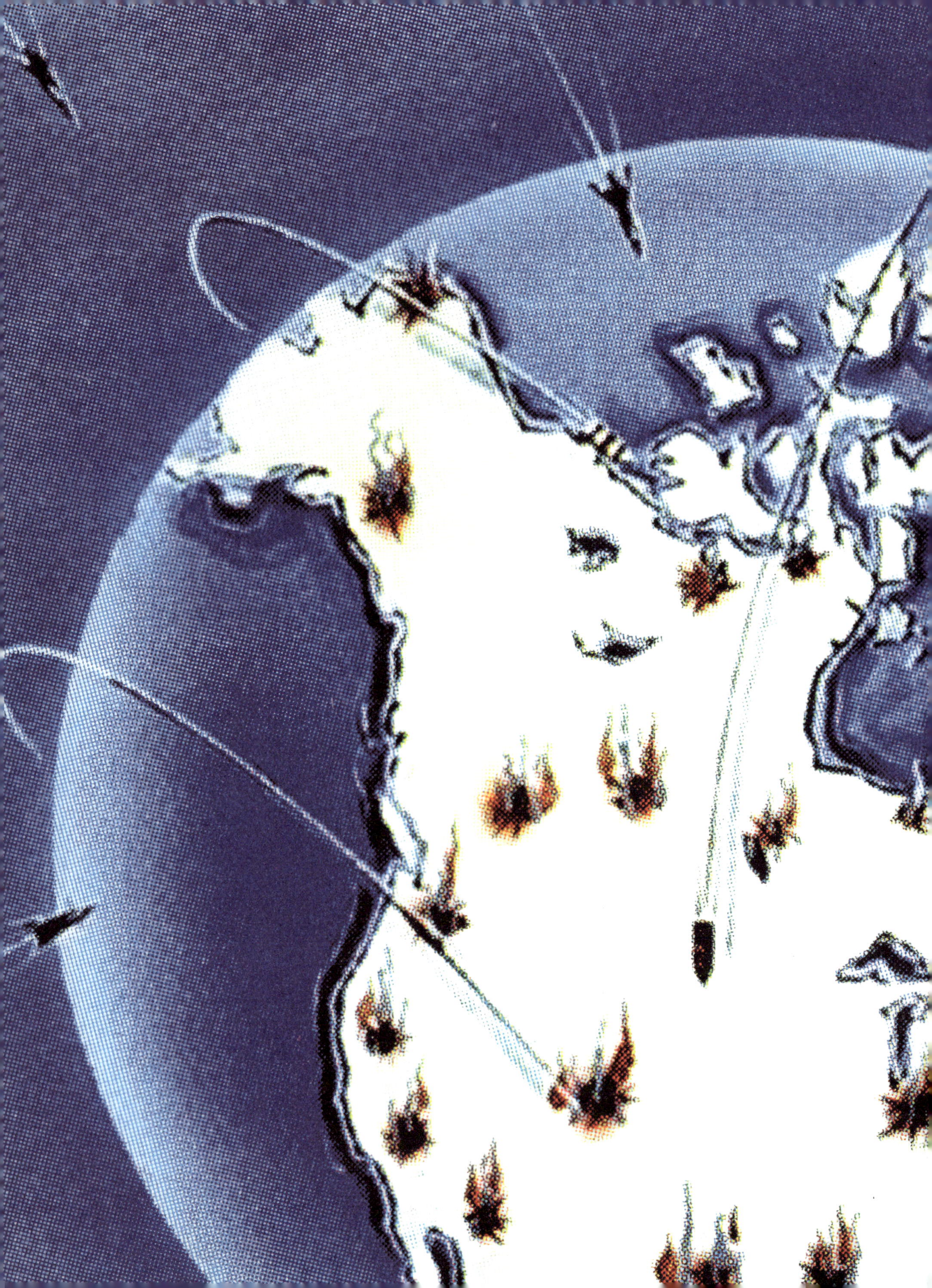

EPILOGUE

Deadpan Pictures

THE US MILITARY NAMED AND PHOTOGRAPHED EVERY ATOMIC BOMB DETOnated at the Nevada Test Site from 1951 to 1962, when atmospheric testing ceased. Moth, Stokes, Climax, Wheeler (one bomb, two bombs, three bombs, four). Smoky, Sugar, Shasta, Diablo (five bombs, six bombs, seven bombs, more).[1] In this roll call of nuclear tests, the military did not give a name to the bomb it dropped on British Columbia on Valentine's Day, 1950, and no images exist of the explosion.[2] Without a visual record, the incident has become a footnote in nuclear history.[3] Ice caused three engines of a US Air Force B-36 bomber, known as the Peacemaker, to fail while it was flying down the BC coast from Alaska en route to Texas. Following the emergency protocol established for US aircraft carrying atomic weapons, Ship 2075 released its nuclear payload over open water. A Mark IV bomb, similar to the one dropped on Nagasaki, was detonated in midair above Queen Charlotte Sound, ninety kilometres northwest of Bella Bella, minus its plutonium core.[4] As the main supplier of uranium to the United States for its weapons program, Canada would have provided some of the uranium used in the bomb.

The device consisted of 5,300 pounds of conventional explosives and 88 pounds of natural uranium casing. The casing dispersed radioactive fallout across the region. The crew had time

Figure E.1: NORAD. *Map of North America with Thermal Fires and Guided Missiles,* reproduced in Douglas Coupland, *Souvenir of Canada.* 2002.

to bail out over Princess Royal Island before the plane turned inland, circled north, and crashed into a mountain in the Kispiox Valley near Smithers. Twelve crew members survived, but five were never found; the last survivor was located hanging upside down in a tree from the harness of his parachute.[5] Despite a massive search, the aircraft was not discovered until 1953, and information about its atomic payload was not made public until 1977, almost twenty-five years later. According to the US military, this nuclear episode was a "Broken Arrow," the code name for a nuclear weapon that has exploded or been lost but without risk of igniting war.[6] It was the first Broken Arrow ever recorded – and perhaps the first dirty bomb as well – making the incident a foundational story in the history of nuclear failure.[7] It is also a foundational story in the history of the Cold War in Canada.[8]

Like all good foundational stories, this particular account of a bomb in the wilderness has symbolic as well as literal dimensions. Douglas Coupland's 2002 book, *Souvenir of Canada,* contains a map of North America designed by NORAD for use in schools, which is pockmarked with miniature thermal fires and guided missiles. [**Figure E.1**] Coupland calls the map a "personal x-ray of my home and native land."[9] A special mark on the map, it seems to me, should be reserved for the bomb detonated over Queen Charlotte Sound. Another mark should be set aside for the mine at Port Radium, which supplied uranium for the first atomic bombs and caused radiation poisoning in the Sahtu Dene people who transported the ore. And further marks should flag Port Hope, where the ore was shipped for processing by Eldorado Mining and Refining and where uranium waste has resulted in a radiogenic town; Chalk River, which sold CANDU nuclear reactors to India and Pakistan that were used in the production of nuclear weapons; and Bruce Power, the largest nuclear facility in the world. With other nuclear organizations in Canada, Bruce Power is supporting the development of a new generation of small modular reactors (SMRs).[10] They require fuel that is more highly enriched than the fuel in CANDU reactors.

Compared to existing reactors, SMRs are designed to produce relatively small amounts of electricity. Some will generate as little as five megawatts, a fraction of the several hundred megawatts produced by the large reactors at Bruce Power. In November 2018, Natural Resources Canada published a "roadmap" for the development of SMRs, which claims there is a market for small nuclear reactors in remote communities in the Canadian North and at mining sites, including the oil sands in Alberta.[11] Aside from the issues of radioactive waste disposal and nuclear risk, policy expert M.V. Ramana writes, the proposal to place SMRs across the North is an economic fantasy: "There are simply not enough remote communities, with adequate purchasing capacity, to be able to drive the manufacture of the thousands of SMRs needed to make them competitive with large reactors, let alone other sources of power."[12] The former head of Westinghouse Electric, Danny Roderick, whose company stopped developing SMRs in 2014, reached a similar conclusion about the viability of the reactors. "The problem I have with SMRs is not the technology, it's not the deployment," Roderick stated, "it's that there are no customers."[13] The only way for the Canadian government to secure SMR customers will be to subsidize them heavily.

Bruce Power is the temporary storage site for almost half of Canada's nuclear fuel waste waiting for permanent storage.[14] Ever since the age of nuclear power began to decline in the 1980s, following the meltdown at Chernobyl, reactors nearing the end of their lives have generally not been replaced. But the problem of managing the radioactive waste produced by the reactors has not gone away; it has become increasingly urgent.[15] "How can this deadly waste be safely contained," asks Robert Jacobs of the Hiroshima Peace Institute, "so that it does not pose a risk to the thousands of generations of human beings for whom it will be an enduring presence?"[16] Western Waste Management Facility operates the temporary radioactive waste storage site at Bruce Power. Forty-eight thousand tonnes of toxic waste were produced by Canada's nuclear reactors from 1950 to 2015, some of which was high-level irradiated waste containing "Iodine-129 (with a half-life of more than 15 million years), Plutonium-239 (with a half-life of 24,000 years, but which decays into Uranium-235 with a half-life of 700 million years) and Neptunium-237 (with a half-life of more than 2 million years)."[17] Invented in 1940, plutonium-239 remains deadly for millennia. By contrast, the temporary concrete and metal containers presently used to store waste are expected to last for no more than a hundred years. The discrepancy between the two time periods is jarring, and it points to the pressing need for a long-term solution. At

| **Figure E.2**: Philip Cheung. *Western Waste Management Facility*. 2015.

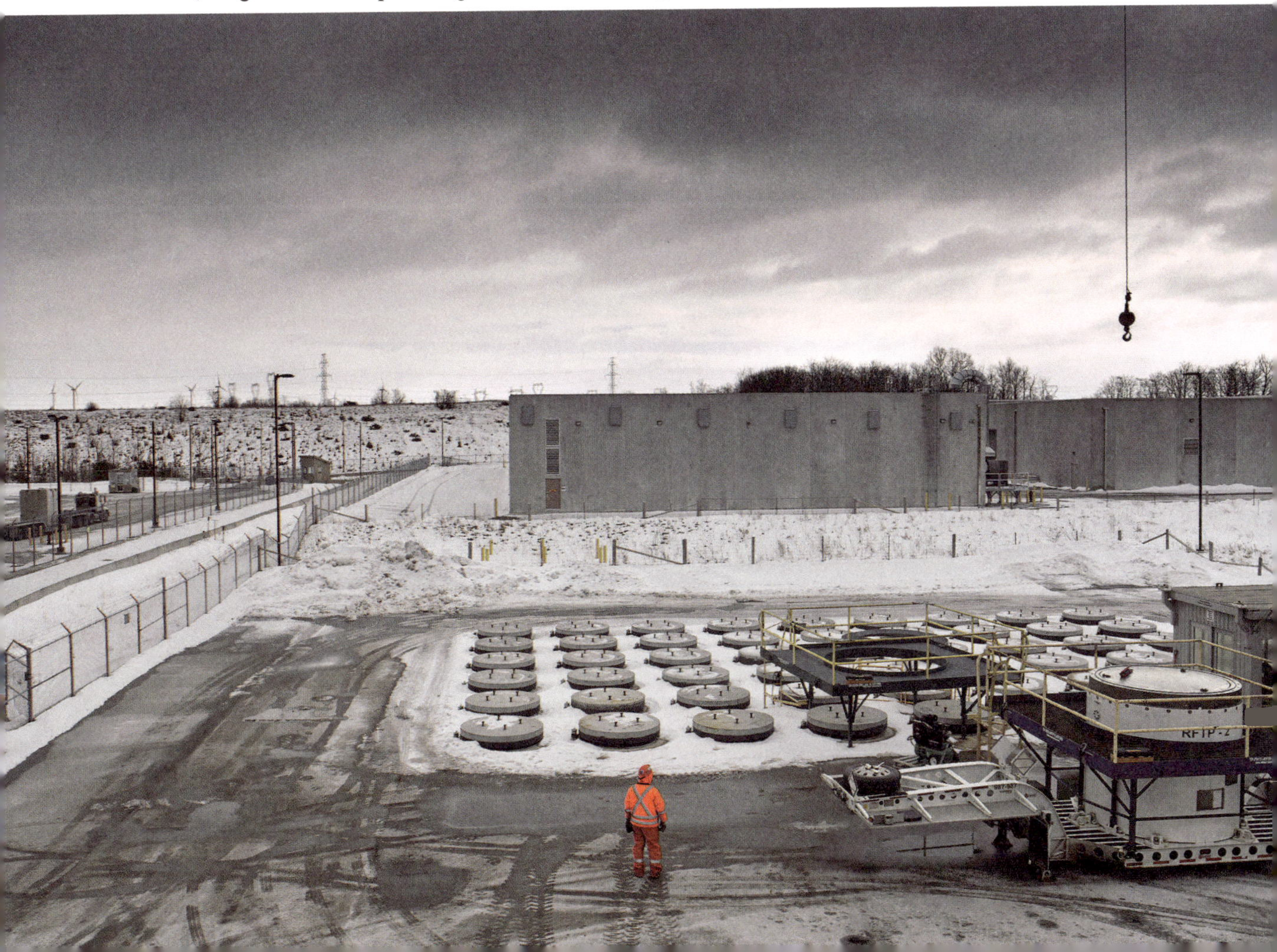

the temporary depot, nuclear waste is lowered into steel-lined containers arranged in a grid; the concrete lids of the receptacles are just visible at ground level. [**Figure E.2**] There is not a lot to see.

Philip Cheung, a freelance photojournalist based in Toronto and Los Angeles, photographed the containers on an overcast winter day in 2015 for an article in *The Globe and Mail Report on Business Magazine.*[18] His representation of the storage site acknowledges that photographs of radioactive storage facilities cannot adequately convey the latent hazards of nuclear waste. The containers are at the centre of his shot, with razor-wire fencing on the left and transportation machinery on the right. A worker dressed in reflective gear stands in the middle foreground, providing an accent of orange against the grey. His back is turned to the camera, his face obscured. The emotional flatness of the picture parallels the apparent ordinariness of the storage site.

For the article, Cheung produced six additional photographs, including one of an organic farmer near the Bruce Power site. All are in the same deadpan style as *Western Waste Management Facility.* Although the farmer is opposed to the continuing production of nuclear waste in the absence of a workable plan for its permanent storage – the waste is "the most dangerous stuff on Earth," he declares – his face is expressionless.[19] Even though a plan exists for the permanent storage of some of Canada's nuclear waste, its workability is still unproven. The scheme calls for the construction of a Deep Geological Repository, a hole bored into rock that is capable of holding steel containers encased in copper, packed with bentonite clay to seal the radioactive waste from leakage, for thousands of years. Two repositories are presently being built in Europe, and more are in the design stage. The repositories under construction are at Forsmark, Sweden, and Onkalo, Finland, and they are being watched closely by other nuclear nations to see if they can overcome problems that occurred at the infamous Waste Isolation Pilot Plant, New Mexico. That facility, which began storing nuclear waste in 1999, has experienced fires, explosions, and structural collapse, not to mention radiation leakage. Canada, meanwhile, is searching for a site that will be "impervious to water penetration, fluctuating ground pressure, earthquakes, climate change, terrorists and human error."[20] That constitutes a tall order. A small number of communities are vying to host it, but the process keeps stalling.

The town of Port Hope will store the radioactive waste generated by its Eldorado factory, the oldest uranium refinery in the world, three kilometres from the plant.[21] The expense of exhuming, trucking, and reinterring the waste is budgeted at $1.28 billion, making it the most costly municipal cleanup in Canadian history. More than a million cubic metres of waste will be buried in a mound seven stories high, located just south of Highway 401, one of the busiest expressways in Canada. At an earlier time, waste from the plant was sent to a dump near the proposed mound. The Welcome Dump has since closed because it was on wetlands that leaked. In 2013, Robert Del Tredici took a photograph of a warning sign at the site. [**Figure E.3**] *Welcome Dump: Keep Out* points to the contradiction between the hospitable name of the

Figure E.3: Robert Del Tredici. *Welcome Dump: Keep Out.* 2013.

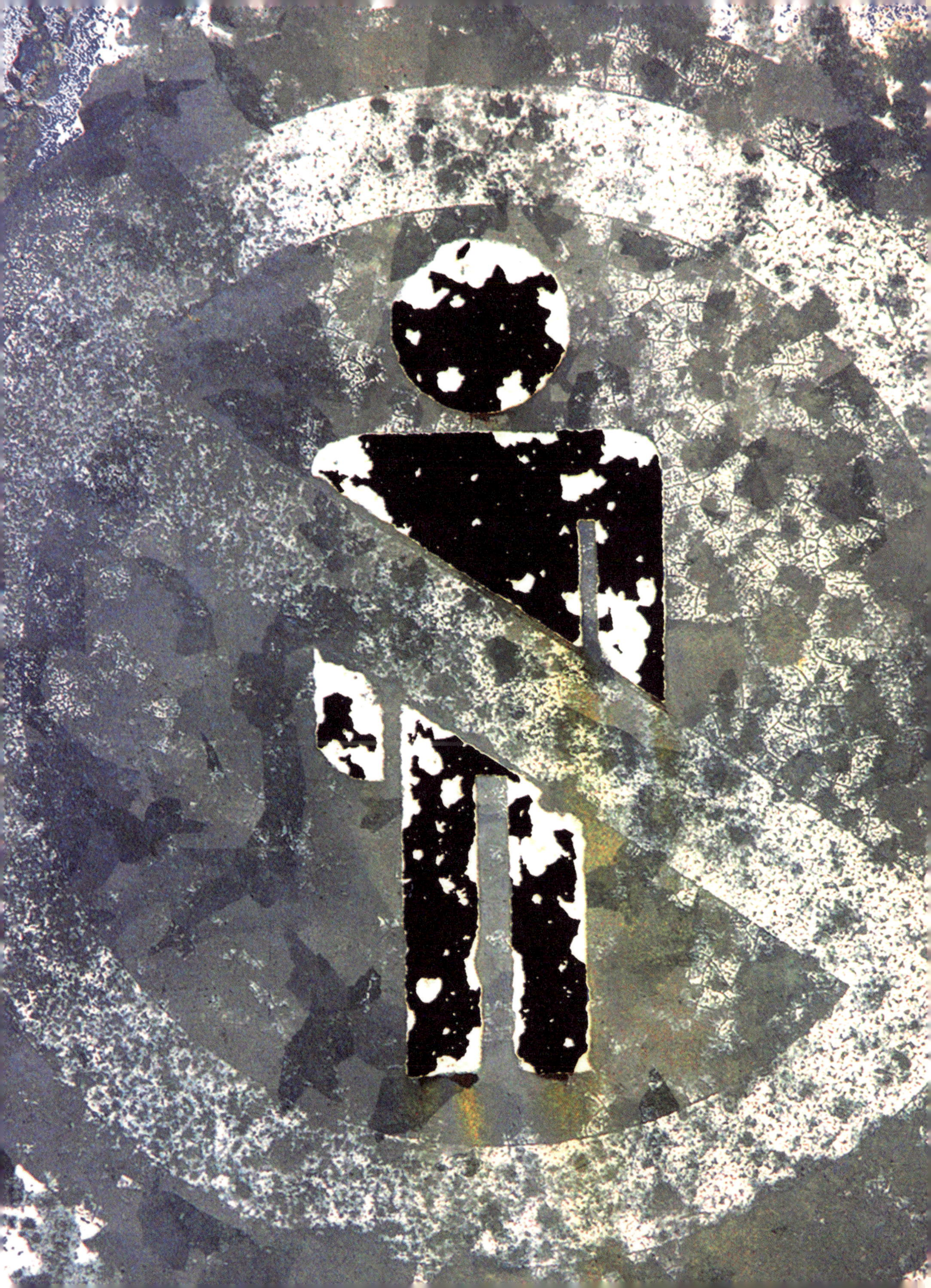

repository and the injunction to stay away. The paint on the sign has peeled, and its metal support has corroded, as if the radioactive contents of the dump might have been responsible for the decay rather than long exposure to weather. The photograph raises this question: How will the radioactive risks posed by the mound, which will contain seventy-five thousand truckloads of waste, be communicated to people now and in the future? How will people hundreds of years from now know to stay clear of the danger?

During my final year of high school in Port Hope, I had a history teacher, A.B. Hodgetts, who told stories and wrote books. The stories had beginnings, middles, and ends – narrative clarity was a virtue. Hodgetts was the author of *Decisive Decades: A History of the Twentieth Century for Canadians,* prepared for high school students in Ontario.[22] I still have my first-edition copy from 1960, which runs to 562 pages. The impetus for Hodgetts to write the book was the ignorance of history displayed by Canadian newspapers during the Suez Crisis in 1956. "Today, as in previous times of crisis," he wrote, "our editors, newscasters and politicians are passionately discussing the tense events of our divided and accusing world without any reference to the long record of the past."[23] In what Hodgetts referred to as a "dwindling world," he hoped to excite a respect for history.

Hodgetts demonstrated what it was like to practise history with passion and rigour. His example had an impact on me. But there are points of difference between us, particularly about how history works and what counts as historical evidence.[24] His past is not my past. Hodgetts did not write history that moved from the local into the world; he wrote history that moved from the world into the local – from the international sphere back into Canada. According to his lights, Canada occupied no significant place in global affairs and even less of a place in nuclear affairs. *Decisive Decades* brushes aside its participation in the Manhattan Project and its role in the development of the atomic bomb, and fails to mention that the Eldorado refinery was located in the author's own backyard. Even though Canada's nuclear programs were developed in three-way collaboration with the United States and Britain during the Second World War, the nuclear story related by *Decisive Decades* unfolds almost entirely under an American flag.

Hodgetts worried that by including too many photographs in *Decisive Decades,* he risked breaking the continuity of history that he was trying to narrate.[25] He feared turning history into spectacle – a justified fear – and was not convinced that photographs could productively collaborate with texts to extend and complicate historical narratives. I have argued otherwise in this investigation; photographs and texts, I have tried to demonstrate, work to complement one another. What is more, history does not evolve in a linear fashion but proceeds erratically. It is made up of discontinuous eruptions in which photography has often played a crucial mediating role.[26] The destruction of Hiroshima and Nagasaki is the primal event of the nuclear era, and it is embedded in collective memory largely because of photographs. Images generated by Little Boy and Fat Man circulate darkly in our minds, just as radioactive fallout generated by the bombs and by nuclear accidents – at Chalk River, Three Mile Island, Chernobyl, and Fukushima Daiichi – circulate in our bodies. Canada was present at the beginning of the nuclear era, and it continues to be present. Its footprint has left a mark.

ACKNOWLEDGMENTS

MY THINKING ABOUT PHOTOGRAPHY AND THE NUCLEAR ERA TOOK SOME unexpected turns while I was writing this book. The discovery of out-of-the-way material required me to abandon some paths of inquiry and pursue others, and ongoing nuclear events caused me to reconsider some of my arguments. The twists and turns slowed my progress on the manuscript. I am grateful to UBC Press for giving me additional time to complete it.

Without the support of universities, particularly in Canada, Japan, and the United States, the book could not have been completed. At the University of British Columbia, the Brenda and David McLean Chair in Canadian Studies gave me the time and resources to investigate questions of nuclear optics in detail; my special thanks to Brenda and David McLean for their support. The McLean chair also provided me with an opportunity to deliver three public lectures and to receive comments back from the audience. I also benefitted from the Orion Lectureship at the University of Victoria, invitations to participate in symposia at the Jarislowsky Institute for Studies in Canadian Art, Concordia University, and the Thakore Award in Human Rights and Peace Studies at Simon Fraser University. In Japan, I was assisted by a visiting research professorship at Ritsumeikan University, Kyoto, as well as by an invitation to lecture at the Hiroshima Peace Institute. At an early stage in my investigations, Harvard University and the University of California, Berkeley, invited me to give lectures and test out ideas.

At the University of British Columbia, I taught seminars on photography and the representation of violence. The give-and-take of the classroom discussions was formative to the development of my thinking. The chapter "Cold War Defence" benefitted from presentations by Karl Fousek and Jeremy Mathers, and the chapter "Nuclear Industry" from a presentation by Blaine Campbell. Other students also contributed to my deliberations, sometimes as research assistants. They include Ashley Belanger, Jesse Birch, Alice Choi, Juan A. Gaitán, Britt Gallpen, Stella Gatto, Ryan Gauvin, Carmella Gray-Cosgrove, Kyoung Yong (Anton) Lee, Klara Manhal, Jeff O'Brien, Kristina Podesva, Daniel Ralston, Robin Simpson, Vanessa Sorensen, Kate Steinmann, Eva Tweedie, Aletheia Wittman, and Sophia Zwiefel. I am grateful to all of them. The artist Andrea Pinheiro generously opened her studio to class visits.

In addition to financial support from universities, I received assistance from the Art Gallery of Ontario, the Social Sciences and Humanities Research Council of Canada, and the Terra Foundation of American Art.

While I made arrangements to look at nuclear photographs and to gain access to archives, my research was facilitated by institutions in half a dozen countries: Albert Einstein Archives, Hebrew University, Jerusalem; Archive of Modern Conflict, London; Arctic Institute of North America, University of Calgary; Art Gallery of Ontario, Toronto; Banff Centre Library and Archives; Canadian Photography Institute, National Gallery of Canada, Ottawa;

Canadian Science and Technology Museum, Ottawa; Canadian War Museum, Ottawa; Center of Photography, Tucson; City of Vancouver Archives; Diefenbunker: Canada's Cold War Museum, Carp, Ontario; Hiroshima Peace Institute; Imperial War Museum, London; International Center of Photography, New York; Kyoto Museum for World Peace; Library and Archives Canada, Ottawa; Library of Congress, Washington; Northwest Territories Archives, Yellowknife; and York University Libraries, Toronto. In addition, I wish to acknowledge the assistance of Shane McCord in locating buried material at Library and Archives Canada; Michael Mao and Jane Young for building a database of nuclear images at the University of British Columbia; Sol Legault for acquiring publication permissions; John Clearwater for helping to locate a wayward image; and James MacNevin, Michelle van der Merwe, Carmen Tiampo, and the rest of the staff at UBC Press for seeing the book through publication.

During many years of conversation focusing on photography and nuclear issues, I particularly recall exchanges with the following artists, colleagues, friends, and family: Grant Arnold, Claudia Beck, Iain Boal, Jeremy Borsos, Paul Boyer, Jessica Bradley, Nick Broderick, Julia Bryan-Wilson, Grant Buday, Ele Carpenter, Richard Cavell, Douglas Coupland, Frank Curzon, David Donaldson, Gordon Edwards, Olivia Fermi, Blake Fitzpatrick, Adam Frank, Lindsey A. Freeman, Carole Gallagher, Daniel Grausam, Bruce Grenville, Andrew Gruft, Serge Guilbaut, Sophie Hackett, James E. Hansen, Ihor Holubizky, Kristan Horton, Robert A. Jacobs, Geoffrey James, Mary Kavanagh, Robert Keziere, Martha Langford, Tim Lee, Akira Mizuta Lippit, Sarah Milroy, Ishiuchi Miyako, Rens van Munster, Ira Nadel, Helen O'Brian, Melanie O'Brian, Peter O'Brian, Helga Pakasaar, David Pantalony, Carol Payne, Giuseppe Penone, Kimberly Phillips, Andrea Pinheiro, M.V. Ramana, Peter Robinson, Mark Ruwedel, Julie Salverson, Eric Sandeen, Elaine Scarry, Susan Schuppli, Joan Schwartz, Geoffrey Sea, the late Allan Sekula, Jim Shedden, Reid Sheir, Trevor Smith, Matthew Speier, Don Stewart, Fred Stockholder, Maia-Mari Sutnik, Bryan C. Taylor, Matthew Teitelbaum, Ann Thomas, Robert Del Tredici, Cheyanne Turions, Eyal Weizman, Peter White, Peter C. van Wyck, and Catherine de Zegher.

None of what follows has appeared in the precise form produced here, but earlier versions of some sections have been published elsewhere. Chapter 4 is adapted from "Nuclear Flowers of Hell," in *Camera Atomica,* a 2015 exhibition catalogue I edited. The final part of Chapter 5 appeared in "Postcard to Moscow," in *Postcards: Ephemeral Histories of Modernity,* edited by Jordana Mendelson and David Proschaska. Chapter 7, with some excisions and additions, appeared as "Amerika Dropt a Bomb on Nevada," in *Open Letter: A Canadian Journal of Writing and Theory,* edited by Ian Rae. And the opening pages of the Epilogue are adapted from "On Photographing a Dirty Bomb," in *The Cultural Work of Photography in Canada,* edited by Carol Payne and Andrea Kunard.

Two anonymous external readers offered perceptive comments on the book, and I am indebted to their recommendations. I am also indebted to the recommendations of three additional readers of the manuscript: Jaclyn Arndt, David Boal, and Robert Del Tredici. Claudette Lauzon, coeditor of *Through Post-Atomic Eyes,* has been an ideal collaborator.

ATOMIC TIMELINE

1942

- United States, Canada, and Britain begin collaborating on the Manhattan Project to build nuclear weapons.
- Eldorado reopens its mine at Port Radium, Northwest Territories, to supply uranium refined at its plant in Port Hope, Ontario, to the Manhattan Project.

1945

- First atomic bomb is tested at Alamogordo, New Mexico (July 16).
- United States uses nuclear weapons against Hiroshima and Nagasaki (August 6 and 9).
- First operational nuclear reactor opened outside the United States at Chalk River, Ontario.
- Soviet atomic spy ring revealed in Canada by Igor Gouzenko, the existence of which is made public the following year.

1946

- Through the United Nations, Canada joins an international effort to control nuclear weapons.

1949

- Soviet Union conducts its first nuclear tests.

1950

- US Air Force bomber detonates a nuclear device off the coast of British Columbia before crashing; the accident is the first Broken Arrow ever recorded.

1952

- United States tests a hydrogen bomb in the Pacific.
- NRX reactor at Chalk River experiences a partial meltdown, the first major nuclear accident in the world.

1954

- United States launches the first nuclear-powered submarine.
- The *Daigo Fukuryū Maru,* a Japanese fishing boat, is contaminated by fallout from a US bomb test at Bikini Atoll, Marshall Islands.

1956

- World's first commercial nuclear power plant opens at Calder Hall, England.
- First reactor exported by Canada to India; the reactor starts operations in 1960.

1957

- Soviet Union launches Sputnik, the first artificial satellite.
- United Nations establishes the International Atomic Energy Agency, a nuclear watchdog.
- First Pugwash Conference, Nova Scotia, brings scientists together to assess the dangers of nuclear weapons.

1958

- Campaign for Nuclear Disarmament is founded in Britain.
- Diefenbaker government agrees to build Bomarc-B missile bases in Canada.

1959

- Diefenbaker government cancels the Avro Arrow, a Canadian-designed and -built interceptor aircraft.

1960

- France conducts its first nuclear test.
- Voice of Women for Peace (VOW) is founded in Canada.
- Canadians protest against Bomarc-B missile bases.

1962

- Cuban Missile Crisis brings the world to the brink of nuclear war.

1963

- United States and Soviet Union sign Partial Test Ban Treaty.
- Bomarc-B missiles delivered to Canadian sites.

1964

- China explodes its first atomic bomb.

1966

- Construction begins on the first of twenty-three commercial CANDU reactors built in Ontario, Quebec, and New Brunswick.

1968

- Treaty signed on the non-proliferation of nuclear weapons.

1971

- Formation of Greenpeace in Vancouver to protest nuclear testing at Amchitka, Alaska.
- Bomarc-B missiles are withdrawn from Canadian sites.

1972

- Richard Nixon and Leonid Brezhnev sign the Strategic Arms Limitation Treaty Accord and Anti-Ballistic Missile Treaty.

1974

- India conducts its first nuclear test.

1979

- Meltdown of Three Mile Island reactor, Pennsylvania.

1981

- Largest-ever anti-nuclear protests occur in North America, Europe, and Japan.

1983

- Ronald Reagan outlines plans for new Star Wars anti-missile defence technology.

1986

- Meltdown at Chernobyl Nuclear Power Plant, Ukraine.
- Ronald Reagan and Mikhail Gorbachev discuss elimination of nuclear weapons.

1989

- Berlin Wall falls, symbolizing the end of the Cold War.

1993

- George H.W. Bush and Boris Yeltsin sign the Strategic Arms Reduction Treaty.

1994

- South Africa completes disarmament of nuclear weapons.

1998

- India and Pakistan conduct nuclear tests within three weeks of each other.

2007

- North Korea conducts a nuclear test.

2009

- Barack Obama calls for the abolition of nuclear weapons.

2010

- Strategic Arms Reduction Treaty signed by the United States and Russia to restrict the number of nuclear warheads.

2011

- Earthquake causes triple meltdown at Fukushima Daiichi Nuclear Power Plant, Japan.

2012

- Bruce Power, Ontario, becomes the largest operating nuclear-generating facility in the world.

2013

- Iran reaches a deal on its nuclear program with six world powers.

2017

- North Korea launches intercontinental ballistic missile capable of reaching North America.

2018

- United States withdraws from the Iran nuclear deal.
- United States terminates treaty with Russia on intermediate-range missiles.

2019

- Kashmir dispute between India and Pakistan heightens risk of nuclear exchange.

NOTES

Preface: Radioactive Footprints

1 Scott Toguri McFarlane, "Un-Ravelling *StoneDGloves* and the Haunt of the Hibakusha," in *All Amazed: For Roy Kiyooka,* ed. John O'Brian, Naomi Sawada, and Scott Watson (Vancouver: Arsenal Pulp Press, Morris and Helen Belkin Art Gallery, Collapse, 2002), 124. Kiyooka refers to Hiroshima in some of his poetry; see Roy Miki, ed., *Pacific Windows: Collected Poems by Roy K. Kiyooka* (Vancouver: Talonbooks, 1997).

2 F.F. Langan, "Peter O'Brian: One of the Last Battle of Britain Pilots," *Globe and Mail,* May 15, 2007; "Group Captain Peter O'Brian," *Daily Telegraph,* May 19, 2007.

3 Kate Harries, "Nuclear Reaction: Accusations of Cancerous Fallout Divide a Small Ontario Town," *Walrus,* March 2008, 40.

4 Philip Pullman, *The Golden Compass* (New York: Random House, 1995).

5 Polonium, radon, and radium are among the radionuclides identified in Port Hope. Gordon Edwards, "Uranium in Quebec: Truth and Consequences," Canadian Coalition for Nuclear Responsibility, February 22, 2016, http://ccnr.org/CCNR_CNSC_BAPE_2016.pdf.

6 Blake Fitzpatrick and Robert Del Tredici, "Port Hope in the Era of Nuclear Waste," in *Through Post-Atomic Eyes,* ed. Claudette Lauzon and John O'Brian (Montreal and Kingston: McGill-Queen's University Press, 2020).

7 Rob Nixon, "Slow Violence, Gender and the Environmentalism of the Poor," *Journal of Commonwealth and Postcolonial Studies* 13, 2 (2006): 14.

8 For a discussion of both Chris Marker's film and Paul Virilio's bunkers, see David Alan Mellor, "On Omaha Beach," in *Conflict*Time*Photography,* ed. Simon Baker and Shoair Mavlian (London: Tate Modern, 2015), 220–21.

9 Several bodies of work by Mary Kelly, such as *Habitus* (2010) and *MIMUS* (2012), draw on her experiences of the Second World War and the Cold War.

10 Hannah Arendt, *On Violence* (New York: Harcourt Brace, 1969), 17.

11 In *The Nuclear Borderlands: The Manhattan Project in Post–Cold War New Mexico* (Princeton: Princeton University Press, 2006), Joseph Masco demonstrates the long-term social consequences of the bomb on American lives.

12 Lauzon and O'Brian, *Through Post-Atomic Eyes.*

13 The contributing artists are Edward Burtynsky, Blaine Campbell, Carole Condé and Karl Beveridge, Robert Del Tredici, Blake Fitzpatrick, Kristan Horton, Mary Kavanagh, Katy McCormick, David McMillan, Andrea Pinheiro, Public Studio, Mark Ruwedel, Susan Schuppli, Erin Siddall, Charles Stankievech, and Donald Weber. The contributing writers are Karen Barad, James Bridle, Eric Cazdyn, Matthew Farish, Lindsey A. Freeman, Derek

Gregory, Kyo Maclear, Joseph Masco, Karla McManus, Julie Salverson and Peter C. van Wyck, Charles Stankievech, and Eyal Weizman.

14 In an article on WikiLeaks and the US National Security Agency, the artist Trevor Paglen makes a similar argument. See Trevor Paglen, "Turnkey Tyranny: Surveillance and the Terror State," *Creative Time Reports,* June 25, 2013, http://creativetimereports.org/2013/06/25/surveillance-and-the-construction-of-a-terror-state.

Introduction: Still Photographs

1 Susan Sontag's *On Photography* (New York: Doubleday, 1977) and her *Regarding the Pain of Others* (New York: Farrar, Straus and Giroux, 2003) address the relationship of photography to historical events.

2 In *The Nuclear Century: Voices of the Hibakusha of the World* (Tokyo: Japan Peace Museum, 1997), the story of the hibakusha is told in a combination of photographs and texts.

3 Some uranium for the bombs was also mined in the Belgian Congo and the United States.

4 In addition to his work in photography and film, Richard Finnie had a career as a writer and lecturer. Peter Geller, *Northern Exposures: Photographing and Filming the Canadian North, 1920–1945* (Vancouver: UBC Press, 2004), 135.

5 Robert Frank, quoted in Sarah Greenough and Philip Brookman, eds., *Robert Frank: Moving Out* (Washington, DC: National Gallery of Art, 1994), 54.

6 Sally Stein, "The Rhetoric of the Colorful and the Colorless: American Photography and Material Culture between the Wars" (PhD diss., Yale University, 1991).

7 Laura Pitkanen and Matthew Farish argue that the United States is heavily overrepresented in the arena of nuclear studies because of its long-standing nuclear weapons program and "the relative accessibility of research sites and data." Laura Pitkanen and Matthew Farish, "Nuclear Landscapes," *Progress in Human Geography* (August 31, 2017): 2, https://doi.org/10.1177/0309132517725808.

8 A search of peer-reviewed sources published between the beginning of 2008 and the end of February 2016 resulted in 109,150 references.

9 Ele Carpenter, ed., *The Nuclear Culture Source Book* (London: Black Dog, 2016); Gabrielle Decamous, *Invisible Colors: The Arts of the Atomic Age* (Cambridge, MA: MIT Press, 2018).

10 Reg Whitaker and Steve Hewitt, *Canada and the Cold War* (Toronto: James Lorimer, 2003); Peter C. van Wyck, *The Highway of the Atom* (Montreal and Kingston: McGill-Queen's University Press, 2010). Books offering a critical Canadian perspective on the nuclear era began to appear in the 1980s. They have continued to appear, though in smaller numbers than the subject warrants. The following works were also valuable in shaping this study: Rosalie Bertell, *No Immediate Danger? Prognosis for a Radioactive Earth* (Toronto: Women's Educational Press, 1985); Lev Bratishenko and Mike Zardini, eds., *It's All Happening So Fast: A Counter-History of the Modern Canadian Environment* (Montreal: Canadian Centre for Architecture, 2016); Robert Del Tredici, *At Work in the Fields of the Bomb*

(Vancouver: Douglas and McIntyre, 1987); Matthew Farish, *The Contours of America's Cold War* (Minneapolis: University of Minnesota Press, 2010); Blake Fitzpatrick and Robert Del Tredici, *The Atomic Photographers Guild: Visibility and Invisibility in the Nuclear Era* (Toronto: Gallery TPW, 2001); Martha Langford and John Langford, *A Cold War Tourist and His Camera* (Montreal and Kingston: McGill-Queen's University Press, 2011); Kyo Maclear, *Beclouded Visions: Hiroshima-Nagasaki and the Art of Witness* (Albany, NY: SUNY Press, 1999); Ian McKay and Jamie Swift, *Warrior Nation: Rebranding Canada in an Age of Anxiety* (Toronto: Between the Lines, 2012); Joyce Nelson, *The Perfect Machine: TV in the Nuclear Age* (Toronto: Between the Lines, 1987); Ernie Regehr and Simon Rosenblum, eds., *Canada and the Nuclear Arms Race* (Toronto: James Lorimer, 1983); Julie Salverson, *Lines of Flight: An Atomic Memoir* (Hamilton, ON: Wolsak and Wynn, 2016); Robert Teigrob, *Warming Up the Cold War: Canada and the United States' Coalition of the Willing, from Hiroshima to Korea* (Toronto: University of Toronto Press, 2009); and Alexander Wilson, *The Culture of Nature: North American Landscape from Disney to the Exxon Valdez* (Toronto: Between the Lines, 1991).

11 Robert Jacobs, "Born Violent: The Origins of Nuclear Power," *Asian Journal of Peacebuilding* 7, 1 (2019): 9–29.

12 Kim Krenz, *Deep Waters: The Ottawa River and Canada's Nuclear Adventure* (Montreal and Kingston: McGill-Queen's University Press, 2004); Chris Saunders, *Whiteshell Laboratories: A Legacy to Nuclear Science and Engineering in Canada* (Pinawa, MB: Pinawa Foundation, 2016), http://online.fliphtml5.com/xzbq/krra.

13 Canadian Press, "Bruce Power Nuclear Unit to Restart after 17 Years," CBC News, March 16, 2012, https://www.cbc.ca/news/canada/toronto/bruce-power-nuclear-unit-to-restart-after-17-years-1.1133921.

14 Van Wyck, *The Highway of the Atom,* 4.

15 Jacques Derrida, "Demeure: Fiction and Testimony," in *Maurice Blanchot, the Instant of My Death* (Stanford: Stanford University Press), 29-30, quoted in van Wyck, *The Highway of the Atom,* 4.

16 Salverson, *Lines of Flight.*

17 Sontag, *On Photography.*

18 Richard Lloyd Parry, "The School beneath the Raging Wave," *Guardian Weekly,* September 29, 2017, discusses the widespread fear of radioactive contamination among the survivors of the Fukushima Daiichi nuclear disaster of March 11, 2011.

19 Judith Butler, "Photography, War, Outrage," *PMLA* 120, 3 (May 2005): 824. See also her *Precarious Life: Powers of Mourning and Violence* (New York: Verso, 2004). On this matter, Eduardo Cadava, *Words of Light: Theses on the Photography of History* (Princeton: Princeton University Press, 1997), concurs with Butler.

20 W.G. Sebald, *On the Natural History of Destruction* (London: Hamish Hamilton, 2003).

21 Christian Scholz, "'But the Written Word Is Not a True Document': A Conversation with W.G. Sebald on Literature and Photography," in *Searching for Sebald: Photography after W.G.*

Sebald, ed. Lise Pratt (Los Angeles: Institute of Critical Inquiry, 2007), 104. The interview was conducted on November 14, 1997.

22 Scholz, "A Conversation with W.G. Sebald," 104. Sebald's approach to photography was idiosyncratic. He claimed to have paid little attention to the aesthetic quality of the prints he found.

23 Scholz, "A Conversation with W.G. Sebald," 104.

24 Geoffrey Batchen is instructive on photography, remembrance, repetition, and modernity. See his *Each Wild Idea: Writing, Photography, History* (Cambridge, MA: MIT Press, 2001) and *Forget Me Not: Photography and Remembrance* (New York: Princeton Architectural Press, 2004).

25 Roland Barthes, *Camera Lucida: Reflections on Photography* (New York: Hill and Wang, 1981).

26 The information is from Peter Beatty O'Brian, author of *The O'Brian Story* (Toronto: privately printed, 2019).

27 Rey Chow, *The Age of the World Target: Self-Referentiality in War, Theory and Comparative Work* (Durham, NC: Duke University Press, 2006).

28 Jacques Derrida, "No Apocalypse, Not Now (Full Speed Ahead, Seven Missiles, Seven Missives)," *Diacritics* 14, 2 (Summer 1984): 24. Derrida first discusses the pharmakon in "Plato's Pharmacy," in his *Dissemination* (Chicago: University of Chicago Press, 1981), 70.

29 "Science Tackles Radiation Peril," *Life,* March 21, 1955, 32–39.

30 Howard Rosenberg, *Atomic Soldiers: American Victims of Nuclear Experiments* (Boston: Beacon, 1980); Steve B. Davis, "Nuclear Guinea Pigs: Canadian and American Troops at Ground Zero," Writings and Ramblings (blog), November 13, 2007, https://stamperdad.wordpress.com/2007/11/13/nuclear-guinea-pigs-canadian-and-american-troops-at-ground-zero/; Eileen Welsome, *The Plutonium Files: America's Secret Medical Experiments in the Cold War* (New York: Dial, 1999), 262–63.

31 "Science Tackles Radiation Peril," 37.

32 During the Cold War, the rhetoric of nuclear politics focused heavily on deterrence, the military doctrine of mutual assured destruction, which argued that full-scale use of nuclear weapons would bring about the complete annihilation of both attackers and defenders.

33 Derrida, "No Apocalypse, Not Now," 24.

34 The archives are listed in the bibliography. In addition, I assembled a personal atomic archive of military photographs, press images, snapshots, nuclear postcards, protest pamphlets, record albums, exhibition catalogues, civil defence pamphlets, corporate brochures, and other items over several years.

35 Walter Benjamin, "On the Concept of History," in *Walter Benjamin: Selected Writings, 1938–1940,* ed. Michael W. Jennings (Cambridge, MA: Harvard University Press, 2003), 4:391.

36 Walter Benjamin, quoted in Eduardo Cadava, "'Lapsus Imaginus': The Image in Ruins," *October* 96 (2001): 42.

37 Maurice Halbwachs, *The Collective Memory* (New York: Colophon, 1980). Halbwachs first introduced his ideas about collective memory during the 1920s.

38 See Jacques Derrida, *The Gift of Death* (Chicago: University of Chicago Press, 1995); Paul Virilio, *War and Cinema: The Logistics of Perception,* trans. Patrick Camiller (London: Verso, 1989); and Rosalyn Deutsche, *Hiroshima after Iraq: Three Studies in Art and War* (New York: Columbia University Press, 2010). In Alain Resnais's 1959 film *Hiroshima mon amour,* a French woman says to her Japanese lover, "I saw everything. Everything," to which he replies, "You saw nothing. Nothing." The notion of "deferred sight" in photography is discussed in Jayne Wilkinson, "Art Documents: The Politics of the Visible in the Work of Taryn Simon and Trevor Paglen" (master's thesis, University of British Columbia, 2014).

39 Akira Mizuta Lippit, *Atomic Light (Shadow Optics)* (Minneapolis: University of Minnesota Press, 2005).

40 *Rutherford Museum of McGill University* (Montreal: McGill University, 1996), 8–9, which has a diagram of an apparatus with a photographic plate, explains how it works. A strip of film used by Rutherford is on display in the museum. See also J.L. Heilbron, *Ernest Rutherford and the Explosion of Atoms* (New York: Oxford University Press, 2003).

41 Peter C. van Wyck, "Highway of the Atom: Recollections along a Route," *Topia* 7 (Spring 2002): 101.

42 J.H. Webb, "The Fogging of Photographic Film by Radioactive Contaminants in Cardboard Packaging Materials," *Physical Review* 76, 3 (August 1, 1949): 375.

43 Peter Kuran, *How to Photograph an Atomic Bomb* (Santa Clarita, CA: VCE, 2006), 7.

44 Rebecca Solnit, *River of Shadows: Eadweard Muybridge and the Technological Wild West* (New York: Viking, 2003), 12.

45 Julia Bryan-Wilson, "Posing by the Cloud: US Nuclear Test Site Photography in Process," in *Camera Atomica,* ed. John O'Brian (London: Black Dog; Toronto: Art Gallery of Ontario, 2015), 119.

46 I am grateful to Blake Fitzpatrick for providing me with an image of the page.

47 See Kaja Silverman, *The Threshold of the Visible World* (New York: Routledge, 1996).

48 Robert A. Jacobs, "The Visible and the Invisible When Considering Northern European Permanent Spent Storage Fuel: Forsmark and Onkalo," *Hiroshima Peace Research Journal* 5 (2019): 15.

49 See Colin S. Gray, *The Second Nuclear Age* (Boulder, CO: Lynne Reinner, 1999); and Bill Keller, "The Second Nuclear Age," *New York Times Magazine,* May 4, 2003, 48–53, 84, 94, 101–2.

50 Canada's engagement is discussed in Richard Cavell, ed., *Love, Hate and Fear in Canada's Cold War* (Toronto: University of Toronto Press, 2004); John M. Clearwater, *Canadian Nuclear Weapons: The Untold Story of Canada's Cold War Arsenal* (Toronto: Dundurn, 1998); Wilfrid Eggleston, *Canada's Nuclear Story* (Toronto: Clarke, Irwin, 1965); James G. Fergusson, *Canada and Ballistic Missile Defence, 1954–2009: Déjà Vu All Over Again* (Vancouver: UBC Press, 2010); Ron Finch, *Exporting Danger: A History of the Canadian Nuclear Energy Export Programme* (Montreal: Black Rose, 1986); Jim Harding, *Canada's Deadly Secret: Saskatchewan Uranium and the Global Nuclear System* (Halifax: Fernwood, 2007); D.M.

LeBourdais, *Canada and the Atomic Revolution* (Toronto: McClelland and Stewart, 1959); Patricia I. McMahon, *Essence of Indecision: Diefenbaker's Nuclear Policy, 1957–1963* (Montreal and Kingston: McGill-Queen's University Press, 2009); and Andrew Richter, *Avoiding Armageddon: Canadian Military Strategy and Nuclear Weapons, 1950–63* (Vancouver: UBC Press, 2002). An important unpublished PhD dissertation is Blake Fitzpatrick, "Making Visible: Photography, Witnessing and the Nuclear Era" (PhD diss., University of Toronto, 2002).

51 Barack Obama, quoted in Robert Fife and Paul Koring, "Obama Details Nuclear Terrorism's Threat," *Globe and Mail,* April 2, 2016.

52 Michael Byers, *Intent for a Nation: What Is Canada For?* (Vancouver: Douglas and McIntyre, 2007), 79–80.

53 John Boyko, *Cold Fire: Kennedy's Northern Front* (Toronto: Alfred A. Knopf, 2016), 235.

54 C.P. Stacey, *Canada and the Age of Conflict: A History of Canadian External Policies* (Toronto: University of Toronto Press, 1981), 2:412.

55 Stacey, *Canada and the Age of Conflict,* 2:412.

56 Commissioned nuclear histories also tend to glide over inconvenient historical evidence. Robert Bothwell has written two of them: *Eldorado: Canada's Nuclear Uranium Company* (Toronto: University of Toronto Press, 1984) and *Nucleus: The History of Atomic Energy of Canada Limited* (Toronto: University of Toronto Press, 1988). *Eldorado* contains a foreword by N.M. Ediger, CEO of Eldorado Nuclear Limited. Krenz's *Deep Waters,* a memoir written by a former employee of Atomic Energy of Canada, exhibits characteristics similar to those of a commissioned history.

57 Kenneth C. Cragg, "Canada Building Plant Near Petawawa Camp to Develop Atom Bomb," *Globe and Mail,* August 7, 1945.

58 Ron Verzuh, "Canada's A-Bomb Secret," Canada's History, July 14, 2015, http://www.canadashistory.ca/explore/science-technology/canada-s-a-bomb-secret.

59 Shane J. Maddock, "Defending the American Way and Containing the Atom: Ideology and U.S. Nuclear Nonproliferation Policy since 1945," in *The Atomic Bomb and American Society: New Perspectives,* ed. Rosemary B. Mariner and G. Kurt Piehler (Knoxville: University of Tennessee Press, 2009), 125.

60 Avner Cohen, *Israel and the Bomb* (New York: Columbia University Press, 1998), 91, 373n58. Ben-Gurion's statement was published in the *Jerusalem Post,* December 22, 1960.

61 Group email sent by Gordon Edwards, subject line "Israel's Nuclear Bomb Factory at Dimona, Thanks to Vanunu," October 10, 2012.

62 In their revised edition of *Canada since 1945,* published in 1989 (and dedicated to C.P. Stacey), Robert Bothwell, Ian Drummond, and John English are more attentive to Canada's role in nuclear affairs. They acknowledge that the country's cooperation with British scientists during the war, in combination with its reserves of uranium, made Canada a global atomic power. See also John Herd Thompson and Stephen J. Randall, *Canada and the United States: Ambivalent*

Allies, 3rd ed. (Montreal and Kingston: McGill-Queen's University Press, 2002), which is also attentive to Canada's nuclear capabilities.

63 Robert Bothwell and J.L. Granatstein, *The Gouzenko Transcripts: The Evidence Presented at the Kellock-Taschereau Royal Commission of 1946* (Ottawa: Deneau, 1982).

64 See Gordon Lunan, *The Making of a Spy: A Political Odyssey* (Montreal: Robert Davies, 1995); and John Sawatsky, *Gouzenko: The Untold Story* (Toronto: Macmillan, 1984).

65 *The Gouzenko Affair: Canada and the Beginnings of Cold War Counter-Espionage,* Library and Archives Canada conference proceedings (Manotick, ON: Penumbra, 2006). One of the foremost historians of the period, Odd Arne Westad, does not address the claim in his book *The Cold War: A World History* (London: Allen Lane, 2017).

66 Igor Gouzenko, *This Was My Choice: Gouzenko's Story* (Toronto: F.M. Dent, 1948); Igor Gouzenko, *The Fall of a Titan* (New York: Norton, 1954).

67 The journalist Drew Pearson conducted the interview in East York, just outside (and now part of) Toronto, on a Sunday morning.

68 The German artist Jürgen Klauke appropriated ninety-six newspaper photographs of masked and hooded figures for an installation called *Antlitze* (Faces, 1972–2000). See Dörte Zbikowski, "Faces," in *CTRL [SPACE]: Rhetorics of Surveillance from Bentham to Big Brother,* ed. Thomas Y. Levin, Ursula Frohne, and Peter Weibel (Cambridge, MA: MIT Press, 2002), 194–99. Gouzenko is not included.

69 W. Eugene Smith, "Photographic Journalism," *Photo Notes,* June 1948, 4.

70 Barthes, *Camera Lucida,* 4.

71 Donald Creighton, *The Story of Canada,* rev. ed. (Toronto: Macmillan, 1971), 263.

72 See van Wyck, *The Highway of the Atom;* and Peter Blow, dir., *Village of Widows* (Peterborough, ON: Lindum Films, 1998), 52 minutes.

73 Paul Baton, quoted in Andrew Nikiforuk, "Echoes of the Atomic Age: Cancer Kills Fourteen Aboriginal Uranium Workers," *Calgary Herald,* March 14, 1998.

74 Cindy Gilday, quoted in Nikiforuk, "Echoes of the Atomic Age."

75 Van Wyck, *The Highway of the Atom,* 38.

76 Van Wyck, "Highway of the Atom"; Nikiforuk, "Echoes of the Atomic Age."

77 Creighton's suggestion that the Precambrian Shield was empty and uninhabited before the Group of Seven went there to paint after the First World War repeats an ongoing colonialist misperception. The misperception is addressed in essays by Jonathan Bordo, Marcia Crosby, Paul Walton, Scott Watson, and other contributors to John O'Brian and Peter White, eds., *Beyond Wilderness: The Group of Seven, Canadian Identity, and Contemporary Art* (Montreal and Kingston: McGill-Queen's University Press, 2007).

78 Ian McKay, "The Liberal Order Framework: A Prospectus for a Reconnaissance of Canadian History," *Canadian Historical Review* 81, 3 (December 2000): 617–45. Ideas in McKay's article figure prominently in Lynda Jessup, Erin Morton, and Kirsty Robertson, eds., *Negotiations in a Vacant Lot: Studying the Visual in Canada* (Montreal and Kingston: McGill-Queen's University Press, 2014).

79 McKay and Swift, *Warrior Nation,* 68.

80 McKay and Swift, *Warrior Nation,* 96.

81 See Thompson and Randall, *Canada and the United States,* 167–68.

82 Government of Canada, Department of Energy Mines and Resources, Nuclear Policy Background Papers, 1981. See also Gordon Edwards, "Fueling the Nuclear Arms Race and How to Stop It," *Ploughshares Monitor* 6, 2 (June 1985), http://www.ccnr.org/non_prolif.html, which discusses what the numbers in the background papers mean. Some Canadian uranium was sold for so-called peaceful purposes. In 1959, for example, Pakistan purchased a CANDU reactor from Canada for peaceful purposes that ultimately helped to advance its nuclear weapons program; by the late 1960s, Prime Minister Zulfikar Ali Bhutto was declaring Pakistan's intention to develop its own nuclear bomb.

83 There is wide-ranging literature on the CANDU reactor and Canada's engagement in nuclear activities. See, for example, Sahadev Basdeo, "Canada's Response to India's Nuclear Detonations in 1974 and 1998: Diversity and Challenge in International Relations," *Indian Journal of Canadian Studies* 13 (2005): 215–38; and Whitaker and Hewitt, *Canada and the Cold War.*

84 David Albright and Andrea Stricker, *Taiwan's Former Nuclear Weapons Program: Nuclear Weapons On-Demand* (Washington, DC: Institute for Science and International Security, 2018).

85 Gordon Edwards, "Canada's Nuclear Industry and the Myth of the Peaceful Atom," in *Canada and the Nuclear Arms Race,* ed. Ernie Regehr and Simon Rosenblum (Toronto: James Lorimer, 1983), 122.

86 Marshall McLuhan, *Understanding Media: The Extensions of Man* (New York: McGraw-Hill, 1964), 339.

Chapter 1: Nuclear Industry

1 Bernard Augustin, "The Sinking of the *Mont-Louis* and Nuclear Safety," *IAEA Bulletin,* Spring 1985, 28.

2 "Atom Wreck Peril," Newspaper Division, Syndication International, London, August 1984.

3 Yves Lenoir, quoted in "Atom Wreck Peril."

4 The *Mont-Blanc* was a French cargo ship, whereas the *Imo* was a Norwegian vessel.

5 Gertrude Pettipas, quoted in Rebecca Solnit, *A Paradise Built in Hell: The Extraordinary Communities That Arise in Disaster* (New York: Penguin, 2009), 73.

6 The term "megadeaths" was coined by the nuclear strategist Herman Kahn in the late 1950s, when he was working at the RAND Corporation, to denote 1 million deaths.

7 Hatsuyo Nakamura, quoted in John Hersey, *Hiroshima* (New York: Alfred A. Knopf, 1946), 12–13.

8 C.R. Blackburn, "Atomic Bomb Rocks Japan," *Globe and Mail,* August 7, 1945.

9 Robert Sanders, "Was First Nuclear Test the Start of New Human-Dominated Epoch, the Anthropocene?" *Berkeley News,* January 16, 2015, http://news.berkeley.edu/2015/01/16/was-first-nuclear-test-dawn-of-new-human-dominated-epoch-the-anthropocene.

10 Peter C. van Wyck, "The Anthropocene's Signature," in Carpenter, *The Nuclear Culture Source Book,* 30.

11 Lucretius, quoted in Anne Michaels, *Fugitive Pieces* (Toronto: McClelland and Stewart, 1996), 21.

12 Richard Rhodes, *Dark Sun: The Making of the Hydrogen Bomb* (New York: Simon and Schuster, 1995), 542.

13 US officials, quoted by the photojournalist Hiromitsu Toyosaki, "Hidden and Forgotten *Hibakusha:* Nuclear Legacy," in O'Brian, *Camera Atomica,* 154.

14 Toyosaki, "Hidden and Forgotten *Hibakusha,*" 151-62.

15 "Bruce Power Tops at US SWAT Event," *Saugeen Times,* May 2008.

16 "Site Security," Canadian Nuclear Association, 2015, https://cna.ca/issues-policy/safety-security/site-security.

17 "Bruce Power Tops."

18 "No Immediate Threat (1986)," C + B, Carole Condé and Karl Beveridge's website, http://condebeveridge.ca/?projects=threat-1986.

19 The term "technoscientific" is taken from Ulrich Beck, *Risk Society: Towards a New Modernity* (London: Sage, 1992). A technoscientific society is one in which the risks of technological innovation and scientific development cannot be managed.

20 Major General Leslie R. Groves, military head of the Manhattan Project, recorded in his journal that Fermi "suddenly offered to take wagers from his fellow scientists on whether or not the bomb would ignite the atmosphere, and if so whether it would merely destroy New Mexico or destroy the world." Groves, quoted in Richard Rhodes, *The Making of the Atomic Bomb* (New York: Simon and Schuster, 1986), 664.

21 Derrida, "No Apocalypse, Not Now," 23.

22 In 2006, Prime Minister Stephen Harper introduced punitive guidelines that prevented scientists from talking about their research. Japan and the United States have also enforced guidelines on scientists. Wendy Palen, "When Canadian Scientists Were Muzzled by Their Government," *New York Times,* February 14, 2017.

23 Rachel Carson, *Silent Spring* (1962; Boston: Houghton Mifflin, 2002), 6. "In this now universal contamination of the environment," Carson writes at the beginning of the book, "chemicals are the sinister and little-recognized partners of radiation in changing the very nature of the world – the very nature of its life. Strontium 90, released through nuclear explosions into the air, comes to earth in rain or drifts down as fallout." In the same year that *Silent Spring* appeared, Norman Cousins published a new and revised edition of *In Place of Folly* (New York: Washington Square Press, 1962), which called for greater controls on nuclear energy in the United States and internationally.

24 In other circles, Carson's message in *Silent Spring* met with more favour. On June 4, 1963, Carson was called to testify before a Senate Subcommittee on Pesticides.

25 Quoted in Linda Lear, *Rachel Carson: Witness for Nature* (Boston: Houghton Mifflin Harcourt, 2009), 434.

26 Del Tredici, Preface to Del Tredici, *At Work in the Fields,* ix.

27 John Roberts, *Photography and Its Violations* (New York: Columbia University Press, 2014), 61.

28 Robert Del Tredici, "Documenting the Dark Period," interview by Michelle Heinemann, *Blackflash,* Spring 1985, 2.

29 James Agee and Walker Evans, *Let Us Now Praise Famous Men* (New York: Houghton Mifflin, 1941).

30 Gary Hall, Introduction to Fitzpatrick and Del Tredici, *The Atomic Photographers Guild,* 2.

31 Robert Del Tredici, email correspondence with the author, October 14, 2018.

32 Ferenc M. Szasz, *Larger Than Life: New Mexico in the Twentieth Century* (Albuquerque: University of New Mexico Press, 2006), 181.

33 See Blake Fitzpatrick, "Atomic Photographs below the Surface," in O'Brian, *Camera Atomica,* 165–79.

34 Robert Del Tredici, "Trinity: Fifty Years After," *Creative Camera,* June–July 1995, 8–9. In the passage, Del Tredici is talking not only about himself, but also about members of the Atomic Photographers Guild.

35 Fitzpatrick, *The Atomic Photographers Guild,* 3.

36 Mike Davis, "The Dead West: Ecocide in Marlboro Country," *New Left Review* 200 (July–August 1993): 9.

37 Blake Fitzpatrick, "The Radioactive Photograph Album," in *Lost in the Archives,* ed. Rebecca Comay (Toronto: Alphabet City, 2002), 734–39. The work is also discussed in John O'Brian, "On Photographing a Dirty Bomb," in *The Cultural Work of Photography in Canada,* ed. Carol Payne and Andrea Kunard (Montreal and Kingston: McGill-Queen's University Press, 2011), 192–99.

38 Jonathan Schell, Introduction to Del Tredici, *At Work in the Fields,* xi.

39 The print run was fifteen thousand copies. The book was supported by the Canada Council for the Arts.

40 Robert Del Tredici, email correspondence with the author, October 14, 2018.

41 James Elkins, "Rapatronic Photographs of Atomic Tests," in James Elkins, *After and Before: Documenting the A-Bomb* (New York: PPP Editions, 2003), n.p.

42 Bryan C. Taylor, "Nuclear Pictures and Metapictures," *American Literary History* 9, 3 (Autumn 1997): 576–78.

43 Yoshito Matsushige, in conversation with Robert Del Tredici, September 5, 1984, in Del Tredici, *At Work in the Fields,* 188.

44 Robert Jay Lifton and Greg Mitchell, *Hiroshima in America: Fifty Years of Denial* (New York: G.P. Putnam's, 1995), 61.

45 Fitzpatrick, "Atomic Photographs below the Surface," 172.

46 For a discussion of Kurt Vonnegut's *Slaughterhouse-Five* (1969), see Simon Baker, "Armageddon in Retrospect," in Baker and Mavlian, *Conflict*Time*Photography,* 195–96.

47 Blake Fitzpatrick, "Atomic Afterimages," *History of Photography* 32, 2 (Summer 2008): 180. In a different but related context, see Georges Didi-Huberman, *Images in Spite of All: Four Photographs from Auschwitz,* trans. Shane B. Lillis (Chicago: University of Chicago Press, 2008).

48 Craig Summers, "Book Review," *Peace Research* 21, 4 (November 1989): 64. *At Work in the Fields of the Bomb* was widely reviewed at the time of its publication. See Samuel H. Day Jr., "Reviews," *Bulletin of the Atomic Scientists* (October 1988): 46–47; Peter Friederici, "Lecture Notes: Robert Del Tredici's Book of the Bomb," *Chicago Reader,* August 3, 1989; Vince Leo, "Book Review," *Afterimage* 15 (May 1988): 18–19; Philip Morrison, "Books," *Scientific American,* November 1988, 148–49; and Patricia Rife, "Book Reviews," *Oral History Review* 19, 1–2 (April 1991): 179.

49 As identified by Taylor, "Nuclear Pictures and Metapictures," 573–75.

50 Del Tredici, *At Work in the Fields,* Plate 15–25 (Ohio workers) and Plate 26 (Mancuso).

51 Del Tredici, *At Work in the Fields,* Plate 26.

52 Del Tredici, *At Work in the Fields,* Plate 97.

53 John O'Brian, "Memory Flash Points," in Allan Sekula, *Geography Lesson: Canadian Notes* (Vancouver: Vancouver Art Gallery; Cambridge, MA: MIT Press, 1997), 79, 79n24.

54 Allan Sekula, *Mining Photographs and Other Pictures, 1948–1968* (Halifax: NSCAD Press, 1983), 99.

55 Richard Misrach, *Violent Legacies: Three Cantos* (New York: Aperture, 1992); Carole Gallagher, *American Ground Zero: The Secret Nuclear War* (New York: Random House, 1993).

56 Robert Del Tredici reviewed Gallagher's photobook in *Technology Review,* January 1994, 69–71, arguing that it contains "a deep and bitter wisdom."

57 H.W. Brands, *The General v. the President: MacArthur and Truman at the Brink of Nuclear War* (New York: Anchor, 2016).

58 John W. Dower, *Embracing Defeat: Japan in the Wake of World War II* (New York: W.W. Norton, 1999), 407.

59 The *New Yorker* published *Hiroshima* in its entirety on August 31, 1946, before it came out as a book. See Michael J. Yavenditti, "John Hersey and the American Conscience: The Reception of 'Hiroshima,'" *Pacifica Historical Review* 43, 1 (February 1974): 24–49.

60 Nicolas Lemann, "The Art of Fact: John Hersey Was a Great Journalist. Why Wasn't That Enough for Him?" *New Yorker,* April 29, 2019, 66.

61 "When Atom Bomb Struck – Uncensored," *Life,* September 29, 1952, 21.

62 See Rupert Jenkins, ed., *Nagasaki Journey: The Photographs of Yosuke Yamahata, August 10, 1945* (San Francisco: Pomegranate Books, 1995), 19.

63 Richard Misrach, quoted on the back cover, *Nagasaki Journey.* Misrach's own photographs are illustrated in Misrach, *Violent Legacies*; and Richard Misrach, *Bravo 20: The Bombing of the American West* (Baltimore: Johns Hopkins University Press, 1990).

64 Masafumi Suzuki, "The Atomized City and the Photograph," in *Nagasaki Journey,* 35.

65 For information on Japanese photobooks, see Ryuichi Kaneko and Ivan Vartanian, *Japanese Photobooks of the 1960s and 1970s* (New York: Aperture, 2009); Martin Parr and Jerry Badger, *The Photobook: A History,* vol. 1 (London: Phaidon, 2004); and Baker, "Armageddon in Retrospect," 195–205.

66 Baker, "Armageddon in Retrospect," discusses many of the photobooks listed.

67 Kenji Higuchi, *Photo Document: Japan's Nuclear Power Plants* (Tokyo: Orijin Shuppan Senta, 1979).

68 Quoted in Parr and Badger, *The Photobook,* 1:269–71.

69 Shot from 1983 to 1986, *The Journey* is the subject of a book by Rachael Rakes and Leo Goldsmith, forthcoming from Verso.

70 Blaine Campbell's residency at TRIUMF took place in 2016–17.

71 Robert Del Tredici, quoted in Fitzpatrick, "Atomic Afterimages," 176.

72 *The Nuclear Machine: Atomic Metabolism on Display,* an exhibition co-curated by Rens van Munster and the author in 2016 at the Danish Institute of International Studies, Copenhagen, investigated the sociopolitical and technoscientific investments required to keep the infrastructural wheels of the nuclear machine grinding.

Chapter 2: Chalk River

1 Spencer R. Weart, *The Rise of Nuclear Fear* (Cambridge, MA: Harvard University Press, 2012), viii.

2 Ernest Rutherford, quoted in W.C.D. Whetham, "Matter and Electricity," *Quarterly Review* 199 (January 1904): 126. Rutherford's letter is also cited in Weart, *The Rise of Nuclear Fear,* 14.

3 Sigmund Freud, quoted in Weart, *The Rise of Nuclear Fear,* 13. The Freud quotation is from *Civilization and Its Discontents.*

4 Steve Ryfle, *Japan's Favorite Mon-Star: The Unauthorized Biography of the Big G* (Toronto: ECW Press, 1998).

5 "Godzilla breath" is a synonym for bad breath. See Urban Dictionary, s.v. "Godzilla Breath," August 27, 2003, http://www.urbandictionary.com/define.php?term=godzilla+breath.

6 Krenz, *Deep Waters.* Krenz's narrative is partially indebted to Eggleston, *Canada's Nuclear Story;* and to Bothwell, *Nucleus.*

7 Energy and Mines Ministers' Conference Report, "Canadian Small Modular Reactor (SMR) Roadmap," 2018, https://www.nrcan.gc.ca/our-natural-resources/energy-sources-distribution/nuclear-energy-and-uranium/canadian-small-modular-reactor-roadmap/21183.

8 Group email sent by Larissa Koculym, subject line "Radiation Perception Survey," July 21, 2009. Koculym's position was research assistant and tutor at Radiological Protection Research and Instrumentation, AECL Chalk River Laboratories. Maurice Yang and Sophie Chang, summer interns at the Deep River Science Academy, assisted with the survey project.

9 The email message did not specify which media were included in the term "information filters."

10 Derrida, "No Apocalypse, Not Now," 24.

11 Scott McQuire, Introduction to Scott McQuire, *Visions of Modernity: Representation, Memory, Time and Space in the Age of Cinema* (London: Sage, 1998), 11.

12 Marjorie Earl, "Who Will Police Our Peacetime Atoms?" *Star Weekly,* November 19, 1960, 11.

13 Quoted in Evan Osnos, "The Fallout: Letter from Fukushima," *New Yorker,* October 17, 2011, 52.

14 Serhii Plokhy, *Chernobyl: History of a Tragedy* (London: Allen Lane, 2018).

15 Daniel Beer, "Revolution from the Brink of Armageddon," *Guardian Weekly,* May 25, 2018, in a review of Plokhy's *Chernobyl: History of a Tragedy.*

16 For a discussion of this image, see Claude Baillargeon, "McMillan's Chernobyl: An Intimation of the Way the World Would End," in David McMillan, *Growth and Decay: Prypiat and the Chernobyl Exclusion Zone* (Göttingen: Steidel, 2018), 254.

17 Octavio Paz, "Order and Accident," in Octavio Paz, *Conjunctions and Disjunctions,* trans. Helen R. Lane (New York: Viking, 1974), 112, quoted in Peter C. van Wyck, *Signs of Danger: Waste, Trauma, and Nuclear Threat* (Minneapolis: University of Minnesota Press, 2005), 13.

18 "They're Called Jumpers, Glowboys and They Go Where No One Else Will," It Won Last (blog), March 15, 2011, http://itwonlast.tumblr.com/post/3890223093/theyre-called-jumpers-glowboys-and-they-go-where.

19 See Christine Conley, "Material, Trace, Trauma: Notes on Some Recent Acquisitions at the Canadian War Museum," *Canadian Military History* 26, 1 (May 26, 2017): 1–19.

20 In 1991, Nordion was acquired by MDS Health Group. Krenz, *Deep Waters,* 90.

21 Max S. Power, *America's Nuclear Wastelands: Politics, Accountability, and Cleanup* (Pullman: Washington State University Press, 2008), 20.

22 Laura Onstot, "The Atomic Reboot," *Seattle Weekly,* March 17–23, 2010, 14. In an Angus Reid poll of 2010, 81 percent of respondents said that waste was their deepest concern.

23 Power, *America's Nuclear Wastelands,* 20.

24 The phenomenon is discussed by Walter Benjamin, "Theses on the Philosophy of History" (1940), in Walter Benjamin, *Illuminations: Essays and Reflections,* ed. Hannah Arendt (New York: Schocken, 1969), 255.

25 Robert Del Tredici, correspondence with the author, October 15, 2018.

26 Cragg, "Canada Building Plant."

27 US Strategic Bombing Survey, *The Effects of Atomic Bombs on Hiroshima and Nagasaki* (Washington, DC: US Government Printing Office, 1946). Useful information is also contained in E.P. Cronkite, "Medical Aspects of Atomic Explosives: Therapy of Radiation Diseases" (lecture transcript, Army Medical Center, Washington, DC, May 19–23, 1947). This is in my personal collection.

28 Louis Slotin, quoted in Martin Zeilig, "Louis Slotin and 'The Invisible Killer,'" *Beaver,* August–September 1995, 24. See also Meeka Walsh, "14 Poems from Bloom," *Border Crossings,* November 2007.

29 Richard L. Miller, *Under a Cloud: The Decades of Nuclear Testing* (Woodlands, TX: Two Sixty, 1991), 69, 77.

30 Thomas P. Ashlock, "Slotin – A Tribute," *Los Alamos Times,* June 14, 1946.

31 Michael Lista, *Bloom* (Toronto: Anansi, 2010).

32 Kate Brown, *Plutopia: Nuclear Families, Atomic Cities, and the Great Soviet and American Plutonium Disasters* (New York: Oxford University Press, 2013); Andrew Blackwell, *Visit Sunny Chernobyl, and Other Adventures in the World's Most Polluted Places* (New York: Rodale, 2012).

33 A.E. Larabee, "Ground Zero: The City, the Bomb, and the End of History," *Canadian Review of American Studies* 22 (1991): 270.

34 For an extended discussion of atomic postcards, see John O'Brian, "Postcard to Moscow," in *Postcards: Ephemeral Histories of Modernity,* ed. Jordana Mendelson and David Proschaska (Philadelphia: Pennsylvania State University Press, 2010), 182–93, 222–24.

35 Even if they were not little Edens, corporate and government authorities put a good face on the mining towns. L. Carson Brown claims that Elliot Lake is "one of Canada's most completely modern, wholly planned and most thoroughly liveable communities ... [It is] a dream town." L. Carson Brown, *Elliot Lake: The World's Uranium Capital,* an undated pamphlet reprinted from the *Canadian Geographical Journal* (October 1967) by the Ontario Ministry of Natural Resources, 3 and 8.

36 "A Town's New Life: Blind River Is Throbbing Capital of Richest Uranium Field," *Life,* August 1, 1955, 78.

37 Lorraine Rekmans, ed., *This Is My Homeland* (Cutler, ON: Serpent River First Nation, 2003), 5. See also Bonnie Devine, dir., *Rooster Rock: The Story of Serpent River* (Toronto: V Tape, 2002).

38 Ruth B. Phillips, "Between Rocks and Hard Places: Indigenous Lands, Settler Art Histories, and the 'Battle for the Woodlands,'" *Journal of Canadian Art History* 37, 1 (2016): 26.

39 Robynne Mellor, "The Radioactive Colonization of Indigenous Bodies," Network in Canadian History and Environment, August 23, 2017, http://niche-canada.org/2017/08/23/the-radioactive-colonization-of-indigenous-bodies.

40 Ara Mooradian, "The Most Exciting Thing in the World Was Atomic Energy, and That's Why I Came Here," interview by Mike English and Sean Mark, *Tamarack Magazine,* June 1995, 24–25.

41 Both are "Real Photo" postcards, c. 1950.

42 Susan Stewart, *On Longing: Narratives of the Miniature, the Gigantic, the Souvenir, the Collection* (Durham, NC: Duke University Press, 1993), 132–39.

43 Gordon Edwards, "Reactor Accidents at Chalk River: The Human Fallout," Canadian Coalition for Nuclear Responsibility, March 27, 1996, http://www.ccnr.org/paulson_legacy.html.

44 Jimmy Carter, quoted in Jane Taber, "Jimmy Neutron: Ex-President Carter Recalls Role in Chalk River Meltdown," *Globe and Mail,* May 13, 2011.

45 Krenz, *Deep Waters,* 95.

Chapter 3: Cold War Defence

1 P. Whitney Lackenbauer and Peter Kikkert, *The Canadian Forces and Arctic Sovereignty: Debating Roles, Interests and Requirements, 1968–1974* (Waterloo: Wilfrid Laurier University Press, 2010), 7.

2 Boyko, *Cold Fire,* 235.

3 National Defence College, Syndicate No. 5, "The Western Alliance: NATO – How Goes NATO," November 5, 1962, quoted in Langford and Langford, *A Cold War Tourist,* 33.

4 Langford and Langford, *A Cold War Tourist,* 7.

5 Leo Heaps, *Operation Morning Light: Terror in Our Skies, the True Story of Cosmos 954* (New York: Paddington Press, 1978).

6 "Kosmos 954," https://en.wikipedia.org/wiki/Kosmos_954.

7 See Daina Augaitis, ed., *Douglas Coupland: Everywhere Is Anywhere Is Anything Is Everything* (Vancouver: Vancouver Art Gallery; London: Black Dog, 2014). Coupland's engagement with the nuclear era is discussed in John O'Brian, "The Bomb, the Group of Seven and Douglas Coupland's G7 Series," in Augaitis, *Douglas Coupland,* 69–71.

8 Douglas Coupland, *Life after God* (New York: Simon and Schuster, 1994), 105.

9 Douglas Coupland, "The Wrong Sun," in O'Brian, *Camera Atomica,* 270–71.

10 *11 Steps to Survival: Blueprint for Survival No. 4* (Ottawa: Emergency Measures Organization, 1961). The Emergency Measures Organization (EMO) was created by the Conservative government of John Diefenbaker in 1957. Each province had its own EMO.

11 John Diefenbaker, Foreword to *Your Basement Fallout Shelter: Blueprint for Survival, No. 1:* (Ottawa: Emergency Measures Organization, 1960), i.

12 John F. Kennedy, "Speech to the Nation," July 25, 1961, quoted in Robert A. Jacobs, *The Dragon's Tail: Americans Face the Atomic Age* (Amherst: University of Massachusetts Press, 2010), 61.

13 Andrew Burtch, *Give Me Shelter: The Failure of Canada's Cold War Civil Defence* (Vancouver: UBC Press, 2012).

14 Kenneth D. Rose, *One Nation Underground: The Fallout Shelter in American Culture* (New York: New York University Press, 2001).

15 *Your Basement Fallout Shelter: Blueprint for Survival No. 1* (Ottawa: Emergency Measures Organization, 1960); *Blast Shelter: Blueprint for Survival No. 6* (Ottawa: Emergency Measures Organization, 1962).

16 Most of the early warning sirens were decommissioned and removed before the end of the Cold War.

17 *Technical and Tactical Study Courses Manual* (London: Home Office Fire Service Department, 1952).

18 *Technical and Tactical Study Courses Manual,* 2.

19 "Casualty Simulation," *National Health and Welfare Defence Bulletin* 43 (February 1955): 3.

20 "Casualty Simulation," 3.

21 *Casualty Simulation* (Ottawa: Information Services Division, Department of National Health and Welfare, 1955), 7. A committee oversaw the publication. C. Ellis Kerr, chief photographer in the Information Services Division, prepared the photographs.

22 The book describes and quantifies the amount of artificial blood. *Casualty Simulation,* 10.

23 *Casualty Simulation,* 7.

24 "Kodachrome," in Paul Simon, *There Goes Rhymin' Simon,* Columbia Warner Brothers, 1973.

25 Carol Payne, *Official Picture: The National Film Board of Canada's Still Photography Division and the Image of Canada, 1941–1971* (Montreal and Kingston: McGill-Queen's University Press, 2013).

26 Joyce Nelson, *The Colonized Eye: Rethinking the Grierson Legend* (Toronto: Between the Lines, 1988).

27 As a category, the term "documentary" is variable and unstable. See Julian Stallabrass, Introduction to Julian Stallabrass, ed., *Documentary: Documents of Contemporary Art* (London: Whitechapel Gallery; Cambridge, MA: MIT Press, 2013), 14, where he asks, "What is documentary?"

28 *Radiological Defense: A Series of Indoctrination Lectures on Atomic Explosion, with Medical Aspects* (Washington, DC: Armed Forces Special Weapons Project, n.d. [1949]), 3:37, Figure 2. The volumes of *Radiological Defense* were not sequentially published. The second volume appeared in 1951, two years after the third, and is subtitled *The Principles of Military Defense against Atomic Weapons.* Its cover is stamped with the Manhattan Project logo. Another scientific report, *The Effects of Atomic Weapons* (Los Alamos: Los Alamos Scientific Laboratory, 1950), is also liberally illustrated. The Canadian photograph of the victim wearing the patterned clothing appears on its page 204 (Figure 6.55b).

29 Anthropologist Hugh Gusterson observes that the photograph tends to generate widely disparate responses. He reports that he recoiled from the image when he first saw it, imagining the pain of the burns, but a nuclear weapons designer to whom he showed the photograph focused strictly on the burn patterns. Hugh Gusterson, "Hiroshima and the Power of Pictures," *Bulletin of the Atomic Scientists* 65, 4 (August 2009): 3.

30 The organizations and their activities in the aftermath of the bombings are discussed in William J. Schull, *Effects of Atomic Radiation: A Half-Century of Studies from Hiroshima and Nagasaki* (New York: Wiley-Liss, 1995); and Frank W. Putnam, "The Atomic Bomb Casualty Commission in Retrospect," *Proceedings of the National Academy of Sciences of the United States of America* 95, 10 (May 12, 1998): 5426–31.

31 *Radiological Defense,* 3:31–32.

32 The first scientific report in Japanese, by Masao Tsuzuki, was not published until 1953. Tsuzuki was the leading Japanese authority on the biological effects of radiation. See M. Susan Lindee, *Suffering Made Real: American Science and the Survivors at Hiroshima* (Chicago: University of Chicago Press, 1994); and Sue Rabbitt Roff, *Hotspots: The Legacy of Hiroshima and Nagasaki* (London: Cassell, 1995).

33 Lifton and Mitchell, *Hiroshima in America,* 41.

34 Lifton and Mitchell, *Hiroshima in America,* 41.

35 *The Nuclear Century,* 54–69, shows photographs of victims in makeshift spaces.

36 Dower, *Embracing Defeat,* 407.

37 Dower, *Embracing Defeat,* 621n21.

38 The photographs were not published in an English-language book, *Nagasaki Journey,* until three decades later.

39 Edward Steichen, ed., *The Family of Man,* exhibition catalogue (New York: Museum of Modern Art and Maco Magazine, 1955).

40 John O'Brian, "The Nuclear Family of Man," *Japan Focus: The Asia-Pacific Journal* 6, 7 (July 2008), https://apjjf.org/-John-O'Brian/2816/article.html.

41 Roland Barthes, "The Great Family of Man," in Roland Barthes, *Mythologies,* trans. Annette Lavers (St. Albans, UK: Picador, 1976), 100–2; Allan Sekula, "The Traffic in Photographs," in Allan Sekula, *Photography against the Grain: Essays and Photo Works, 1973–1983* (Halifax: Press of the Nova Scotia College of Art and Design, 1984), 77–101.

42 Steichen, *The Family of Man,* 179.

43 For more information on the Carp shelter, see the website of Diefenbunker: Canada's Cold War Museum, https://diefenbunker.ca/en/.

44 Fitzpatrick, "Atomic Photographs below the Surface," 168. Fitzpatrick refers to the air vents as "tombstones," markers of the vault built below.

45 James Bridle, "Big Data, Nein Danke" (paper presented at "Through Post-Atomic Eyes" conference, OCAD University and University of Toronto, October 23–24, 2015).

Chapter 4: Cloud Shadows

1 Robert A. Jacobs, "Good Bomb / Bad Bomb: Talking about Atomic Tests in Nevada," *Interdisciplinary Humanities* 1 (Spring 2007): 65.

2 Paul Wagner, "Interview with Paul Wagner, Public Relations Manager for the Department of Energy at Pantex, 10 August 1982," interview by Robert Del Tredici, in Del Tredici, *At Work in the Fields of the Bomb,* 135. In the Introduction to the book, Jonathan Schell also quotes Wagner, xi.

3 Coupland visited the site before 9/11, when tours could be readily arranged. Douglas Coupland, in conversation with the author, West Vancouver, September 25, 2013.

4 The development of this chapter was helped by ideas in Didi-Huberman, *Images in Spite of All;* and Robin Kelsey, "Making Photography into Public Memory" (paper presented at "The 'Public Life' of Photographs" symposium, Toronto, May 9–11, 2013).

5 Scott Kirsch, "Watching the Bombs Go Off: Photography, Nuclear Landscapes, and Spectator Democracy," *Antipode* 29, 3 (1997): 227–55.

6 Del Tredici, Preface to Del Tredici, *At Work in the Fields,* ix.

7 See also van Wyck, *The Highway of the Atom.*

8 Oppenheimer was quoting from the Bhagavad Gita. The film is *The Town That Never Was,* a sixteen-minute documentary on Los Alamos and the Manhattan Project. It is regularly screened at the Bradbury Science Museum, Los Alamos. See Peggy Rosenthal, "The Nuclear Mushroom Cloud as Cultural Image," *American Literary History* 3, 1 (Spring 1991): 73.

9 Photographs, statues, and busts of Oppenheimer are standard issue in atomic museums like the Bradbury, despite his later fall from grace for speaking out on the dangers of nuclear weapons.

10 William Broad, "The Bomb Chroniclers," *New York Times,* September 13, 2010.

11 See Kuran, *How to Photograph an Atomic Bomb;* and Susan Schuppli, "Radical Contact Prints," in O'Brian, *Camera Atomica,* 280.

12 Virilio, *War and Cinema,* back cover.

13 Schuppli, "Radical Contact Prints," 281. In *Atomic Light (Shadow Optics),* Lippit also writes about radiographic surfaces and visibility and invisibility at Hiroshima and Nagasaki.

14 Father John A. Siemes, quoted in James P. Delgado, *Nuclear Dawn: The Atomic Bomb from the Manhattan Project to the Cold War* (Oxford: Osprey, 2009), 92.

15 "Within weeks of the United States dropping an atomic bomb," historian Tom Engelhardt observes, "Americans were already immersed in new scenarios of nuclear destruction. As the late Paul Boyer so vividly describes in his classic book *By the Bomb's Early Light,* it took no time at all – at a moment when no other nation had such destructive weaponry – for a triumphant America to begin to imagine itself in ruins and for its newspapers and magazines to start drawing concentric circles of death and destruction around American cities while consigning their future country to the stewardship of the roaches." Tom Engelhardt, "Best of TomDispatch: Noam Chomsky, 'The Most Dangerous Moment,'" TomDispatch.com, March 22, 2014, http://www.tomdispatch.com/post/175821/best_of_tomdispatch%3A_noam_chomsky%2C_%22the_most_dangerous_moment%22. See also Paul Boyer, By the Bomb's Early Light: American Thought and Culture at the Dawn of the Atomic Age (New York: Pantheon, 1985).

16 The Nagasaki explosion was also recorded on film, with colour footage shot from the window of an observation plane accompanying *Bockscar,* the B-29 bomber carrying the bomb. The recently discovered film is unedited and begins with the preparation and loading of the bomb on Tinian Island. "Haunting Unedited Footage of the Bombing of Nagasaki (1945)," Open Culture, February 7, 2014, http://www.openculture.com/2014/02/haunting-unedited-footage-of-the-bombing-of-nagasaki-1945.html.

17 William Carlos Williams, "Asphodel, That Greeny Flower," in William Carlos Williams, *Journey to Love* (New York: Random House, 1955). For a commentary on the poem, see J. Hillis Miller, *Poets of Reality: Six Twentieth-Century Writers* (Cambridge, MA: Harvard University Press, 1965).

18 Williams's poem relates to Michael Lista's volume of poetry *Bloom,* which centres on the Canadian physicist Louis Slotin and his death following his botched experiment with a plutonium core at Los Alamos in 1946. When plutonium goes critical, it releases what scientists call a "bloom."

19 For example, a Google image search for "Hiroshima bombing" will bring up several versions of the Nagasaki mushroom cloud image.

20 Ariella Azoulay, *The Civil Contract of Photography* (New York: Zone Books, 2008).

21 Kyungah Ham's large-scale textile diptych *Nagasaki and Hiroshima Mushroom Clouds* (2008) was produced using North Korean labour unauthorized by the regime. It speaks not only to the Cold War and the partitioning of Korea but also to the present sociopolitical situation where North Korea's nuclear capabilities threaten the region. Sumi Kang, "The Hypertext of Some Art," *Art Monthly* 2 (2008): 136–41.

22 Coupland, *Life after God;* Douglas Coupland, *Generation X: Tales for an Accelerated Culture* (New York: St. Martin's, 1991).

23 Coupland, "The Wrong Sun," in Coupland, *Life after God,* 97.

24 Coupland, "The Wrong Sun," 97.

25 As a doctor measuring radioactive contamination, David Bradley was present at Bikini for the first tests in 1946. His influential book *No Place to Hide* (Boston: Little, Brown, 1948) concludes that nuclear energy has "made us less free, more interdependent, and has threatened to enslave us all in a frantic and foredoomed race for military power" (dustjacket).

26 Michael Light, *100 Suns: 1945–1962* (New York: Alfred A. Knopf, 2003).

27 Office of the Historian, Joint Task Force One, *Operation Crossroads: The Official Pictorial Record* (New York: Wm. H. Wise, 1946), 41.

28 "Look Pleasant Please," in Office of the Historian, *Operation Crossroads,* 72–73; Schuppli, "Radical Contact Prints," 280.

29 Whereas some early reports of the explosion described the cloud as having a cauliflower form, others stated that it looked like a tree or a cabbage or, more abstractly, like a colossus.

30 "Base Surge, Lagoon Level," in Office of the Historian, *Operation Crossroads,* 202. The book contains 228 images, many "reclassified from secret files," which bear lengthy captions.

31 A comprehensive glossary of nuclear terms, such as "kill radius" and "radioactive decay," can be found in Eric Semler, James Benjamin, and Adam Gross, *The Language of Nuclear War: The Intelligent Citizen's Dictionary* (New York: Harper and Row, 1987).

32 The United States also conducted extensive radiation experiments on American citizens. Eileen Welsome reveals their scope in *The Plutonium Files,* an exposé of the secret studies. Pregnant women and children were given radioactive iron (children got it in lemonade); prison inmates were targeted with X-rays to see what doses they could take before becoming sterile; and "radiation bombs" were tossed from military aircraft into civilian areas to measure the distances that radiation could travel. See also Catherine Caulfield, *Multiple Exposures: Chronicles of the Radiation Age* (London: Secker and Warburg, 1989); Philip J. Hilts, "Fallout Risk Near Atom Tests Was Known, Documents Show," *New York Times,* March 15, 1995; and John LaForge, "Contaminated Nation: Inhuman Radiation Experiments," *Counterpunch,* April 12–14, 2013.

33 Clarence H. White, ed., *Operation Sandstone: The Story of Joint Task Force Seven* (Washington, DC: Infantry Journal Press, 1949).

34 The website Image Atlas (http://imageatlas.org), created by Taryn Simon and Aaron Shwartz, provides evidence that *Mushroom Cloud, Test Baker* has itself recently been supplanted in the collective memory. Image Atlas allows users to compare top image results in various countries according to subject matter. If one enters "mushroom cloud" into its search engine, *Mushroom Cloud, Test Baker,* does not appear. Most images that do are of hydrogen bomb explosions reproduced in colour.

35 For an extended account of the photograph, see "The Atomic Cake Controversy of 1946," Conelrad Adjacent (blog), September 7, 2010, http://conelrad.blogspot.ca/2010/09/atomic-cake-controversy-of-1946.html.

36 "The Atomic Cake Controversy."

37 The Inco ad appeared in the March 22, 1954, issue of *Time,* on page 109.

38 This advertisement is in the October 1, 1954, issue of *Collier's,* on page 55.

39 "Atomic Five-Year Plan," *Time,* March 22, 1954, 68. Four additional articles on nuclear matters appear in the same issue: "The Atom: Five Hundred Hiroshimas," 21; "The Cold War: Facing the Facts," 31; "Atomic Diagnosis," 59; and "Time Clock: First Full-Scale Atomic Power Plant," 101.

40 As discussed in more detail in Chapter 1, the explosive yield of Castle Bravo was fifteen megatons, double what was expected, and the fallout from the detonation fell on the islanders of Rongelap and Utrik as well as on the crew of a Japanese fishing boat, the *Daigo Fukuryū Maru,* with deadly results.

41 Retort (Iain Boal, T.J. Clark, Joseph Matthews, Michael Watts), *Afflicted Powers: Capital and Spectacle in a New Age of War* (London: Verso, 2005).

42 US Department of Defense, quoted in Lawrence Robert Aronsen, *American Security and Economic Relations, 1945–1954* (Westport, CT: Praeger, 1997), 128.

43 Inco Annual Report for 1953, quoted in *Inco Triangle,* April 1954, 2.

Chapter 5: Atomic Soldiers

1 Several Canadian eyewitness accounts are given in Dave Pugliese, "What the Hell Is This Thing?" *Ottawa Citizen,* November 8, 2007. See also Dick Hodgson, "They Dropped the A-Bomb on Me: Atom Blast Minus Four Miles," *Men,* November 1952, 11–15, 48–49. The blast must have been a "spectacular sight for the photographers," an eyewitness told Hodgson.

2 Woodrow Whitefield Coward, *Connections: A Story of the Family* (Vancouver: Ellenwood, 1999), 141–42. I am grateful to Erin Coward for bringing this memoir to my attention.

3 Bertha Röntgen, quoted in Gottfried Landwehr, "Wilhelm Röntgen and the Beginning of Modern Physics," in *Röntgen Centennial: X-Rays in Natural and Life Sciences,* ed. Axel Haase, Gottfried Landwehr, and Eberhard Umbach (Singapore: World Scientific, 1997), 7–8. In his last book, written while mourning the death of his mother, Roland Barthes insisted that *all* photographs traffic in the "what-has-been" and therefore in death. "Death is the *eidos* of [the] photograph," he wrote in Barthes, *Camera Lucida,* 15.

4 In Chapter 2 of *Atomic Light (Shadow Optics),* Lippit offers a compelling psychoanalytical account of the X-ray in relation to cinema.

5 John Bodnar, *Remaking America: Public Memory, Commemoration, and Patriotism in the Twentieth Century* (Princeton: Princeton University Press, 1994), 13–14. See also Matthew Farish, "Below the Bombs," in Lauzon and O'Brian, *Through Post-Atomic Eyes.*

6 Stein, "The Rhetoric of the Colorful and the Colorless."

7 Peter B. Hales, "Imagining the Atomic Age: Life and the Atom," in *Looking at Life Magazine,* ed. Erika Doss (Washington, DC: Smithsonian Institution Press, 2001), 105.

8 Pugliese, "What the Hell Is This Thing?" Steve B. Davis also reports on Canadian participation in the tests, in "Nuclear Guinea Pigs."

9 See Michael Uhl and Tod Ensign, *GI Guinea Pigs: How the Pentagon Exposed Our Troops to Dangers More Deadly Than War* (New York: Wideview Books, 1980); Thomas H. Saffer and Orville E. Kelly, *Countdown Zero: GI Victims of U.S. Atomic Bombing* (New York: Penguin, 1982); and Jacobs, "Good Bomb / Bad Bomb," 65–81.

10 Rosenberg, *Atomic Soldiers.*

11 Davis, "Nuclear Guinea Pigs."

12 J.R. McNeill and Corinna R. Unger, Introduction to *Environmental Histories of the Cold War,* ed. J.R. McNeill and Corinna R. Unger (New York: Cambridge University Press, 2010).

13 "The Big Picture (TV Series)," Wikipedia, updated March 27, 2018, https://en.wikipedia.org/wiki/The_Big_Picture_(TV_series). The quotation is from Gallagher, *American Ground Zero,* xii.

14 Gallagher, *American Ground Zero,* xii.

15 Michael Light, "A Note on the Photographs," in Light, *100 Suns,* n.p.

16 Jeffrey Jones, "Canada Offers to Compensate Nuclear Test Veterans," Reuters, September 2, 2008, http://www.reuters.com/article/healthNews/idUSN0248391320080902. See also Jason Fekete, "Canadian Atomic Testing Veteran Compensation: $24,000," *Calgary Herald,* September 2, 2008.

17 *La Conquête du Monde* was released by Miro in France and *Risk* by Parker Brothers in North America.

18 "Feel Like RISKing Everything?" *Dragon Magazine,* February 1980.

19 Herman Kahn, *Thinking about the Unthinkable* (New York: Horizon, 1962). See also Fred Kaplan, *The Wizards of Armageddon* (New York: Simon and Schuster, 1983); and Sharon Ghamari-Tabrizi, *The Worlds of Herman Kahn: The Intuitive Science of Thermonuclear War* (Cambridge, MA: Harvard University Press, 2005).

20 Derrida, "No Apocalypse, Not Now," 21.

21 Ghamari-Tabrizi, *The Worlds of Herman Kahn,* 275.

22 Kahn, *Thinking about the Unthinkable,* 27.

23 Stanley Kubrick, quoted in Loudon Wainwright, "The Strange Case of Strangelove," *Life,* March 13, 1964, 15.

24 Sylvia Plath deployed the same technique, "treating the horrific with comic irreverence." Robin Peel, *Writing Back: Sylvia Plath and Cold War Politics* (Madison, NJ: Fairleigh Dickinson University Press, 2002), 42.

25 Kuran, *How to Photograph an Atomic Bomb,* 7.

26 George Yoshitake, interviewed by Peter Kuran, in *How to Photograph an Atomic Bomb,* 46.

27 Yoshitake, interviewed by Kuran, 46.

28 George Yoshitake, "George Yoshitake, Nuclear Test Photographer, Recalls Filming Nuclear Blast 55 Years Ago," interview by Timothy Stenovec, *Huffington Post,* July 20, 2012, quoted in Bryan-Wilson, "Posing by the Cloud," 119.

29 The nickname Genie was given to the missile by its manufacturer, Douglas Aircraft Company.

30 "Operation Plumbbob," US Defense Threat Reduction Agency, Fact Sheet, 2017, www.dtra.mil/Portals/61/Documents/NTPR/1-Fact_Sheets/19_PLUMBBOB.pdf.

31 See Robert F. Dorr, *McDonnell F-101 Voodoo* (London: Osprey, 1987); and Clearwater, *Canadian Nuclear Weapons.*

32 Robert McIntyre, *CF-101 Voodoo: Canadian Profile* (Ottawa: SMS, 1984).

33 Atomic Energy Commission, quoted in Howard Ball, *Justice Downwind: America's Atomic Testing Program in the 1950s* (New York: Oxford University Press, 1986), 77. The quotation is from the *Desert News,* May 30, 1957.

34 Gallagher, *American Ground Zero.*

35 Valerie L. Kuletz, *The Tainted Desert: Environmental and Social Ruin in the American West* (New York: Routledge, 1998), xv.

36 Henry Jackson, quoted in Joseph C. Harsch, "John Foster Dulles: A Very Complicated Man," *Harper's Magazine,* September 1956, 27.

37 John Foster Dulles, quoted in Stephen E. Ambrose, *Rise to Globalism: American Foreign Policy since 1938* (New York: Penguin, 1988), 148.

38 The United States also contemplated using nuclear weapons in Korea in 1950. When President Harold Truman was asked by Merriman Smith of the United Press whether the atomic bomb was under consideration, he replied, "Always has been. It is one of our weapons." Quoted in Paul Boyer, *Fallout: A Historian Reflects on America's Half-Century Encounter with Nuclear Weapons* (Columbus: Ohio State University Press, 1998), 35.

39 Dwight D. Eisenhower, quoted in Ambrose, *Rise to Globalism,* 147. Information in this paragraph is also drawn from John Wilson Lewis and Xue Litai, *China Builds the Bomb* (Stanford, CA: Stanford University Press, 1988).

40 Robert Frank denied permission for the reproduction of *Hoover Dam, Nevada,* in this volume. The photograph has been reproduced elsewhere, including in O'Brian, *Camera Atomica,* 70. It is also widely available on the Internet.

41 For Joan Didion, the Hoover Dam has an emotional resonance that other dams lack. She refers to "a plaque dedicated to the 96 men who died building the first of the great high dams." The plaque states, "They died to make the desert bloom." Joan Didion, *The White Album* (New York: Simon and Schuster, 1979), 198.

42 Robert Frank, *The Americans* (New York: Scalo, 1995). The book was first published in 1958 in France and then a year later in the United States.

43 Sarah Greenough, "Fragments That Make a Whole: Meaning in Photographic Sequences," in Greenough and Brookman, *Robert Frank,* 96–97.

44 Nathan Lyons, ed., *Toward a Social Landscape* (New York: Horizon, 1966).

45 Robert Frank, quoted in Sarah Greenough, "Transforming Destiny into Awareness: The Americans," in Sarah Greenough, ed., *Looking In: Robert Frank's The Americans* (Washington, DC: National Gallery of Art, 2009), 177.

46 *Hoover Dam, Nevada,* was published in Robert Frank, *The Lines of My Hand* (New York: Lustrom, 1972).

47 Even though the archive and its plural representations are, as Pierre Nora says, "the deliberate and calculated secretion of lost memory ... a secondary memory, a prosthesis-memory," I am conscious of Allan Sekula's warning that archival memory insistently reproduces dominant cultural norms, even when broken up and reordered. Pierre Nora, "Between Memory and History," *Representations* 26 (Spring 1989), 8; Allan Sekula, "The Body and the Archive," *October* 39 (Winter 1986): 3–64.

48 Sandra S. Phillips, "To Subdue the Continent: Photographs of the Developing West," in Sandra S. Phillips, ed., *Crossing the Frontier: Photographs of the Developing West, 1949 to the Present* (San Francisco: San Francisco Museum of Modern Art, 1997), 39.

49 Peter B. Hales, "The Atomic Sublime," *American Studies* 32 (Spring 1991): 5–31. Hales does not refer to *Hoover Dam, Nevada,* directly, but he is certainly aware of it. He draws on Vincent Leo, "The Mushroom Cloud Photograph: From Fact to Symbol," *Afterimage* 13 (Summer 1985): 6–12, which devotes several paragraphs to it.

50 Leo, "The Mushroom Cloud Photograph," 11.

51 W.T. Lhamon Jr., *Deliberate Speed: The Origins of a Cultural Style in the American 1950s* (Washington, DC: Smithsonian Institution, 1990), 125. Lhamon stops short of transforming the plywood shelf into a theatrical stage, as I am inclined to do.

52 Written in 1949, *Waiting for Godot* had its French premiere in Paris in 1953. The 1955 English premiere was in London.

53 Jacques Derrida, *The Post Card: From Socrates to Freud and Beyond,* trans. Alan Bass (Chicago: University of Chicago Press, 1987).

54 Walker Evans, "Test Exposures: Six Photographs from a Film Manufacturer's Files," *Fortune,* July 1954, 80.

55 Stein, "The Rhetoric of the Colorful and the Colorless."

56 Nelson, *The Perfect Machine,* 118.

57 Nelson, *The Perfect Machine,* 134.

58 Stephen Hilgartner, Rory O'Connor, and Richard Bell, *Nukespeak: Nuclear Language, Visions, and Mindset* (San Francisco: Sierra Club, 1982), 75.

59 Light, *100 Suns,* n.p.

Chapter 6: Mass Protest

1 Bob Hunter and Robert Keziere, *Greenpeace* (Toronto: McClelland and Stewart, 1972). See also Jim Bohlen, *Making Waves: The Origins and Future of Greenpeace* (Montreal: Black Rose, 2001); Bob Hunter, *The Greenpeace to Amchitka: An Environmental Odyssey* (Vancouver: Arsenal Pulp Press, 2004); Dean W. Kohlhoff, *Amchitka and the Bomb: Nuclear Testing in Alaska* (Seattle: University of Washington Press, 2002); and Rex Wyler, *Greenpeace: How a Group of Ecologists, Journalists, and Visionaries Changed the World* (Vancouver: Raincoast, 2004).

2 Hunter and Keziere, *Greenpeace,* n.p. On October 9, 1971, for example, three weeks after the *Greenpeace* departed from Vancouver, Hunter's *Vancouver Sun* column was accompanied by two photographs, *Protest Vessel* and *Peace Signs Flashing.*

3 Lawrence S. Wittner, *Confronting the Bomb: A Short History of the World Nuclear Disarmament Movement* (Stanford, CA: Stanford University Press, 2009), 117.

4 Wittner, *Confronting the Bomb,* 116.

5 There was a strong anti-nuclear movement in France before the sinking of the *Rainbow Warrior* and the cancellation of French testing. See Dorothy Nelkin and Michael Pollak, *The Atom Besieged: Anti-Nuclear Movements in France and Germany* (Cambridge, MA: MIT Press, 1982).

6 David Lange, *Nuclear Free: The New Zealand Way* (Auckland: Penguin, 1990).

7 David Levi Strauss, "The Documentary Debate: Aesthetic or Anaesthetic? Or What's So Funny about Peace, Love, Understanding, and Social Documentary Photography?" in David Levi Strauss, *Between the Eyes: Essays on Photography and Politics* (New York: Aperture, 2003), 6. The essay focuses on the work of Sebastião Salgado.

8 Sontag, *Regarding the Pain of Others,* 117; Susan Sontag, "Regarding the Torture of Others," *New York Times Magazine,* May 23, 2004, 42.

9 Lifton and Mitchell, *Hiroshima in America,* 44.

10 Donald Rumsfeld, speech to the Council on Foreign Relations, February 17, 2006, quoted in Retort, "All Quiet on the Eastern Front," *New Left Review* 41 (September–October 2006), https://newleftreview.org/issues/II41/articles/retort-all-quiet-on-the-eastern-front.

11 Roberts, *Photography and Its Violations,* 10.

12 George W. Bush, quoted in Sontag, "Regarding the Torture," 26.

13 Wittner, *Confronting the Bomb,* 53. The petition was called the Suganami Appeal.

14 McMahon, *Essence of Indecision,* 64.

15 Ursula M. Franklin, *The Real World of Technology* (Toronto: Anansi, 1990).

16 Martin Luther King Jr., interview by John Freedom, *Face to Face,* BBC Two, October 19, 1961.

17 Hugh Brock, "Marching to Aldermaston – Ten Years Ago!" *Sanity,* April 20, 1962.

18 Wyler, *Greenpeace,* 71, 455.

19 Becki L. Ross, quoted in the exhibition *City on Edge: A Century of Vancouver Activism,* Museum of Vancouver, 2017. See also Kate Bird, *City on Edge: A Rebellious Century of Vancouver Protests, Riots, and Strikes* (Vancouver: Greystone, 2017).

20 Whitaker and Hewitt, *Canada and the Cold War,* 165.

21 McMahon, *Essence of Indecision,* 125–30. In addition to petitions, new publications such as *Our Generation against Nuclear War* began to circulate. The first issue of *Our Generation against Nuclear War* appeared in 1961.

22 The Diefenbaker government first entered into negotiations to acquire nuclear weapons in January 1958. Boyko, *Cold Fire,* 61.

23 On the Voice of Women movement, see Braden Hutchinson, "Fighting the War at Home: Voice of Women and War Toy Activism in Postwar Canada," in *Worth Fighting For: Canada's Tradition of War Resistance from 1812 to the War on Terror,* ed. Lara Campbell, Michael Dawson, and Catherine Gidney (Toronto: Between the Lines, 2015), 147–68.

24 Boyko, *Cold Fire,* 61.

25 Whitaker and Hewitt, *Canada and the Cold War,* 152.

26 The American pollster Lou Harris was engaged to undertake the survey. See Boyko, *Cold Fire,* 212.

27 For an account of the protest at La Macaza, see Bruce Douville, "Project La Macaza: A Study of Two Canadian Peace Protests in the 1960s," in Campbell, Dawson, and Gidney, *Worth Fighting For,* 161–71.

28 Pierre Trudeau, quoted in John English, *The Worldly Years: The Life of Lester B. Pearson, 1949–1972* (Toronto: Alfred A. Knopf, 1992), 238.

29 Ryan M. Touhey, *Conflicting Visions: Canada and India in the Cold War World, 1946–76* (Vancouver: UBC Press, 2015), 150.

30 "Canada and India: The Nuclear Genie, 40 Years On," *Globe and Mail,* April 15, 2015.

31 Touhey, *Conflicting Visions,* 98–101.

32 Robert S. Anderson, *Nucleus and Nation: Scientists, International Networks, and Power in India* (Chicago: University of Chicago Press, 2010).

33 Eugene Boyko, dir., *Juggernaut,* National Film Board of Canada, 1968, 27 minutes, https://www.nfb.ca/film/juggernaut/.

34 Manmohan Singh, quoted in Bill Curry, "Canada Signs Nuclear Deal with India," *Globe and Mail,* June 28, 2010.

35 Ernie Regehr, "The Canada-India Civilian Nuclear Cooperation Deal," *Centre for International Governance Intervention,* June 29, 2010.

36 "Cancelled Nuclear Units Ordered in the United States," July 24, 2011, http://www.rogerwitherspoon.com/docs/cancellednukeplants.pdf.

37 Wyler, *Greenpeace;* Carole Roy, "The Original Raging Grannies: Using Creative and Humorous Protests for Political Education," Raging Grannies International, http://raginggrannies.org/herstory.

38 Whitaker and Hewitt, *Canada and the Cold War,* 221.

39 Useful accounts of the cruise missile controversy include Allan Antliff, *Only a Beginning: An Anarchist Anthology* (Vancouver: Arsenal Pulp Press, 2004); John M. Clearwater, *Just*

Dummies: Cruise Missile Testing in Canada (Calgary: University of Calgary Press, 2006); F.H. Knelman, *America, God and the Bomb: The Legacy of Ronald Reagan* (Vancouver: New Star, 1987); and Simon Rosenblum, *Misguided Missiles: Canada, the Cruise and Star Wars* (Toronto: James Lorimer, 1985).

40 Bridle, "Big Data, Nein Danke."

41 Clearwater, *Just Dummies,* 99.

42 Krzysztof Wodiczko, "Public Projections," *Canadian Journal of Political and Social Theory* 7, 1–2 (Winter–Spring 1983): 186.

43 Krzysztof Wodiczko, quoted in Peter Boswell, *Public Address: Krzysztof Wodiczko* (Minneapolis: Walker Art Center, 1992), 103.

Chapter 7: Camera Targets

1 George Bowering, "Taking Pictures" (1982), quoted in Ian Rae, *From Cohen to Carson: The Poet's Novel in Canada* (Montreal and Kingston: McGill-Queen's University Press, 2008), 143.

2 Virilio, *War and Cinema,* 1.

3 George Bowering, *The Man in Yellow Boots / El Hombre de las Botas Amarillas* was published in English and Spanish as a special issue of *El Corno Emplumado / The Plumed Horn* 16 (Mexico City: October 1965), edited by Sergio Mondragón and Margaret Randall, http://opendoor.northwestern.edu/archive/files/original/3e8471258a33391b9a132aee2afbac48.pdf.

4 Over the years, Bowering has collaborated with a number of visual artists, as discussed in Eva-Marie Kröller, *George Bowering: Bright Circles of Colour* (Vancouver: Talonbooks, 1992). See also W.F. Garrett-Petts and Donald Lawrence, *PhotoGraphic Encounters: The Edges and Edginess of Reading Prose Pictures and Visual Fictions* (Edmonton: University of Alberta, 2000), Chapter 3, on some of the collaborations.

5 Bowering later commented on Kiyooka's serial practices in his essay "Roy Kiyooka's Poetry (an Appreciation)," in *Roy K. Kiyooka: 25 Years* (Vancouver: Vancouver Art Gallery, 1975), n.p. He also wrote about Kiyooka in the poem "Fill Our Houses," first published in *Brick* 48 (Spring 1994); and republished in George Bowering, *Blonds on Bikes* (Vancouver: Talonbooks, 1997).

6 For more information on the hoarfrost series, see John O'Brian, *White Paint, Hoarfrost, and the Cold Shoulder of Neglect* (Vancouver: UBC Fine Arts Gallery, 1992), an essay published to accompany the exhibition *Roy Kiyooka: The Hoarfrost Paintings,* which I also curated.

7 Kröller, *George Bowering,* 35.

8 Kröller, *George Bowering,* 35.

9 George Bowering, "Old Time Photo of the Present," in Bowering, *The Man in Yellow Boots,* 56.

10 Bowering comments on some of the censorship battles appear in letters published in the back pages of Bowering, *The Man in Yellow Boots,* 97–101. The letters are printed on yellow paper.

11 Bowering, "Old Time Photo," 56.

12 Bowering's remark about Alberta is on page 99 of *The Man in Yellow Boots.*

13 Roy Kiyooka, "Intersections," interview by Chris Varley, in *Roy K. Kiyooka: 25 Years,* n.p. As Varley comments, "If I haven't told you before, you should know that [Kiyooka] wrote every word of 'our' interview in the exhibition catalogue. I must have seen it before publication, but it ain't mine. It's just Roy leading the witness." Chris Varley, email to the author, July 25, 2010.

14 This strategy corresponds to what Roberts, *Photography and Its Violations,* calls upon photography to do.

15 Roy Kiyooka, conversation with the author, 1991. At the time, we were collaborating on the exhibition *Roy Kiyooka: The Hoarfrost Paintings,* UBC Fine Arts Gallery, Vancouver, 1992.

16 Kröller, *George Bowering,* 36.

17 George Bowering, "Her Act Was a Bomb," in Bowering, *The Man in Yellow Boots,* 38.

18 Mark Ruwedel published *Lili's Wedding* in an edition of three. I am grateful to him for permission to reproduce images from the book. Ruwedel is an American who taught photography at Concordia University, Montreal, from 1984 to 2000, before returning to the United States.

19 "Obituary: Lili St. Cyr," *Independent* (UK), February 8, 1999.

20 James Martin, "Sex and the City," *McGill Reporter,* September 22, 2005, 1–2.

21 Although most fully embraced by Montreal, St. Cyr also performed in the United States and elsewhere in Canada, notably in Vancouver. As Becki L. Ross observes in *Showgirls, Sex, and Sin in Postwar Vancouver* (Toronto: University of Toronto Press, 2009), Vancouver was called "Vegas North" in some quarters for its hospitality to strippers and striptease acts. They greased the city economy. Ross's book is a significant contribution to the so-called sex war debates between radical and libertarian feminists that have focused on pornography, sex workers, and discrimination since the 1980s.

22 After the wedding took place, *Time* reported the essential details in "Milestones," March 7, 1955, 67: "Lili St. Cyr, 36, blond striptease, and Ted Jordan, 28, Hollywood and Broadway bit actor ... she for the fifth time, he for the third, in Las Vegas, Nev."

23 Information on Test Moth is from Light, *100 Suns,* n.p.

24 Canadair's "Communism's Technical Ability" ad appeared in the September 17, 1955, issue of *Maclean's.* The new Soviet bomber was the T-37.

25 James Opp, "Picturing Communism: Yousuf Karsh, Canadair, and Cold War Advertising," in Payne and Kunard, *The Cultural Work of Photography,* 120–33. Opp examines all eight ads in the Canadair campaign, as well as its implications.

26 Lewis and Litai, *China Builds the Bomb,* 188.

27 Bowering, *The Man in Yellow Boots,* 98.

28 George Bowering, "The Day before the Chinese A-Bomb," in Bowering, *The Man in Yellow Boots,* 36.

29 Richard Harrington, *Canadian Photographer* (Edmonds, WA: Alaska Northwest, 1998).

30 The three nuclear photos exist only as single, unique prints. On behalf of the Richard Harrington estate, they were sold by Stephen Bulger Gallery, Toronto, to a private collector.

31 Wu Hufan's scroll is titled *Celebrate the Success of Our Atomic Bomb Explosion.* It is reproduced in Julia Frances Andrews, *Painters and Politics in the People's Republic of China, 1949–1979* (Berkeley: University of California Press, 1994), Figure 108.

32 Stanley Kubrick, quoted in Lawrence H. Suid, *Guts and Glory: The Making of the American Military Image in Film* (Lexington: University of Kentucky Press, 2002), 232.

33 Michael Herr, a friend and sometime collaborator of the director, reports that Kubrick "once asked me if I'd mind moving with my family to Vancouver for a year to check it out for him." Michael Herr, *Kubrick* (New York: Grove, 2000), 50.

34 War is the dominant theme in half of Kubrick's films, as discussed in Philip Kuberski, *Kubrick's Total Cinema: Philosophical Themes and Formal Qualities* (London: Continuum, 2012). Kuberski divides the war films into four categories: classical war (*Spartacus* and *Barry Lyndon*); modern war (*Paths of Glory, Fear and Desire,* and the unmade *Napoleon*); guerilla war *(Full Metal Jacket);* and, of course, nuclear war *(Dr. Strangelove).* Some commentators argue that Kubrick is an anti-war filmmaker, but this is to misunderstand his dialectical cast of mind and his refusal to moralize on the subject. His films are about the anatomy of war, about creative and destructive forces released in times of human conflict, more than they are about ethical matters of right and wrong. Like the seventeenth-century philosopher Thomas Hobbes in *Leviathan,* or Gilles Deleuze and Félix Guattari in *Nomadology: The War Machine* (New York: Semiotext(e), 1986), Kubrick focuses on the logic of war in an unsentimental manner. He is interested in the codifications of war and the apparatus of the state with which such codifications are entwined. If *Paths of Glory* is about mechanized killing and its deadly rationalizations during the First World War, *Dr. Strangelove* is about planetary weaponization and the endgame consequences for humanity.

35 Two other photographers, Bob Penn and Nicolas Tikhomiroff, were also hired by Columbia Pictures to take publicity stills for the film. The tabloid photographer Weegee also took pictures on the set, but he was paid directly by Kubrick. John O'Brian, *Strangelove's Weegee* (North Vancouver: Presentation House Gallery, 2013), 15.

36 "Dr. Strangelove: A Continuity Transcript," The Kubrick Site, http://www.visual-memory.co.uk/amk/doc/0055.html.

37 Kristan Horton, *Dr. Strangelove Dr. Strangelove* (Toronto: Art Gallery of York University, 2007).

38 George Bowering, "The Good Prospects," in Bowering, *The Man in Yellow Boots,* 52.

39 See Margot A. Henriksen, *Dr. Strangelove's America: Society and Culture in the Atomic Age* (Berkeley: University of California Press, 1997), 319–20, for an extended discussion of the exaggerated names of the characters in the film.

40 Bowering, *The Man in Yellow Boots,* 99.

41 Barry Goldwater, "Goldwater 1964 Acceptance Speech," YouTube video, 48:08, posted by SystemSoldat, July 29, 2015, https://youtu.be/RFSiyueal7Q.

42 Made for the Democratic Party by DDB Worldwide, the Daisy Ad aired on Monday, September 7, 1964, on NBC. It is widely available on the Internet. DDB made other presidential campaign spots for Johnson in 1964 about nuclear risk, notably "Ice Cream" and "Merely Another Weapon."

Epilogue: Deadpan Pictures

1 See Andrea Pinheiro, *The Bomb Book* (Vancouver: Presentation House Gallery and Publication Studio, 2013), a set of twelve volumes naming every nuclear test detonation that has occurred in the world in chronological order. Each page carries the name of a single nuclear detonation; where a test has no name, the page is left blank. The total number of pages is 2,450.

2 There are conflicting reports of the exact time the bomb was released. Dirk Septer states it was just before midnight on February 13, whereas Norman Leach reports that it was just after midnight. Dirk Septer, *Lost Nuke: The Last Flight of Bomber 075* (Toronto: Heritage House, 2012), 40–41; Norman S. Leach, *Broken Arrow: America's First Lost Nuclear Weapon* (Calgary: Red Deer Press, 2008), 2.

3 The accident is recounted in detail in Septer, *Lost Nuke;* and Leach, *Broken Arrow.* Other accounts can be found in John M. Clearwater, "The First One to Get Away," *Bulletin of the Atomic Scientists* 60, 6 (November–December 2004): 22–27; and John M. Clearwater, *U.S. Nuclear Weapons in Canada* (Toronto: Dundurn Group, 1999), 97–103. A documentary film on the incident, called *Lost Nuke,* was made by Michael Jorgensen for the Discovery Channel in 2004. See also John Mackie, "Canadian Army Interested in Old Nuke That May Have Been Found off Haida Gwaii," *Vancouver Sun,* November 3, 2016.

4 These are the facts provided in the official account of the accident, which was released under the US Freedom of Information Act in 1977. In *Lost Nuke,* Septer supplies evidence to suggest that the bomb may have been dropped further south in the Inside Passage, near the spot where the *Queen of the North* ferry sank off Gil Island in 2006.

5 Sergeant Vitale Trippodi, radioman-gunner on the B-36 bomber, spoke to reporters about the ordeal from his hospital bed at McChord Air Force Base, Washington: "I was hanging there in that tree, head down, with a foot caught in the 'chute strap ... I felt like I was dying." "I Felt Like I Was Dying," *Vancouver Daily Province,* February 17, 1950, front page.

6 The accident posed no risk of starting a nuclear war. Had it presented a risk, it would have been categorized as a "Nucflash." In descending order of seriousness, the names of the four coded categories of nuclear accident are Nucflash, Broken Arrow, Bent Spear, and Dull Sword.

7 News of the accident was immediately picked up by the media. The *Vancouver Daily Province* put it front page ("Storm Hides Fate of B-36, 17 Men") on February 14, 1950, the same day the plane went down. The *Globe and Mail* ran it as a front-page story the following day ("Blazing Six-Motor Bomber Down in Sea, 17 Aboard"). The media were not informed that Ship 2075 had been carrying an atomic weapon or that the crew had detonated it before bailing out. The US Air Force controlled the release of all information, stating only that the aircraft was engaged in a routine training flight.

8 During the Cold War, US military aircraft armed with nuclear warheads flew regularly over Canada. Also in 1950, an American B-50 exploded another nuclear bomb above the St. Lawrence River after experiencing engine trouble. As in the Queen Charlotte Sound detonation, it was not a fission explosion. Still, the blast was felt forty kilometres away, and it scattered uranium throughout the area.

9 Douglas Coupland, *Souvenir of Canada* (Vancouver: Douglas and McIntyre, 2002), 25. The map appears with other nuclearized maps.

10 I am indebted to a discussion on SMRs with M.V. Ramana on February 27, 2019, in Vancouver, and to his article "Are Thousands of New Nuclear Generators in Canada's Future?" *The Tyee*, November 7, 2018, https://thetyee.ca/Opinion/2018/11/07/Nuclear-Generators-Canada-Future/. See also Gordon Edwards, "Small Nuclear Reactors – Media Conference in Ottawa," Canadian Coalition for Nuclear Responsibility press release, October 25, 2018.

11 Canadian Nuclear Laboratories wants to produce a demonstration SMR by 2026. "Canadian Utilities Build SMR Alliances as Developers Tackle Licensing," Nuclear Energy Insider, January 16, 2019, https://www.terrestrialenergy.com/2019/01/nuclear-energy-insider-canadian-utilities-build-smr-alliances-as-developers-tackle-licensing/.

12 Ramana, "Are Thousands of New."

13 Danny Roderick, quoted in Ramana, "Are Thousands of New."

14 See Charles Wilkins, "The Deep: Canada's Nuclear Waste Will Be Deadly for 400,000 Years," *The Globe and Mail Report on Business Magazine*, March 2015, 57.

15 Fitzpatrick and Del Tredici, "Port Hope in the Era."

16 Jacobs, "The Visible and the Invisible," 13.

17 The list of isotopes is from Joanna Macy and Chris Johnstone, *Active Hope: How to Face the Mess We're in without Going Crazy* (Novato: CA: New World Library, 2012), 148.

18 Wilkins, "The Deep," 58.

19 Tony McQuail, quoted in Wilkins, "The Deep," 56.

20 Wilkins, "The Deep," 58.

21 The information in this section is from Matt Flowers, "What You Need to Know about the Port Hope Area Radioactive Waste Cleanup," Lake Ontario Waterkeeper (blog), November 9, 2016, http://www.waterkeeper.ca/blog/2016/11/8/what-you-need-to-know-about-the-port-hope-area-radioactive-waste-cleanup; and from Fitzpatrick and Del Tredici, "Port Hope in the Era."

22 A.B. Hodgetts, *Decisive Decades: A History of the Twentieth Century for Canadians* (Toronto: Thomas Nelson and Sons, 1960).

23 Hodgetts, *Decisive Decades*, xi.

24 Hodgetts was in agreement with his contemporary E.H. Carr, who rejected notions of historical contingency in his 1961 book *What Is History?* (London: Penguin, 1961).

25 Hodgetts, *Decisive Decades*, xii.

26 See Benjamin, "Theses on the Philosophy of History," 255.

SELECTED BIBLIOGRAPHY

Archival Sources

Albert Einstein Archives, Hebrew University, Jerusalem
Archive of Modern Conflict, London
Arctic Institute of North America, University of Calgary
Art Gallery of Ontario, Toronto
Banff Centre Library and Archives
Canadian Photography Institute, National Gallery of Canada, Ottawa
Canadian Science and Technology Museum, Ottawa
Canadian War Museum, Ottawa
Center for Creative Photography, Tucson
City of Vancouver Archives
Diefenbunker: Canada's Cold War Museum, Carp, Ontario
Hiroshima Peace Institute
Imperial War Museum, London
International Center of Photography, New York
Kyoto Museum for World Peace
Library and Archives Canada, Ottawa
Library of Congress, Washington
Northwest Territories Archives, Yellowknife
York University Libraries, Toronto

Secondary Sources

Albright, David, and Andrea Stricker. *Taiwan's Former Nuclear Weapons Program: Nuclear Weapons On-Demand.* Washington, DC: Institute for Science and International Security, 2018.

Arendt, Hannah. *On Violence.* New York: Harcourt Brace, 1969.

Augustin, Bernard. "The Sinking of the *Mont-Louis* and Nuclear Safety." *IAEA Bulletin,* Spring 1985.

Azoulay, Ariella. *The Civil Contract of Photography.* New York: Zone Books, 2008.

Baker, Simon, and Shoair Mavlian, eds. *Conflict*Time*Photography.* London: Tate Modern, 2015.

Ball, Howard. *Justice Downwind: America's Atomic Testing Program in the 1950s.* New York: Oxford University Press, 1986.

Barthes, Roland. *Camera Lucida: Reflections on Photography.* New York: Hill and Wang, 1981.

Basdeo, Sahadev. "Canada's Response to India's Nuclear Detonations in 1974 and 1998: Diversity and Challenge in International Relations." *Indian Journal of Canadian Studies* 13 (2005).

Benjamin, Walter. *Illuminations: Essays and Reflections.* Edited by Hannah Arendt. New York: Schocken, 1969.

Bertell, Rosalie. *No Immediate Danger? Prognosis for a Radioactive Earth.* Toronto: Women's Educational Press, 1985.

Blow, Peter, dir. *Village of Widows.* Peterborough, ON: Lindum Films, 1998.

Blueprint for Survival: 11 Steps to Survival. Foreword by John Diefenbaker. Ottawa: Emergency Measures Organization, 1961.

Bohlen, Jim. *Making Waves: The Origins and Future of Greenpeace.* Montreal: Black Rose, 2001.

Bothwell, Robert. *Eldorado: Canada's Nuclear Uranium Company.* Toronto: University of Toronto Press, 1984.

–. *Nucleus: The History of Atomic Energy of Canada Limited.* Toronto: University of Toronto Press, 1988.

Bothwell, Robert, Ian Drummond, and John English. *Canada since 1945.* Rev. ed. Toronto: University of Toronto Press, 1989.

Bothwell, Robert, and J.L. Granatstein. *The Gouzenko Transcripts: The Evidence Presented at the Kellock-Taschereau Royal Commission of 1946.* Ottawa: Deneau, 1982.

Bowering, George. *The Man in Yellow Boots / El Hombre de las Botas Amarillas.* Special issue, *El Corno Emplumado / The Plumed Horn* 16 (October 1965). Edited by Sergio Mondragón and Margaret Randall. http://opendoor.northwestern.edu/archive/files/original/3e8471258a-33391b9a132aee2afbac48.pdf.

Boyer, Paul. *By the Bomb's Early Light: American Thought and Culture at the Dawn of the Atomic Age.* New York: Pantheon, 1985.

–. *Fallout: A Historian Reflects on America's Half-Century Encounter with Nuclear Weapons.* Columbus: Ohio State University Press, 1998.

Boyko, John. *Cold Fire: Kennedy's Northern Front.* Toronto: Alfred A. Knopf, 2016.

Bradley, David. *No Place to Hide.* Boston: Little, Brown, 1948.

Bratishenko, Lev, and Mike Zardini, eds. *It's All Happening So Fast: A Counter-History of the Modern Canadian Environment.* Montreal: Canadian Centre for Architecture, 2016.

Broad, William. "The Bomb Chroniclers." *New York Times,* September 13, 2010.

Brown, Kate. *Plutopia: Nuclear Families, Atomic Cities, and the Great Soviet and American Plutonium Disasters.* New York: Oxford University Press, 2013.

Burtch, Andrew. *Give Me Shelter: The Failure of Canada's Cold War Civil Defence.* Vancouver: UBC Press, 2012.

Butler, Judith. "Photography, War, Outrage." *PMLA* 120, 3 (May 2005).

–. *Precarious Life: Powers of Mourning and Violence.* New York: Verso, 2004.

Byers, Michael. *Intent for a Nation: What Is Canada For?* Vancouver: Douglas and McIntyre, 2007.

Cadava, Eduardo. *Words of Light: Theses on the Photography of History.* Princeton: Princeton University Press, 1997.

Campbell, Lara, Michael Dawson, and Catherine Gidney, eds. *Worth Fighting For: Canada's Tradition of War Resistance from 1812 to the War on Terror.* Toronto: Between the Lines, 2015.

Carpenter, Ele, ed. *The Nuclear Culture Source Book.* London: Black Dog, 2016.

–. "Shifting the Nuclear Imaginary." In *Cold War Legacies: Systems, Theories, Aesthetics,* ed. John Beck and Ryan Bishop. Edinburgh: Edinburgh University Press, 2016.

Carson, Rachel. *Silent Spring.* 1962; Boston: Houghton Mifflin, 2002.

Cavell, Richard, ed. *Love, Hate and Fear in Canada's Cold War.* Toronto: University of Toronto Press, 2004.

Chow, Rey. *The Age of the World Target: Self-Referentiality in War, Theory and Comparative Work.* Durham, NC: Duke University Press, 2006.

Clearwater, John M. *Canadian Nuclear Weapons: The Untold Story of Canada's Cold War Arsenal.* Toronto: Dundurn, 1998.

–. "The First One to Get Away." *Bulletin of the Atomic Scientists* 60, 6 (November–December 2004).

–. *Just Dummies: Cruise Missile Testing in Canada.* Calgary: University of Calgary Press, 2006.

–. *U.S. Nuclear Weapons in Canada.* Toronto: Dundurn Group, 1999.

Coupland, Douglas. *Life after God.* New York: Simon and Schuster, 1994.

–. *Souvenir of Canada.* Vancouver: Douglas and McIntyre, 2002.

Cragg, Kenneth C. "Canada Building Plant Near Petawawa Camp to Develop Atom Bomb." *Globe and Mail,* August 7, 1945.

Creighton, Donald. *The Story of Canada.* Rev. ed. Toronto: Macmillan, 1971.

Davis, Mike. "The Dead West: Ecocide in Marlboro Country." *New Left Review* 200 (July–August 1993).

Decamous, Gabrielle. *Invisible Colors: The Arts of the Atomic Age.* Cambridge, MA: MIT Press, 2018.

Del Tredici, Robert. *At Work in the Fields of the Bomb.* Vancouver: Douglas and McIntyre, 1987.

–. "Documenting the Dark Period." Interview by Michelle Heinemann. *Blackflash,* Spring 1985.

–. *The Nuclear Map of Canada Project.* Charlottetown: Confederation Center Art Gallery and Museum, 1998.

Derrida, Jacques. *Dissemination.* Chicago: University of Chicago Press, 1981.

–. "No Apocalypse, Not Now (Full Speed Ahead, Seven Missiles, Seven Missives)." *Diacritics* 14, 2 (Summer 1984).

Deutsche, Rosalyn. *Hiroshima after Iraq: Three Studies in Art and War.* New York: Columbia University Press, 2010.

Devine, Bonnie, dir. *Rooster Rock: The Story of Serpent River.* Toronto: V Tape, 2002.

Diack, Heather. "Hand over Fist: A Chronicle of Cold War Photography," *Visual Studies* 30, 2 (2015).

Didi-Huberman, Georges. *Images in Spite of All: Four Photographs from Auschwitz.* Translated by Shane B. Lillis. Chicago: University of Chicago Press, 2008.

Dower, John W. *Embracing Defeat: Japan in the Wake of World War II.* New York: W.W. Norton, 1999.

Edwards, Gordon. "Canada's Nuclear Industry and the Myth of the Peaceful Atom." In *Canada and the Nuclear Arms Race,* ed. Ernie Regehr and Simon Rosenblum. Toronto: James Lorimer, 1983.

–. "Fueling the Nuclear Arms Race and How to Stop It." *Ploughshares Monitor* 6, 2 (June 1985). http://www.ccnr.org/non_prolif.html.

–. "Reactor Accidents at Chalk River: The Human Fallout." Canadian Coalition for Nuclear Responsibility, March 27, 1996. http://www.ccnr.org/paulson_legacy.html.

The Effects of Atomic Weapons. Los Alamos: Los Alamos Scientific Laboratory, 1950.

Eggleston, Wilfrid. *Canada's Nuclear Story.* Toronto: Clarke, Irwin, 1965.

Elkins, James. "Rapatronic Photographs of Atomic Tests." In James Elkins, *After and Before: Documenting the A-Bomb.* New York: PPP Editions, 2003.

Evans, Walker. "Test Exposures: Six Photographs from a Film Manufacturer's Files." *Fortune,* July 1954.

Farish, Matthew. "Below the Bombs." In *Through Post-Atomic Eyes,* ed. Claudette Lauzon and John O'Brian. Montreal and Kingston: McGill-Queen's University Press, 2020.

–. *The Contours of America's Cold War.* Minneapolis: University of Minnesota Press, 2010.

Finch, Ron. *Exporting Danger: A History of the Canadian Nuclear Energy Export Programme.* Montreal: Black Rose, 1986.

Fitzpatrick, Blake. "Atomic Afterimages." *History of Photography* 32, 2 (Summer 2008).

–. "The Radioactive Photograph Album." In *Lost in the Archives,* ed. Rebecca Comay. Toronto: Alphabet City, 2002.

Fitzpatrick, Blake, and Robert Del Tredici. *The Atomic Photographers Guild: Visibility and Invisibility in the Nuclear Era.* Toronto: Gallery TPW, 2001.

Frank, Robert. *The Americans.* New York: Scalo, 1995.

–. *The Lines of My Hand.* New York: Lustrom, 1972.

Franklin, Ursula M. *The Real World of Technology.* Toronto: Anansi, 1990.

Freeman, Lindsey A. *Longing for the Bomb: Oak Ridge and Atomic Nostalgia.* Chapel Hill: University of North Carolina Press, 2015.

Gallagher, Carole. *American Ground Zero: The Secret Nuclear War.* New York: Random House, 1993.

Geller, Peter. *Northern Exposures: Photographing and Filming the Canadian North, 1920–1945.* Vancouver: UBC Press, 2004.

Ghamari-Tabrizi, Sharon. *The Worlds of Herman Kahn: The Intuitive Science of Thermonuclear War.* Cambridge, MA: Harvard University Press, 2005.

Gouzenko, Igor. *This Was My Choice: Gouzenko's Story.* Toronto: F.M. Dent, 1948.

Gusterson, Hugh. "Hiroshima and the Power of Pictures." *Bulletin of the Atomic Scientists* 65, 4 (August 2009).

Halbwachs, Maurice. *The Collective Memory.* New York: Colophon, 1980.

Hales, Peter B. "The Atomic Sublime." *American Studies* 32 (Spring 1991).

–. "Imagining the Atomic Age: Life and the Atom." In *Looking at Life Magazine*, ed. Erika Doss, Washington, DC: Smithsonian Institution Press, 2001.

Harding, Jim. *Canada's Deadly Secret: Saskatchewan Uranium and the Global Nuclear System.* Halifax: Fernwood, 2007.

Harries, Kate. "Nuclear Reaction: Accusations of Cancerous Fallout Divide a Small Ontario Town." *Walrus*, March 2008.

Heilbron, J.L. *Ernest Rutherford and the Explosion of Atoms.* New York: Oxford University Press, 2003.

Henriksen, Margot A. *Dr. Strangelove's America: Society and Culture in the Atomic Age.* Berkeley: University of California Press, 1997.

Hersey, John. *Hiroshima.* New York: Alfred A. Knopf, 1946.

Higuchi, Kenji. *Photo Document: Japan's Nuclear Power Plants.* Tokyo: Orijin Shuppan Senta, 1979.

Hodgetts, A.B. *Decisive Decades: A History of the Twentieth Century for Canadians.* Toronto: Thomas Nelson and Sons, 1960.

Hunter, Bob. *The Greenpeace to Amchitka: An Environmental Odyssey.* Vancouver: Arsenal Pulp Press, 2004.

Hunter, Bob, and Robert Keziere. *Greenpeace.* Toronto: McClelland and Stewart, 1972.

Jacobs, Robert A. *The Dragon's Tail: Americans Face the Atomic Age.* Amherst: University of Massachusetts Press, 2010.

–. "Good Bomb / Bad Bomb: Talking about Atomic Tests in Nevada." *Interdisciplinary Humanities* 1 (Spring 2007).

–. "The Visible and the Invisible When Considering Northern European Permanent Spent Fuel Storage: Forsmark and Onkalo." *Hiroshima Peace Research Journal* 5 (2019).

Jenkins, Rupert, ed. *Nagasaki Journey: The Photographs of Yosuke Yamahata, August 10, 1945.* San Francisco: Pomegranate Books, 1995.

Johnson, Genevieve Fuji. *Deliberative Democracy for the Future: The Case of Nuclear Waste Management in Canada.* Toronto: University Toronto Press, 2008.

Kahn, Herman. *Thinking about the Unthinkable.* New York: Horizon, 1962.

Kaneko, Ryuichi, and Ivan Vartanian. *Japanese Photobooks of the 1960s and 1970s.* New York: Aperture, 2009.

Kaplan, Fred. *The Wizards of Armageddon.* New York: Simon and Schuster, 1983.

Keller, Bill. "The Second Nuclear Age." *New York Times Magazine*, May 4, 2003.

Kirsch, Scott. "Watching the Bombs Go Off: Photography, Nuclear Landscapes, and Spectator Democracy." *Antipode* 29, 3 (1997).

Kiyooka, Roy. *Roy K. Kiyooka: 25 Years.* Vancouver: Vancouver Art Gallery, 1975.

Knelman, F.H. *America, God and the Bomb: The Legacy of Ronald Reagan.* Vancouver: New Star, 1987.

Kohlhoff, Dean W. *Amchitka and the Bomb: Nuclear Testing in Alaska.* Seattle: University of Washington Press, 2002.

Krenz, Kim. *Deep Waters: The Ottawa River and Canada's Nuclear Adventure.* Montreal and Kingston: McGill-Queen's University Press, 2004.

Kröller, Eva-Marie. *George Bowering: Bright Circles of Colour.* Vancouver: Talonbooks, 1992.

Kuletz, Valerie L. *The Tainted Desert: Environmental and Social Ruin in the American West.* New York: Routledge, 1998.

Kuran, Peter. *How to Photograph an Atomic Bomb.* Santa Clarita, CA: VCE, 2006.

Langford, Martha, and John Langford. *A Cold War Tourist and His Camera.* Montreal and Kingston: McGill-Queen's University Press, 2011.

Lauzon, Claudette, and John O'Brian, eds. *Through Post-Atomic Eyes.* Montreal and Kingston: McGill-Queen's University Press, 2020.

Leach, Norman S. *Broken Arrow: America's First Lost Nuclear Weapon.* Calgary: Red Deer Press, 2008.

LeBourdais, D.M. *Canada and the Atomic Revolution.* Toronto: McClelland and Stewart, 1959.

Leo, Vincent. "The Mushroom Cloud Photograph: From Fact to Symbol." *Afterimage* 13 (Summer 1985).

Lewis, John Wilson, and Xue Litai. *China Builds the Bomb.* Stanford, CA: Stanford University Press, 1988.

Lifton, Robert Jay, and Greg Mitchell. *Hiroshima in America: Fifty Years of Denial.* New York: G.P. Putnam's, 1995.

Light, Michael. *100 Suns: 1945–1962.* New York: Alfred A. Knopf, 2003.

Lindee, M. Susan. *Suffering Made Real: American Science and the Survivors at Hiroshima.* Chicago: University of Chicago Press, 1994.

Lippit, Akira Mizuta. *Atomic Light (Shadow Optics).* Minneapolis: University of Minnesota Press, 2005.

Lista, Michael. *Bloom.* Toronto: Anansi, 2010.

Maclear, Kyo. *Beclouded Visions: Hiroshima-Nagasaki and the Art of Witness.* Albany, NY: SUNY Press, 1999.

Masco, Joseph. *The Nuclear Borderlands: The Manhattan Project in Post–Cold War New Mexico.* Princeton: Princeton University Press, 2006.

McKay, Ian, and Jamie Swift. *Warrior Nation: Rebranding Canada in an Age of Anxiety.* Toronto: Between the Lines, 2012.

McLuhan, Marshall. *Understanding Media: The Extensions of Man.* New York: McGraw-Hill, 1964.

McMahon, Patricia I. *Essence of Indecision: Diefenbaker's Nuclear Policy, 1957–1963.* Montreal and Kingston: McGill-Queen's University Press, 2009.

McMillan, David. *Growth and Decay: Prypiat and the Chernobyl Exclusion.* With an essay by Claude Baillargeon. Göttingen: Steidel, 2018.

Mellor, Robynne. "The Radioactive Colonization of Indigenous Bodies." Network in Canadian History and Environment, August 23, 2017. http://niche-canada.org/2017/08/23/the-radioactive-colonization-of-indigenous-bodies.

Misrach, Richard. *Bravo 20: The Bombing of the American West.* Baltimore: Johns Hopkins University Press, 1990.

Nelkin, Dorothy, and Michael Pollak. *The Atom Besieged: Anti-Nuclear Movements in France and Germany.* Cambridge, MA: MIT Press, 1982.

Nelson, Joyce. *The Perfect Machine: TV in the Nuclear Age.* Toronto: Between the Lines, 1987.

Nikiforuk, Andrew. "Echoes of the Atomic Age: Cancer Kills Fourteen Aboriginal Uranium Workers." *Calgary Herald,* March 14, 1998.

Nixon, Rob. *Slow Violence and the Environmentalism of the Poor.* Cambridge, MA: Harvard University Press, 2011.

Nora, Pierre. "Between Memory and History." *Representations* 26 (Spring 1989).

The Nuclear Century: Voices of the Hibakusha of the World. Tokyo: Japan Peace Museum, 1997.

O'Brian, John, ed. *Camera Atomica.* London: Black Dog; Toronto: Art Gallery of Ontario, 2015.

–. "The Nuclear Family of Man." *Japan Focus: The Asia-Pacific Journal* 6, 7 (July 2008). https://apjjf.org/-John-O'Brian/2816/article.html.

–. "Postcard to Moscow." In *Postcards: Ephemeral Histories of Modernity,* ed. Jordana Mendelson and David Proschaska. Philadelphia: Pennsylvania State University Press, 2010.

–. *Strangelove's Weegee.* North Vancouver: Presentation House Gallery, 2013.

O'Brian, John, and Jeremy Borsos. *Atomic Postcards: Radioactive Message from the Cold War.* Bristol: Intellect Books, 2011.

Office of the Historian, Joint Task Force One. *Operation Crossroads: The Official Pictorial Record.* New York: Wm. H. Wise, 1946.

Palen, Wendy. "When Canadian Scientists Were Muzzled by Their Government." *New York Times,* February 14, 2017.

Parsons, Sarah. "Sontag's Lament: Emotion, Ethics, and Photography." *Photography and Culture* 2, 3 (November 2009).

Payne, Carol. *Official Picture: The National Film Board of Canada's Still Photography Division and the Image of Canada, 1941–1971.* Montreal and Kingston: McGill-Queen's University Press, 2013.

Payne, Carol, and Andrea Kunard, eds. *The Cultural Work of Photography in Canada.* Montreal and Kingston: McGill-Queen's University Press, 2011.

Phillips, Ruth B. "Between Rocks and Hard Places: Indigenous Lands, Settler Art Histories, and the 'Battle for the Woodlands.'" *Journal of Canadian Art History* 37, 1 (2016).

Pinheiro, Andrea. *The Bomb Book.* Vancouver: Presentation House Gallery and Publication Studio, 2013.

Pitkanen, Laura, and Matthew Farish. "Nuclear Landscapes." *Progress in Human Geography* (August 31, 2017). https://doi.org/10.1177/0309132517725808.

Plokhy, Serhii. *Chernobyl: History of a Tragedy.* London: Allen Lane, 2018.

Power, Max S. *America's Nuclear Wastelands: Politics, Accountability, and Cleanup.* Pullman: Washington State University Press, 2008.

Ramana, M.V. "Are Thousands of New Nuclear Generators in Canada's Future?" *The Tyee,* November 7, 2018. https://thetyee.ca/Opinion/2018/11/07 Nuclear-Generators-Canada-Future/.

Rayner-Canham, Marelene F., and Geoffrey W. Rayner-Canham. *Harriet Brooks: Pioneer Nuclear Scientist.* Montreal and Kingston: McGill-Queen's University Press, 1992.

Regehr, Ernie, and Simon Rosenblum, eds. *Canada and the Nuclear Arms Race.* Toronto: James Lorimer, 1983.

Rekmans, Lorraine, ed. *This Is My Homeland.* Cutler, ON: Serpent River First Nation, 2003.

Retort (Iain Boal, T.J. Clark, Joseph Matthews, Michael Watts). *Afflicted Powers: Capital and Spectacle in a New Age of War.* London: Verso, 2005.

Rhodes, Richard. *Dark Sun: The Making of the Hydrogen Bomb.* New York: Simon and Schuster, 1995.

–. *The Making of the Atomic Bomb.* New York: Simon and Schuster, 1986.

Richter, Andrew. *Avoiding Armageddon: Canadian Military Strategy and Nuclear Weapons, 1950–63.* Vancouver: UBC Press, 2002.

Roberts, John. *Photography and Its Violations.* New York: Columbia University Press, 2014.

Rose, Kenneth D. *One Nation Underground: The Fallout Shelter in American Culture.* New York: New York University Press, 2001.

Rosenberg, Howard. *Atomic Soldiers: American Victims of Nuclear Experiments.* Boston: Beacon, 1980.

Rosenblum, Simon. *Misguided Missiles: Canada, the Cruise and Star Wars.* Toronto: James Lorimer, 1985.

Rosenthal, Peggy. "The Nuclear Mushroom Cloud as Cultural Image." *American Literary History* 3, 1 (Spring 1991).

Salverson, Julie. *Lines of Flight: An Atomic Memoir.* Hamilton, ON: Wolsak and Wynn, 2016.

Saunders, Chris. *Whiteshell Laboratories: A Legacy to Nuclear Science and Engineering in Canada.* Pinawa, MB: Pinawa Foundation, 2016. http://online.fliphtml5.com/xzbq/krra.

Sekula, Allan. *Geography Lesson: Canadian Notes.* Vancouver: Vancouver Art Gallery; Cambridge, MA: MIT Press, 1997.

–. *Photography against the Grain: Essays and Photo Works, 1973–1983.* Halifax: Press of the Nova Scotia College of Art and Design, 1984.

Septer, Dirk. *Lost Nuke: The Last Flight of Bomber 075.* Toronto: Heritage House, 2012.

Silverman, Kaja. *The Threshold of the Visible World.* New York: Routledge, 1996.

Smith, W. Eugene. "Photographic Journalism." *Photo Notes,* June 1948.

Solnit, Rebecca. *A Paradise Built in Hell: The Extraordinary Communities That Arise in Disaster.* New York: Penguin, 2009.

Sontag, Susan. *On Photography.* New York: Doubleday, 1977.

–. *Regarding the Pain of Others.* New York: Farrar, Straus and Giroux, 2003.

Stacey, C.P. *Canada and the Age of Conflict: A History of Canadian External Policies.* Vol. 2. Toronto: University of Toronto Press, 1981.

Stallabrass, Julian, ed. *Documentary: Documents of Contemporary Art*. London: Whitechapel Gallery; Cambridge, MA: MIT Press, 2013.

Steichen, Edward, ed. *The Family of Man*. New York: Museum of Modern Art and Maco Magazine, 1955.

Stein, Sally. *The Rhetoric of the Colorful and the Colorless: American Photography and Material Culture between the Wars*. PhD diss., Yale University, 1991.

Stewart, Susan. *On Longing: Narratives of the Miniature, the Gigantic, the Souvenir, the Collection*. Durham, NC: Duke University Press, 1993.

Stimson, Blake. *The Pivot of the World: Photography and Its Nation*. Cambridge, MA: MIT Press, 2006.

Szasz, Ferenc M. *Larger Than Life: New Mexico in the Twentieth Century*. Albuquerque: University of New Mexico Press, 2006.

Taylor, Bryan C. "Nuclear Pictures and Metapictures." *American Literary History* 9, 3 (Autumn 1997).

Teigrob, Robert. *Warming Up the Cold War: Canada and the United States' Coalition of the Willing, from Hiroshima to Korea*. Toronto: University of Toronto Press, 2009.

Thompson, John Herd, and Stephen J. Randall. *Canada and the United States: Ambivalent Allies*. 3rd ed. Montreal and Kingston: McGill-Queen's University Press, 2002.

Toguri McFarlane, Scott. "Un-Ravelling *StoneDGloves* and the Haunt of the Hibakusha." In *All Amazed: For Roy Kiyooka*, ed. John O'Brian, Naomi Sawada, and Scott Watson. Vancouver: Arsenal Pulp Press, Morris and Helen Belkin Art Gallery, Collapse, 2002.

Touhey, Ryan M. *Conflicting Visions: Canada and India in the Cold War World, 1946–76*. Vancouver: UBC Press, 2015.

Van Wyck, Peter C. *The Highway of the Atom*. Montreal and Kingston: McGill-Queen's University Press, 2010.

–. "Highway of the Atom: Recollections along a Route." *Topia* 7 (Spring 2002).

–. *Signs of Danger: Waste, Trauma, and Nuclear Threat*. Minneapolis: University of Minnesota Press, 2005.

Virilio, Paul. *War and Cinema: The Logistics of Perception*. Translated by Patrick Camiller. London: Verso, 1989.

Weart, Spencer R. *The Rise of Nuclear Fear*. Cambridge, MA: Harvard University Press, 2012.

Westad, Odd Arne. *The Cold War: A World History*. London: Allen Lane, 2017.

Whitaker, Reg, and Gary Marcuse. *Cold War Canada: The Making of the National Insecurity State, 1945–1957*. Toronto: University of Toronto Press, 1994.

Whitaker, Reg, and Steve Hewitt. *Canada and the Cold War*. Toronto: James Lorimer, 2003.

Wilkins, Charles. "The Deep: Canada's Nuclear Waste Will Be Deadly for 400,000 Years." *Globe and Mail Report on Business Magazine*, March 2015.

Williams, William Carlos. *Journey to Love*. New York: Random House, 1955.

Wilson, Alexander. *The Culture of Nature: North American Landscape from Disney to the Exxon Valdez*. Toronto: Between the Lines, 1991.

Wittner, Lawrence S. *Confronting the Bomb: A Short History of the World Nuclear Disarmament Movement.* Stanford, CA: Stanford University Press, 2009.

Wodiczko, Krzysztof. "Public Projections." *Canadian Journal of Political and Social Theory* 7, 1–2 (Winter–Spring 1983).

Wyler, Rex. *Greenpeace: How a Group of Ecologists, Journalists, and Visionaries Changed the World.* Vancouver: Raincoast, 2004.

ILLUSTRATIONS

Unless otherwise noted, illustrations are gelatin silver prints.

Cover

Robert Del Tredici. *Stanrock Tailings Wall, Elliot Lake, Ontario,* August 25, 1985. Courtesy of the artist.

Frontispiece

Philippe Halsman. *American Physicist J. Robert Oppenheimer,* 1958. Philippe Halsman / Magnum Photos. Students from Canadian universities were drawn to Oppenheimer, who taught at Berkeley and was director of the Manhattan Project. Three students from the University of British Columbia alone undertook their doctoral studies with him: Robert Christy, Shuichi Kusaka, and George Volkoff.

Table of Contents

US Army Air Force. *Hiroshima Ruins,* August 1945. JO'B Atomic Archive.

Preface: Radioactive Footprints

Introduction: Still Photographs

Chapter 1: Nuclear Industry

Chapter 2: Chalk River

Chapter 3: Cold War Defence

Chapter 4: Cloud Shadows

Chapter 5: Atomic Soldiers

Chapter 6: Mass Protest

Chapter 7: Camera Targets

Epilogue: Deadpan Pictures

INDEX

Note: "(i)" after a page number indicates an illustration

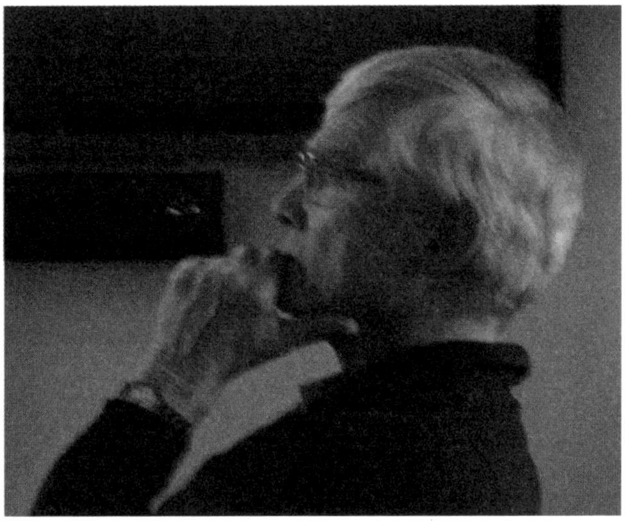

Photo by Philippe Guillaume

ABOUT THE AUTHOR

JOHN O'BRIAN IS AN ART HISTORIAN, WRITER, AND CURATOR. Until 2017, he taught at the University of British Columbia. He has authored or edited twenty books, including *Clement Greenberg: The Collected Essays and Criticism* – one of *the New York Times'* "Notable Books of the Year" – and *Ruthless Hedonism,* which won an award from the Association of American University Presses. His publications on nuclear photography include *Strangelove's Weegee, Camera Atomica, Through Post-Atomic Eyes*, and *Atomic Postcards: Radioactive Messages from the Cold War.* He has organized five exhibitions on nuclear photography, in Copenhagen, London, Toronto, and Vancouver, and is a recipient of the Thakore Award in Human Rights and Peace Studies from Simon Fraser University.